Geotechnical Laboratory Measurements for Engineers

Geotechnical Laboratory Measurements for Engineers

John T. Germaine and Amy V. Germaine

WILEY

John Wiley & Sons, Inc.

Copyright © 2009 by John Wiley & Sons, Inc. All rights reserved

Published by John Wiley & Sons, Inc., Hoboken, New Jersey
Published simultaneously in Canada

For general information about our other products and services, please contact our Customer Care Department within the United States at (800) 762-2974, outside the United States at (317) 572-3993 or fax (317) 572-4002.

Wiley also publishes its books in a variety of electronic formats. Some content that appears in print may not be available in electronic books. For more information about Wiley products, visit our web site at www.wiley.com.

Library of Congress Cataloging-in-Publication Data:

ISBN: 978-0-470-15093-1
Germaine, John T.
    Geotechnical laboratory measurements for engineers / John T. Germaine and Amy V. Germaine.
     p. cm.
    Includes index.
    ISBN 978-0-470-15093-1 (paper/website) 1. Soil dynamics. 2. Soils—Testing. 3. Soils—Composition. 4. Soils—Density—Measurement. I. Germaine, Amy V. II. Title.
    TA711.G48 2009
    624.1'510287—dc22
                                             2009007439

Printed in the United States of America
10 9 8 7 6 5 4 3 2 1

Contents

PREFACE   XI

ACKNOWLEDGEMENTS   XIII

PART I

Chapter 1   **Background Information for Part I   3**
Scope, 3
Laboratory Safety, 4
Terminology, 5
Standardization, 6
Evaluation of Test Methods, 8
Precision and Bias Statements, 9
Laboratory Accreditation, 11
Proficiency Testing, 12
Technician Certification, 12
Unit Convention, 12
Significant Digits, 13
Test Specification, 14
Sampling, 15
Processing Bulk Material, 17
Test Documentation, 19
Spreadsheets, 20
Reporting Test Results, 20
Typical Values, 21
Further Reading and Other References, 21
References, 22

Chapter 2   **Phase Relationships   24**
Scope and Summary, 24
Typical Materials, 25
Background, 25

Typical Values, 33
Calibration, 33
Specimen Preparation, 35
Procedure, 35
Precision, 37
Detecting Problems with Results, 38
Reference Procedures, 38
References, 38

Chapter 3   **Specific Gravity   39**
Scope and Summary, 39
Typical Materials, 39
Background, 39
Typical Values, 44
Calibration, 45
Specimen Preparation, 45
Procedure, 45
Precision, 50
Detecting Problems with Results, 51
Reference Procedures, 51
References, 51

Chapter 4   **Maximum Density, Minimum Density   52**
Scope and Summary, 52
Typical Materials, 52
Background, 52
Typical Values, 55
Calibration, 55
Specimen Preparation, 57
Procedure, 57
Precision, 58
Detecting Problems with Results, 58
Reference Procedures, 59
References, 59

Chapter 5   **Calcite Equivalent   60**
Scope and Summary, 60
Typical Materials, 60
Background, 60
Typical Values, 62
Calibration, 63
Specimen Preparation, 65
Procedure, 65
Precision, 66
Detecting Problems with Results, 66
Reference Procedures, 66
References, 66

Chapter 6   **pH and Salinity   68**
Scope and Summary, 68
Typical Materials, 68
Background, 68
Typical Values, 74
Calibration, 75
Specimen Preparation, 76
Procedure, 76

Precision, 78
Detecting Problems with Results, 78
Reference Procedures, 79
References, 79

Chapter 7   **Organic Content   80**
Scope and Summary, 80
Typical Materials, 80
Background, 80
Typical Values, 82
Calibration, 82
Specimen Preparation, 82
Procedure, 82
Precision, 83
Detecting Problems with Results, 83
Reference Procedures, 83
References, 83

Chapter 8   **Grain Size Analysis   84**
Scope and Summary, 84
Typical Materials, 84
Background, 84
Typical Values, 107
Calibration, 109
Specimen Preparation, 111
Procedure, 112
Precision, 115
Detecting Problems with Results, 116
Reference Procedures, 116
References, 116

Chapter 9   **Atterberg Limits   117**
Scope and Summary, 117
Typical Materials, 117
Background, 117
Typical Values, 130
Calibration, 132
Specimen Preparation, 134
Procedure, 135
Precision, 137
Detecting Problems with Results, 138
Reference Procedures, 138
References, 138

Chapter 10   **Soil Classification and Description   140**
Scope and Summary, 140
Typical Materials, 140
Background, 141
Calibration, 156
Specimen Preparation, 157
Procedure, 157
Precision, 160
Detecting Problems with Results, 160
Reference Procedures, 160
References, 160

PART II

Chapter 11   **Background Information for Part II   163**
Scope and Summary, 163
Intact Sampling, 164
Processing Intact Samples, 169
Reconstituting Samples, 184
Transducers, 189
Data Collection and Processing, 203
References, 209

Chapter 12   **Compaction Test Using Standard Effort   210**
Scope and Summary, 210
Typical Materials, 210
Background, 211
Typical Values, 219
Calibration, 220
Specimen Preparation, 220
Procedure, 220
Precision, 221
Detecting Problems with Results, 222
Reference Procedures, 222
References, 222

Chapter 13   **Hydraulic Conductivity: Cohesionless Materials   223**
Scope and Summary, 223
Typical Materials, 223
Background, 223
Typical Values, 233
Calibration, 234
Specimen Preparation, 235
Procedure, 235
Precision, 237
Detecting Problems with Results, 238
Reference Procedures, 238
References, 238

Chapter 14   **Direct Shear   239**
Scope and Summary, 239
Typical Materials, 239
Background, 240
Typical Values, 248
Calibration, 248
Specimen Preparation, 250
Procedure, 251
Precision, 254
Detecting Problems with Results, 254
Reference Procedures, 255
References, 255

Chapter 15   **Strength Index of Cohesive Materials   256**
Scope and Summary, 256
Typical Materials, 256
Background, 256
Typical Values, 264

Calibration, 264
Specimen Preparation, 266
Procedure, 267
Precision, 272
Detecting Problems with Results, 272
Reference Procedures, 273
References, 273

Chapter 16   **Unconsolidated-Undrained Triaxial Compression   275**
Scope and Summary, 275
Typical Materials, 276
Background, 276
Typical Values, 286
Calibration, 286
Specimen Preparation, 289
Procedure, 289
Precision, 292
Detecting Problems with Results, 292
Reference Procedures, 293
References, 293

Chapter 17   **Incremental Consolidation By Oedometer   294**
Scope and Summary, 294
Typical Materials, 294
Background, 294
Typical Values, 324
Calibration, 324
Specimen Preparation, 325
Procedure, 326
Precision, 331
Detecting Problems with Results, 331
Reference Procedures, 332
References, 332

APPENDICES

Appendix A   **Constants And Unit Conversions 334**

Appendix B   **Physical Properties Of Pure Water 338**

Appendix C   **Calculation Adjustments For Salt 340**

INDEX   345

Preface

This textbook is divided into two parts, according to a general division of test result characteristics and level of background knowledge necessary to perform the tests. Part I focuses on relatively simple tests that are used to characterize the nature of soils and can be performed on bulk materials. Part II increases the level of testing complexity, places more emphasis on engineering properties, and requires a larger investment in laboratory equipment. These topics may be covered in an undergraduate civil or geotechnical engineering laboratory course.

An ancillary web site has been created for this textbook. The web site (www.wiley.com/college/germaine) is divided by chapter and includes data sheets, spreadsheets, and example data sets. In addition, there are online resources for instructors that provide template data sheets with embedded data reduction formulas.

There are a large number of tests that are performed on geo-materials. This book is not intended to be all inclusive, but rather covers a selection of the most common and essential tests, while maintaining a broad cross-section of methods and devices. In general, testing of geo-materials is a slow process according to "clock time." This is mostly due to the need for pore water to come to equilibrium. Cost-effective, high-quality testing is possible by understanding the important factors and working with nature to use labor wisely. A goal of this text is to provide guidance for efficient testing without sacrificing the quality of results. Efficiency can be achieved by tailoring techniques to individual circumstances and understanding when shortcuts are feasible.

The background chapter to each of the two parts provides general information that applies to the chapters that follow. Test-specific information is included in the pertinent chapter. Each testing chapter provides background information to understand the concepts and objectives of the method, a discussion of important factors useful for professional practice, a list of the minimum equipment requirements, detailed procedures and guidance for performing the test, and the calculations required to produce the results. The procedures are provided for specific situations and would be most useful for instructional purposes. These instructions could be modified for commercial application to increase productivity and efficiency.

The text is set up to allow instructors to choose which laboratories to include in their courses. The knowledge gained through individual tests tends to be cumulative as the text progresses. However, it is not intended that all of the laboratories would be taught, one after the other, as part of a single course.

Although this book is well suited to teaching a geotechnical laboratory course, the practicing geotechnical engineer should find this text useful as a reference on the important details relative to testing. This is especially important when designing sophisticated subsurface characterization and corresponding advanced laboratory testing programs. The information presented is essential to the geotechnical engineer. The text helps develop a working knowledge of laboratory capabilities and testing methods. Laboratory testing is also a large part of geotechnical research. Perhaps the most valuable experience is that the knowledge gained by performing the laboratories reinforces the understanding of soil behavior.

It is hoped that the practicing engineer will understand the following motto: *Only perform the tests you need. But if you need to do it, do it correctly.*

Acknowledgements

Several individuals in the geotechnical field have had tremendous impacts on the authors, both professionally and as family friends.

Stephen Rudolph is the machinist and designer responsible for the equipment modifications necessary to accomplish the experiments that appear in many chapters of this work. His dedication and skills are admired and appreciated.

Charles Ladd has been a mentor and colleague to both of us since we entered MIT as students. His guidance in understanding soil behavior is treasured.

Richard Ladd has been a valued colleague and mentor to both of us through the years. Richard is always available and ready to discuss the finer points of testing, and his feedback is valued dearly.

Several consulting clients deserve recognition for providing interesting work and for probing into the details relative to geotechnical testing. Two of those individuals are Demetrious Koutsoftas and Richard Reynolds. We can never thank them enough.

Jack's students of the past, present, and future are to be commended for asking questions and pushing the boundaries of geotechnical testing. They keep the job of teaching fresh, challenging, and interesting.

Finally, we wish to thank Bill DeGroff and Fugro Consultants, Inc., for affording Amy the flexible schedule that allowed for the writing of this textbook.

Part I

Chapter 1

Background Information for Part I

Experimental investigation requires an appreciation for more than the individual test method. In order to perform the tests effectively, interpret the measurements properly, and understand the results, background information is required on a variety of general topics. This chapter provides general information important to the overall operation of a laboratory, evaluation of a test method and test result, and handling of disturbed materials. Some of the individual topics are ASTM International, Interlaboratory Test Programs, Precision and Bias, Sampling, Bulk Material Processing, and Test Documentation.

The tests covered in Part I are normally performed on disturbed material and are used to characterize the nature of soils. There are a vast number of specific tests used to characterize particles, the pore fluid, and also the combination of both. Part I contains a variety of the most essential test methods used in geotechnical engineering to quantify the properties of a particular soil as well as providing exposure to a range of experimental techniques. The test methods in Part I are:

- Phase Relations
- Specific Gravity
- Maximum Density, Minimum Density

- Calcite Equivalent
- pH, Salinity
- Organic Content
- Grain Size Analysis
- Atterberg Limits
- Soil Classification and Description

These tests are generally referred to as index and physical property tests. These tests are performed in large numbers for most projects because they provide an economical method to quantify the spatial distribution of material types for the site investigation. The results of these tests are also useful in combination with empirical correlations to make first estimates of engineering properties.

LABORATORY SAFETY

Laboratories have numerous elements that can cause injury, even if an individual is merely present in a laboratory as opposed to actively engaged in testing. Some of the most significant dangers for a typical geotechnical laboratory are listed here. Significant applies to either most harmful or most common. Most of these dangers are entirely preventable with education, some preparation, and common sense. The most dangerous items are listed first. Unlimited supply means that once an event initiates, someone must intervene to stop it. Sometimes a person besides the afflicted individual has to step in, such as with electrocution. Limited supply means that once an event initiates it only occurs once, such as a mass falling from a bench onto someone's toe.

- Electricity (equipment, power supplies, transducers)—unlimited supply, no warning, could result in death. Observe appropriate electrical shut off and lock off procedures. Allow only professionals to perform electrical work. Dispose of equipment with damaged electrical cords rather than attempting to repair them. Do not expose electricity to water, and use Ground Fault Interrupters (GFIs) when working near water. Master proper grounding techniques.
- Fire (Bunsen burner, oven, electrical)—unlimited supply, some warning, could result in death, injury, and significant loss of property. Do not allow burnable objects or flammable liquids near Bunsen burners. Do not place flammable substances in the oven. Dispose of equipment with damaged electrical cords rather than attempting to repair them. Review evacuation procedures and post them in a visible, designated place in the laboratory.
- Chemical reaction (acids mixed with water, mercury, explosions)—large supply, little warning, could result in death or illness. Proper training and personal protection is essential when working with or around any chemicals in a laboratory. Procedures for storage, manipulation, mixing, and disposal must be addressed. Mercury was once used in laboratories (such as in thermometers and mercury pressure pots) but is slowly being replaced with other, less harmful methods.
- Blood (HIV, hepatitis)—contact could result in illness or death. Proper personal protection measures, such as gloves, are required, as well as preventing the other accidents described herein.
- Pressure (triaxial cells, containers under vacuum)—air can have a large supply, little warning, could result in significant injury. Open valves under pressure or vacuum carefully. Inspect containing devices for any defects, such as cracks, which will cause explosion or implosion at a smaller pressure than specified by the manufacturer.
- Power tools and machinery (motors, gears, circular saw, drill)—can have large supply if no safety shutoff, could result in significant injury and release of

blood. Proper procedures, protective gear, and attire are required, as well as common sense.

- Heat (oven racks, tares)—limited supply, could result in burn. Use protective gloves specifically designed for heat, as well as tongs to manipulate hot objects. Arrange procedures so that reaches are not required over or near an open flame.
- Sharp objects (razor blades, broken glass)—limited supply, dangers should be obvious, could result in injury and release of blood. Dispose of sharp objects using a sharps container.
- Mass (heavy pieces of equipment that fall)—limited supply but dangers can blend into background, could result in injury. Do not store heavy or breakable objects up high.
- Tripping, slipping, and falling hazards—limited supply but can blend into background, could result in injury. Do not stretch when trying to reach objects on shelves; instead, reposition to avoid overextension. Maintain a clear path in the laboratories. Put tools, equipment, and boxes away when finished. Clean up spills immediately and put up signage to indicate wet floors when necessary.
- Particulates (silica dust, cement dust)—unlimited supply, could result in serious long-term illness. Use dust masks when working with dry soils and cement. Note that other considerations may be required, such as ventilation.
- Noise (sieve shaker, compressor, compaction hammer)—unlimited supply, could result in damage in the long-term. Use ear protection when presence is absolutely required near a noisy object, such as a compressor. A better solution is to have this type of equipment enclosed in a sound barrier or placed in another designated room away from people. Note that other considerations may be required for the machinery, such as ventilation.

Laboratories require safety training to prevent accidents from happening, and to provide instruction on how to minimize damage should these events occur. Proper attire must be insisted upon. The laboratory must also provide safety equipment, such as eye protection; ear protection; latex, vinyl, or other gloves; and dust masks. A designated chair of authority is essential to facilitating an effective laboratory safety program.

Any person entering a laboratory must be made aware of the dangers lurking. In addition, it must be impressed upon persons working in the laboratory that organization and cleanliness are paramount to preventing unnecessary injuries.

TERMINOLOGY

Terminology is a source of confusion in any profession. Imprecise language can lead to misinterpretation and cause errors. Definitions of several very important material conditions terms follow, along with a discussion of appropriate and intended use. These terms are generally consistent with those found in the ASTM D653 Standard Terminology Relating to Soil, Rock, and Contained Fluids. ASTM International is discussed in the next section.

In situ

"In situ" describes rock or soil as it occurs in the ground. This applies to water content, density, stress, temperature, chemical composition, and all other conditions that comprise the importance characteristics of the material.

Sample versus Specimen

Throughout this text, soil will be discussed in terms of both samples and specimens. The two terms are frequently misused in practice. In reality, the two refer to different entities. A sample is a portion of material selected and obtained from the ground or other source by some specified process. Ideally, the sample is representative of the whole. A specimen is a subset of a sample and is the specific soil prepared for and used for a test. A specimen is generally manipulated or altered due to the test process.

Undisturbed versus Intact

"Undisturbed" is a very specific condition that signifies the in situ state of the soil. Literally taken, the adjective encompasses everything from temperature to stress to strain to chemistry. In concept, it can be used to describe samples or specimens, but as a practical matter it is impossible to remove material from the ground without causing some measurable disturbance. "Intact" is the preferred adjective to sample or specimen to signify that the material has been collected using state of the practice methods to preserve its in situ conditions commensurate with the testing to be performed. Describing material as "intact" acknowledges the fact that some disturbance has occurred during the sampling operation. This level of disturbance will depend on the method used to obtain the sample and the level of care used in the sampling operation.

Remolded versus Reconstituted

"Remolded" signifies modifying soil by shear distortion (such as kneading) to a limiting destructured condition without significantly changing the water content and density. A remolded sample is completely uniform and has no preferential particle structure. The mechanical properties at this limiting state are dependent on water content and void ratio. This is a terminal condition and from a practical perspective the completeness of remolding will depend on the method used to remold the material. "Reconstituted" describes soil that has been formed in the laboratory to prescribed conditions by a specified procedure. The fabric, uniformity, and properties of a reconstituted sample will depend on the method and specific details used to make the sample.

STANDARDIZATION

Commercial testing is not an arbitrary process. At the very least, each test method must have a specific procedure, defined characteristics of the equipment, and method of preparing the material. This is essential for a number of reasons. It provides consistency over time. It allows comparison of results from different materials. But most importantly, it allows others to perform the test with the expectation of obtaining similar results. There are many levels of formalization for this information. It may reside in an individual's laboratory notebook, be an informal document for a company laboratory, or be a formalized document available to the general public. Obviously, the level of effort, scrutiny, and value increase with the level of availability and formalization.

There are several standardization organizations, including the International Standardization Office (ISO), American Association of State Highway and Transportation Officials (AASHTO), British Standards (BS), and ASTM International (ASTM). The authors both do extensive volunteer work for ASTM and that experience is heavily represented in this book.

ASTM is a not-for-profit volunteer standardization organization, formerly known as American Society for Testing and Materials. ASTM documents are referred to as "standards" to accentuate the fact that they are products of the consensus balloting process. ASTM produces standard test methods, guides, practices, specifications, classifications, and terminology documents. The criteria for each of these terms as given in ASTM documentation (2008) is presented below:

> **standard**, *n*—as used in ASTM International, a document that has been developed and established within the consensus principles of the Society and that meets the approval requirements of ASTM procedures and regulations.
>
> DISCUSSION—The term "standard" serves in ASTM International as a nominative adjective in the title of documents, such as test methods or specifications, to connote specified consensus and approval. The various types of standard documents are based on the needs and usages as prescribed by the technical committees of the Society.
>
> **classification**, *n*—a systematic arrangement or division of materials, products, systems, or services into groups based on similar characteristics such as origin, composition, properties, or use.

guide, *n*—a compendium of information or series of options that does not recommend a specific course of action.

DISCUSSION—A guide increases the awareness of information and approaches in a given subject area.

practice, *n*—a definitive set of instructions for performing one or more specific operations that does not produce a test result.

DISCUSSION—Examples of practices include, but are not limited to, application, assessment, cleaning, collection, decontamination, inspection, installation, preparation, sampling, screening, and training.

specification, *n*—an explicit set of requirements to be satisfied by a material, product, system, or service.

DISCUSSION—Examples of specifications include, but are not limited to, requirements for physical, mechanical, or chemical properties, and safety, quality, or performance criteria. A specification identifies the test methods for determining whether each of the requirements is satisfied.

terminology standard, *n*—a document comprising definitions of terms; explanations of symbols, abbreviations, or acronyms.

test method, *n*—a definitive procedure that produces a test result.

DISCUSSION—Examples of test methods include, but are not limited to, identification, measurement, and evaluation of one or more qualities, characteristics, or properties. A precision and bias statement shall be reported at the end of a test method.

ASTM does not write the documents, but rather manages the development process and distribution of the resulting products. This is a very important distinction. The information contained in the document is generated by, and is approved by, the volunteer membership through a consensus process. It is essential to recognize that the very nature of the consensus process results in the standard establishing minimum requirements to perform the test method. An expert in the method will be able to make improvements to the method.

ASTM has over 200 Main Committees, including Steel, Concrete, and Soil and Rock. Main Committees are generally divided by technical interest but a particular profession may have interest in several committees. Each Main Committee is divided into subcommittees according to technical or administrative specialization.

ASTM has over 30,000 members, who are volunteers from practice, government, research, and academia. ASTM is an all-inclusive organization. ASTM has no particular membership qualification requirements and everyone with professional interest in a discipline is encouraged to join. Within each committee, there are specific requirements on the distribution of member types that have a vote as well as the restriction that each organization is limited to one vote. This is done so that manufacturers cannot sway the operation of the committees for financial gain.

Committee D18 is the Soil and Rock committee. It is divided into twenty technical subcommittees and seven administrative subcommittees. The committee meets twice per year for three days to conduct business in concurrent meetings of the subcommittees followed by a final Main Committee wrap-up.

ASTM mandates that every standard stays up to date. Each standard is reviewed every five years and placed on a subcommittee ballot. If any negative votes are cast and found persuasive by the subcommittee with jurisdiction, that negative vote must be accommodated. Comments must be considered as well, and if any technical changes are made to the document, it must be sent back to subcommittee ballot. Once the document makes it through subcommittee balloting without persuasive negatives and without technical changes, the item is put on a Main Committee ballot. Similarly, the document must proceed through the process at the Main Committee level without persuasive negatives or any required technical changes. The item is then published with any editorial changes resulting from the process. Any technical changes or persuasive negatives require that the item be sent back to subcommittee-level balloting. If successful ballot action has not been completed at both levels after seven years, the standard is removed from publication.

Each standard has a template format with required sections. This makes the tandard easy to use once familiar with the format but it also makes for uninteresting reading.

Standards are used extensively in all types of laboratory testing from very simple manual classification procedures to complicated engineering tests. In short, standardization provides a means of maintaining consistency of testing equipment and test methods across testing organizations. ASTM standards are the reference standards wherever possible in this textbook.

ASTM International publishes their standards in over seventy-five volumes. The volumes can be obtained individually or as various sets, and are published in three formats: print, compact disc, or online subscriptions. Libraries and organizations may have full sets of the ASTM volumes. Individual members are able to choose one volume a year as part of their membership fee. Annual membership dues are relatively small as compared to other professional organizations. Standards under the jurisdiction of D18 the Soil and Rock Committee are published in two volumes: 04.08 and 04.09.

ASTM also offers student memberships and has an educational program where professors can choose up to ten standards to use as part of their curriculum. This package is made available to students for a nominal fee. For more information, refer to ASTM's web site at www.astm.org. Navigate to the "ASTM Campus" area for student memberships, and educational products and programs.

EVALUATION OF TEST METHODS

How good is a test result? This is a very important question and one that has been very difficult to answer relative to testing geo-materials. Conventional wisdom holds that the natural variability of geo-materials is so large that any two results using the same method are "just as likely to be different because of material variability as due to the variation in performing the test." This line of thinking has had a serious negative impact on the advancement of quality testing. Within the last two decades there have been several attempts to improve the quality of testing. However, the cost of testing, the number of test methods, and the variability of geo-materials make this a difficult task.

Several terms are used to express the quality of a measurement such as accuracy, bias, precision, and uncertainty. ASTM has chosen to quantify the goodness of a test method in terms of two quantities: precision and bias. In fact, Precision and Bias is a mandatory section of every ASTM test method. Precision and bias are two separate measures that replace what one might typically consider "accuracy." Bias quantifies the difference between a measured quantity and the *true* value. Precision quantifies the scatter in measurements around an average value. Refer to Figure 1.1 for a schematic depiction of precision and bias.

Precision is especially useful in testing geo-materials because one can quantify the variability in measuring a rather arbitrary quantity. A good example of these concepts is the liquid limit test. The liquid limit is defined by the test method and is not

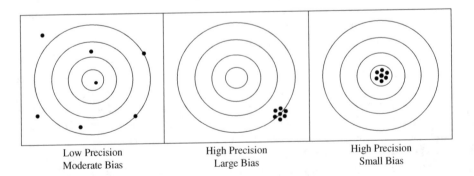

| Low Precision | High Precision | High Precision |
| Moderate Bias | Large Bias | Small Bias |

Figure 1.1 Schematic depiction of precision and bias. (Adapted from Germaine and Ladd, 1988).

an absolute quantity. Therefore, there can not be bias for this test result. On the other hand, we could run many tests and compute the standard deviation of the results. This would be a measure of the scatter in the test method or the precision.

The framework (or standard method) for determining the quality of an ASTM test method is prescribed by E691 Standard Practice for Conducting an Interlaboratory Study to Determine the Precision of a Test Method. ASTM E691 defines the process that must be followed to develop a numerical precision statement for a specific test method. The practice also specifies the minimum requirements for the process to be valid. At least six independent laboratories must return results of triplicate testing on a single material. In addition, the test program should include ruggedness testing. Since standard test methods are normally written as generally as possible, there will be a range of acceptable parameters that satisfy the method specification. The test program must include the range of conditions, procedures, and equipment allowed in the standard test method. Finally, the test program should include a range of soils. One can easily see the practical difficulty in performing an all-inclusive program.

Either a round robin testing program or an interlaboratory testing program can be used to obtain the necessary test results to develop the numerical precision statement. A round robin program uses one specimen which is sent around to the laboratories participating in the program. Each laboratory performs the three test measurements and then sends the specimen to the next laboratory. Round robin testing programs are appropriate for nondestructive test methods. The use of one specimen eliminates scatter associated with specimen variability.

If the testing alters or destroys the specimen, such as in most geotechnical testing, a round robin program would not be appropriate. An interlaboratory test program uses a uniform source material and distributes a different sample to each laboratory. The source material must be homogenized by blending and then pretested prior to distribution. The laboratory then prepares the test specimen and performs the three tests. This method is used most frequently in soils testing. Interlaboratory test programs add a component of variability due to the fact that each sample is unique.

An important component of variability in the test results arises from individual interpretation of the standard test method. For this reason, each laboratory participating in the program is reviewed by the team conducting the study to be confident that the testing is conducted in accordance with the method.

PRECISION AND BIAS STATEMENTS

Once the interlaboratory test program is complete and the results are returned, they are analyzed by the team conducting the study. The test documentation is first reviewed to be sure the assigned procedures were followed and the data set is complete.

Statistics are performed on the final data set to develop the repeatability and reproducibility statements for the test method. "Repeatability" is a measure of the variability of independent test results using the same method on identical specimens in the same laboratory by the same operator with the same equipment within short intervals of time. "Reproducibility" is a measure of the variability of independent test results using the same method on identical specimens, but in different laboratories, different operators, and different equipment.

Using basically the same terminology as E691, the statistics are calculated as follows. The average of the test results are calculated for each laboratory using Equation 1.1:

$$\bar{x}_j = \sum_{i=1}^{n} x_{i,j} / n \qquad (1.1)$$

Where:

$\bar{x}_j$ = the average of the test results for one laboratory
$x_{i,j}$ = the individual test results for one laboratory, j
n = the number of test results for one laboratory

The standard deviation is calculated using Equation 1.2:

$$s_j = \sqrt{\sum_{i=1}^{n}(x_{i,j} - \bar{x}_j)^2/(n-1)}$$ (1.2)

Where:

s_j = standard deviation of the test results for one laboratory

Both the average and the standard deviation calculations are those used in most calculators. However, since some will use "n" in the denominator of the standard deviation
calculation in place of "n−1," it must be verified that the calculator is using the correct
denominator shown above.

The results for each laboratory are then used to calculate the average and standard
deviation of the results for all laboratories. The average value for all laboratories is calculated using Equation 1.3:

$$\bar{\bar{x}} = \sum_{j=1}^{p}\bar{x}_j/p$$ (1.3)

Where:

$\bar{x}_j$ = the average of the test results for one material
p = the number of participating laboratories

The standard deviation of the average of the test results for one material is calculated using Equation 1.4:

$$s_{\bar{x}} = \sqrt{\sum_{j=1}^{p}(\bar{x}_j - \bar{\bar{x}})^2/(p-1)}$$ (1.4)

Where:

$s_{\bar{x}}$ = standard deviation of the average results of all participating laboratories

The repeatability standard deviation and the reproducibility standard deviation are
calculated as Equation 1.5 and Equation 1.6, respectively:

$$s_r = \sqrt{\sum_{j=1}^{p}s_j^2/p}$$ (1.5)

Where:

s_r = repeatability standard deviation

$$s_R = \sqrt{(s_{\bar{x}})^2 + (s_r)^2(n-1)/n}$$ (1.6)

Where:

s_R = reproducibility standard deviation (minimum value of s_r)

Finally, the 95 percent repeatability and reproducibility limits are calculated using
Equation 1.7 and Equation 1.8, respectively:

$$r = 2.8 \cdot s_r$$ (1.7)

$$R = 2.8 \cdot s_R$$ (1.8)

Where:

r = 95 percent repeatability limit
R = 95 percent reproducibility limit

E691 also provides for the removal of outlier results. These outliers are removed
from the data set prior to performing the final statistics to obtain the precision statement.

It is important to realize that even under the best of circumstances, occasionally a test result will simply be unacceptable.

The final results are referenced in the test method in the form of a precision statement. The details of the interlaboratory study and the results generated are archived by ASTM in the form of a research report.

Precision statements can be extremely useful. Assuming that the measurement errors are random, the precision values can be used to compare two individual measurements. There is a 95 percent probability that the two measurements will be within this range, provided the tests were performed properly. This is essentially the acceptable difference between the measurements. The precision values can be used to compare the results for different laboratories and can be used to evaluate the relative importance of measurements in a single test program.

Bias is defined in ASTM as the difference between the expected test results and a reference value. Bias applies to most manufactured products, but is not relevant for naturally occurring materials such as soil. Therefore, most of the standards in ASTM Soil and Rock Committee will not have numerical bias statements.

LABORATORY ACCREDITATION

Accreditation provides a means for assuring that laboratories meet minimum requirements for testing. There are many individual accreditation programs, each of which has different criteria, levels of inspection, frequency of visits by the accrediting body, proficiency testing requirements, and fees. Specific accreditation may be required by an organization to perform work for a client or to bid on a job. Many accreditation bodies exist that are required to work in certain geographic areas. Trends in the practice are such that eventually a centralized, international body may exist for accreditation. Two nationally recognized accreditation programs are described in this section; however, there are numerous others.

ASTM International does not provide accreditation. It does, however, have a standard titled D3740 Standard Practice for Minimum Requirements for Agencies Engaged in the Testing and/or Inspection of Soil and Rock as Used in Engineering Design and Construction. The document provides guidance on the basic technical requirements for performing geotechnical testing including record keeping, training, and staff positions. Other agencies that do provide accreditation are described below.

American Association of State Highway and Transportation Officials

The American Association of State Highway and Transportation Officials (AASHTO) operates an accreditation program. The program has several requirements ranging from paying application and site assessment fees, to developing a quality management system that meets the requirements in the AASHTO R18 manual, to an on-site assessment where the AASHTO inspector observes technicians performing tests, and enrollment in the appropriate proficiency testing program. On-site assessments are performed every eighteen to twenty-four months, and must be completed to maintain accreditation.

AASHTO accreditation establishes the ability to run certain tests. The laboratory will receive an AASHTO accreditation certificate listing the specific tests for which it is accredited. In addition, AASHTO accreditation allows the laboratory to choose to be accredited for the AASHTO or ASTM version of a particular test method, or both. AASHTO requires enrollment in their proficiency testing program. The soils proficiency program is managed by the Material Reference Laboratory (AMRL), while for concrete products, the program is run by the Cement and Concrete Reference Laboratory (CCRL).

American Association for Laboratory Accreditation

American Association for Laboratory Accreditation (A2LA) works in a manner very similar to AASHTO with a few exceptions. There is no on-site assessment for A2LA accreditation. The guidance document for the certification is International Organization for Standardization (ISO) 17025 General Requirements for the Competence of Testing and Calibration Laboratories. Finally, the proficiency testing program is not operated by A2LA, but rather the laboratory must choose from an approved list of accredited proficiency testing providers.

PROFICIENCY TESTING

Proficiency testing is a useful tool to evaluate lab procedures, as well as being required as part of some laboratory accreditation programs. Proficiency programs are conducted by an agency that sends out uniform, controlled materials to the participating laboratories at a specified, regular frequency.

Individual details of the proficiency programs vary according to the material of interest and the requirements of the accreditation program. In most cases, the labs perform the required tests and return the results to the managing agency within a specified timeframe. The results of all the participating laboratories are compiled, and the participating laboratories are sent the overall results along with information on where their laboratory fell within the results. Laboratories with outlier results must respond with a report outlining the cause of their poor results. Soils proficiency samples are sent out at a regular frequency.

Laboratories can purchase samples of the reference soils used for the interlaboratory study (ILS) conducted by the ASTM Reference Soils and Testing Program on several test methods. Five-gallon buckets of sand, lean clay, fat clay, and silt can be purchased from Durham Geo Enterprises (Durham Geo web site, 2008). These samples were produced for uniformity testing in the ASTM ILS and are an invaluable resource for teaching students, as well as qualifying technicians in commercial laboratories. The bucket samples come with the summary information and testing results used to develop precision statements for six ASTM test methods. The poorly graded sand bucket samples include the summary analysis sheets for D854 (Specific Gravity), D1140 (Percent Finer than the No. 200 Sieve), D4253 (Maximum Index Density), and D4254 (Minimum Index Density). The silt, lean clay, and fat clay bucket samples include the summary analysis sheets for D854, D1140, D698 (Standard Effort Compaction), and D4318 (Liquid Limit, Plastic Limit, and Plasticity Index).

TECHNICIAN CERTIFICATION

Various regions and agencies have technician certification programs for laboratory and field technicians, as well as a combination of both. The concrete industry has a certification program managed by the ACI (American Concrete Institute).

One national technician certification program that includes soil technicians is National Institute for Certification in Engineering Technologies (NICET). The NICET program was developed by the National Society of Professional Engineers. There are four levels of certification corresponding to levels of skill and responsibility. The individual applies to take a written exam, and if a passing grade is achieved, the individual is given a NICET certification for that level.

UNIT CONVENTION

Hopefully, it is not surprising to find an introductory section focused on the selection and application of units. From a purely academic perspective this is a rather boring topic, but consistency in units has enormous implications for the application of calculations to practice. One of the most public unit-caused mistakes resulted in the Mars Climate Orbiter being lost in space in 1999 (Mishap Investigation Board, 1999). The message is clear: always state the units you are working with, and be sure to use the correct unit conversions in all your calculations.

There are many different systems of units in use around the world and it appears that the United States uses them all. You will find different measures for stress depending on company, region, and country. This is not inherently wrong, but does require more care in documentation of test results.

One should develop good habits relative to calculations and documentation of unit specific information. All equations, tables, and graphs should be properly labeled with the designated units. Conversions between various units will always be necessary. Conversion constants should be carried to at least two more significant digits than the associated measurement. Appendix A contains conversion constants for commonly used parameters in geotechnical practice. A far more general list of conversions can be found online or in various textbooks, such as the *CRC Handbook of Chemistry and Physics* (Lide, 2008).

The choice of units for a specific project can be a difficult decision. Two absolute rules must be followed. While in the laboratory, one must use the local units of measure to record data. This is an absolute rule even if it results in working with mixed units while in the laboratory. Never make an observation (say in inches), convert to another unit (inches to cm), and then record the result (cm) on a data sheet. This practice encourages confusion, invites round-off errors, and causes outright mistakes. The second rule is always to provide final results (tables, graphs, example calculations, and the like) in the client's units of choice. This is because individuals (the client in this case) develop a sense of comfort (or a feel) with one particular set of units. It is generally good practice to make use of this "engineering judgment" For quality control. As a result, it is common practice to post-process the data from the "lab" units to the "client" units as the last step in the testing process.

A commonly used collection of measurement units comprises a system. Every system has a set of base units and a series of derived units. There are many systems and even variations of systems, leading to a laundry list of terms. The two systems most commonly used in engineering practice today are the SI system and the British system. For the SI system (and limiting attention to geotechnical practice), the base units are meters, kilograms, and seconds. Unfortunately there are two British systems, the absolute and the gravitational. The British Absolute system is based on the foot, pound mass (lbm), and second. The British gravitational system (also called the U.S. Customary System) is based on the foot, slug, and second. All of these systems make use of a unique and consistent collection of terminology.

Past engineering practice has caused problems relative to the specification of force and mass when working with the British systems. Force is a derived unit ($F = ma$). In the absolute system, force is reported in poundals. In the gravitational system, the unit of force is a pound. The situation is exacerbated by the fact that at standard gravity, 1 lbm results in a force of roughly 32 poundals and a mass of 1 slug generates a force of roughly 32 lbf. Since there are about 32 lbm in one slug, it is understandable how pound became interchangeable for mass and force. Making matters even worse, the same casual reference was applied to the kilogram.

In the laboratory, the mass is obtained, not the weight. Weight is a force. In this text, the SI system is used wherever practical. The system is clean, easy to use, and avoids most of the confusion between mass and force.

In geotechnical practice, compression is positive and extension is negative, unless indicated otherwise. This is contrary to the practice in structural engineering.

SIGNIFICANT DIGITS

It is important to report measurements and calculated results to the appropriate significant digit. The individual performing the test calculations is normally in the best position to make the decision as to how many significant digits are appropriate to report for a particular measurement. Reporting too many digits is poor practice because it misleads the user of the results by conveying a false sense of accuracy. On the other hand, at times it can be a challenge to determine the appropriate number of digits to report. In geotechnical testing, five factors must be considered when determining the least significant digit of a number: the mathematical operation, the rules of rounding, the resolution of the measurement, the size of the specimen, and in some cases, the practice associated with the test method.

Determination of the number of significant digits in the result of a specific calculation depends on the mathematical operation. There are several variations on the best practice, and the degree of precision depends on the operation. For addition and subtraction, the final result is reported to the position of the least precise number in the calculation. For multiplication and division, the final result is reported to the same number of significant digits as in the least significant input. Other operations, such as exponentials, logarithms, and trigonometry functions need to be evaluated individually but can be conservatively assumed to be the same as the input. Intermediate calculations are performed using one additional significant digit. Constants can contain two more significant digits than the least significant measurement to be sure the constant does not control the precision of the calculation.

It will often be necessary to round off a calculation to the appropriate significant digit. The most common rules for rounding are to round up if the next digit to the right is above 5 and to round down if the digit to the right is below 5. Uncertainty arises when dealing with situations when the digit to the right is exactly 5. Calculators will round numbers up in this situation, which introduces a systematic bias to all calculations. The more appropriate rule is to round up if the digit to the left of the 5 is odd, and round down if it is even.

The resolution of a measuring device sets one limit on significant digits. When using electronic devices (e.g., a digital scale), the resolution is automatically set as the smallest increment of the display. When using manual devices, the situation is less clear. A pressure gage will have numbered calibration markings and smaller "minor" unnumbered tick marks. The minor tick marks are clearly considered significant numbers. It is often necessary to estimate readings between these minor tick marks. This measurement is an estimate and can be made to the nearest half, fifth, or tenth of a division, depending on the particular device. This estimate is generally recorded as a superscript and should be used with caution in the calculations.

The specimen size also contributes to the significant digit consideration. This is simply a matter of keeping with the calculation rules mentioned in the previous paragraphs. It is an important consideration when working in the laboratory. The size of the specimen and the resolution of the measuring device are both used to determine the significant digits of the result. While this may seem unfair, all other factors being equal, there is a loss of one significant digit in the reported water content if the dry mass of a specimen drops from 100.0 g to 99.9 g. Being aware of such factors can be important when comparing data from different programs.

The final consideration comes for the standard test method. In geotechnical practice, some of the results have prescribed reporting resolutions, independent of the calculations. For example, the Atterberg Limits are reported to the nearest whole number. This seemingly arbitrary rule considers the natural variability of soils as well as application of the result. ASTM D6026 Standard Practice for Using Significant Digits in Geotechnical Data provides a summary of reporting expectations for a number of test methods.

TEST SPECIFICATION

Individual test specification is part of the larger task of a site characterization program. Developing such a program is an advanced skill. Mastering the knowledge required to test the soil is a first step, which this textbook will help to accomplish. However, eventually a geotechnical engineer must specify individual tests in the context of the project as a whole. Designing a site characterization and testing program while balancing project needs, budget, and schedule is a task requiring skill and knowledge. A paper titled "Recommended Practice for Soft Ground Site Characterization: Arthur Casagrande Lecture" written by Charles C. Ladd and Don J. DeGroot (2003, rev. 2004) is an excellent resource providing information and recommendations for testing programs. Analysis-specific testing recommendations are also provided in this paper. Although this paper specifically addresses cohesive soils, many of the principles of planning are similar for granular soils.

There are two general, complementary categories of soil characteristics: index properties and engineering properties.

Index tests are typically less expensive, quick, easy to run, and provide a general indication of behavior. The value of index properties is many-fold: index properties can define an area of interest, delineate significant strata, indicate problem soils where further investigation is needed, and estimate material variability. They can be used to approximate engineering properties using more or less empirical correlations. There is a tremendous amount of data in the literature to establish correlations and trends. The most common index tests are covered in the first part of this book, such as water content, particle size distribution, Atterberg Limits, soil classification, and so on.

Engineering properties, on the other hand, provide numbers for analysis. These tests generally simulate specific boundary conditions, cost more, and take longer to perform. They typically require more sophisticated equipment, and the scale of error is equipment dependent. Engineering testing includes strength, compressibility, hydraulic conductivity, and damping and fatigue behavior, among others. The compaction characteristics of a material fall into an odd category. Compaction is not an index property, nor does it provide numbers for an analysis. However, determining the level of compaction is used as an extremely important quality-control measure.

A properly engineered site characterization program must achieve a balance of index and engineering properties testing. More index tests are usually assigned to characterize the materials at a site. The results are then used to select a typical material or critical condition. These materials or locations are then targeted for detailed engineering testing.

Once a program has been established, individual tests are assigned on specific samples. To avoid a waste of time, resources, and budget, the tests must be consistent with the project objectives, whether that is characterization, determining engineering properties, or a combination of both. Test specification should be done by the project engineer or someone familiar with the project objectives and the technical capabilities of the laboratory. In addition to general test specification, details including, but not limited to, sample location, specimen preparation criteria, stress level, and loading schedule, may need to be provided, depending on test type.

The testing program can not be so rigid as to prevent changes as new information unfolds during the investigation. Rarely does a test program run on "autopilot." The results must be evaluated as they become available, and rational changes to the program made based on the new findings. As experience develops, the radical changes in a testing program will not occur as often.

SAMPLING

Field sampling methods can have a significant impact on the scope of a testing program as well as on the quality of the final results of laboratory testing. The sampling methods to be used for a site investigation must be aligned with the type of soils to be sampled, the field conditions, and the quality of specimen needed for the specific tests. Sampling technology is an extensive topic and beyond the scope of this textbook. A brief discussion of some of the most important (and often overlooked) aspects of sampling is included in this section and in Chapter 11, "Background Information for Part II." The reader is referred to other literature (such as the U.S. Army Corps of Engineers manual *Geotechnical Investigations: EM 1110-1-1804*) for further information on sampling methods.

Field sampling can be divided into two general categories: disturbed methods and intact methods. As the name implies, disturbed methods are used to collect a quantity of material without particular concern for the condition of the material. Sometimes preservation of the water content is important but the primary concern is to collect a representative sample of the soil found in the field. Intact methods are designed to collect a quantity of material and, at the same time, preserve the in situ conditions to the extent practical. Changes to the in situ conditions (disturbance) will always happen. The magnitude of the disturbance depends on soil condition, sampling method, and expertise.

Intact sampling normally recovers much less material, requires more time, and more specialized sampling tools. When working with intact samples, it is always important to preserve the water content, to limit exposure to vibrations, and to limit the temperature variations. When maintaining moisture is a priority, the samples must be properly sealed immediately upon collection and stored on site at reasonable temperatures. Intact samples should be transported in containers with vibration isolation and under reasonable temperature control. ASTM D4220 Standard Practices for Preserving and Transporting Soil Samples provides a very good description of the technical requirements when working with either intact or disturbed samples.

Disturbed Sampling

A test pit is an excavated hole in the ground. A very shallow test pit can be excavated by hand with a shovel. A backhoe bucket is normally used, however, which has an upper limit of about 8 to 10 m achievable depth, depending on the design of the backhoe. Soil is removed and set aside while the exposed subsurface information (soil strata, saturated interface, buried structures) is recorded, photos taken, and samples obtained from target strata. Usually, grab samples are collected at representative locations and preserved in glass jars, plastic or burlap bags, or plastic buckets. Each sample container must be labeled with project, date, initials, exploration number, depth, and target strata at a minimum. At the completion of these activities, the test pit is backfilled using the backhoe.

Disturbed sampling is very common when evaluating materials for various post-processing operations. Typical examples are borrow pit deposits being using for roadway construction, drainage culverts, sand and aggregate for concrete production, mining operations, and a myriad of industrial applications. Grab samples are generally collected in plastic buckets or even small truckloads. The sampling focus is to collect representative materials with little concern for in situ conditions.

Auger sampling is accomplished by rotating an auger into the ground. Hand augers can be used for shallow soundings (up to about 3 m). Augers attached to a drilling rig can be used up to about 30 meters. Soil is rotated back up to the surface as the auger is rotated to advance the hole. This sampling technique gives only a rough correlation of strata with depth and returns homogenized samples to the surface. Since layers are mixed together, the method has limited suitability for determining stratigraphy. In addition, the larger particles may be pushed aside by the auger rather than traveling up the flights to the surface. The location of the water table can also be approximated with auger methods. Samples are normally much smaller due to the limited access and are stored in glass jars or plastic bags. A typical sample might be 1 to 2 kg.

Split spoon sampling involves attaching the sampler to a drill string (hollow steel rods) and driving the assembly into the ground. This is done intermittently at the bottom of a boring, which is created by augering or wash boring. Split spoon sampling is usually combined with the standard penetration test (SPT) (ASTM D1586 Standard Test Method for Penetration Test and Split-Barrel Sampling of Soils) where a specified mass (63.5 kg [140 lb]) is dropped a standard distance (0.76 m [30 in.]) and the number of drops (blows) is recorded for 6 inches of penetration. The blow counts provide a measure of material consistency in addition to providing a disturbed sample for examination. The sampler is driven a total of 24 inches. The middle two number of blows (number of blows to drive the split spoon sampler 12 inches) are added to give the N-value. Numerous correlations between N-value and soil properties exist. The SPT test and split spoon sample combined provide a valuable profiling tool as well as providing material for classification and index tests. The small inside diameter of the split spoon sampler automatically limits the maximum collectable particle size. Split spoon samples are typically placed in a jar (usually referred to as jar samples) and labeled with project name or number, exploration number, sample number, initials, and date at a minimum. Sometimes other information, such as blow counts and group symbol, are included as well.

Sampling Method	Samples per day	Coverage	Sample Size
Hand excavation	8 to 10	1 m depth 5 to 10 m spacing	Up to 5 gallon bucket
Test pit	10 to 15	10 m depth 5 to 10 m spacing	Depends on max particle
Borrow pit	10 to 15	1 m depth 5 to 10 m spacing	Depends on max particle
Auger returns	20+	Up to 1.5 m intervals 40 m depth	Up to 2 kg
Split spoon	20+	Up to 1.5 m intervals 40 m depth	Less than 1 kg

Disturbed methods are useful for profiling the deposit, approximately locating the water table and obtaining samples for measuring physical properties and classification of soils. The borehole methods can also be used to advance the hole for in situ tests, observation wells, intact samples, or for installing monitoring instrumentation. The sampling operations are typically fast and relatively cheap. Disturbed methods are especially useful when combined with interspersed intact sampling. Table 1.1 provides an overview of the attributes of the various disturbed sampling methods.

Intact Sampling

Intact samples can be collected near the ground surface or exposed face of an excavation using hand techniques and are referred to as block samples. More commonly, intact samples are collected from boreholes using a variety of specialized sampling tools. Sampling is generally limited to soils that are classified as fine-grained soils with a small maximum particle size. If the deposit contains a few randomly located particles, the maximum size can be nearly as large as the sampler. When the large particles are more persistent, sample quality will suffer as the maximum size approaches 4.75 mm in diameter (No. 4 sieve).

Intact samples are collected to observe in situ layering and to supply material for engineering tests. Characterization and index tests can be performed on intact samples, but the added cost and effort required to collect intact samples are typically only justified when performing engineering tests as well. There are specific techniques involved in controlling the intact sampling operation to preserve these properties. These sampling details, along with processing of intact samples, are addressed in Chapter 11, "Background Information for Part II."

PROCESSING BULK MATERIAL

Bulk material is considered any sample that arrives at the laboratory as a disturbed sample or portions of intact samples that will be used for index testing. Disturbed samples are normally in loose form and transported by dump truck, 5-gallon bucket, and gallon-size sealable bags. A laboratory usually receives a much larger amount of material than needed for the specified tests. Even if just enough soil is received, it may need to be manipulated so multiple tests can be run on matching samples. Furthermore, many tests have limiting specifications and require specific processing of a fraction of the sample. As a result, materials must be processed prior to testing.

Three generic processing methods are available to manipulate the material. They are blending, splitting, and separating. Each has well-defined objectives and can be performed using a variety of techniques and devices.

Independent of the method used to process the bulk sample, consideration must be given to the quantity of material required to maintain a representative sample. This topic is discussed in detail in Chapter 8, "Grain Size Analysis." One possible criterion is to consider the impact of removing the largest particle from the sample. If the goal was to limit the impact to less than 1 percent, the minimum sample size would be 100 times the mass of the largest particle. Using this criterion leads to the values presented in Table 1.2.

Blending

It is very common for bulk samples to segregate during transport. Vibration is a very effective technique to separate particles by size. Blending is the process of making a sample homogeneous by mixing in a controlled manner. This can be done through hand mixing, V-blenders, tumble mixers, and the like. Fine-grained materials will not segregate during mixing. Blending fine-grained soils is easily performed on dry material (with proper dust control), or on wet materials. When mixing coarse-grained materials, separation of sizes is a significant problem. The best approach is to process the materials when moist (i.e., at a moisture content between 2 and 5 percent). The water provides surface tension, giving the fine particles adhesive forces to stick to the larger particles.

Blending is relatively easy when working with small quantities. Hand mixing can be done on a glass plate with a spatula or even on the floor with a shovel. For large quantities, based on the largest quantity that fits in a mixer, the material must be mixed in portions and in sequential blending operations. Figure 1.2 provides a schematic of this operation for a sample that is four times larger than the available blender. The material is first divided (it does not matter how carefully) into 4 portions labeled 1, 2, 3, and 4. Each of these portions is blended using the appropriate process. Each blended portion is then carefully split into equal quarters labeled a, b, c, and d. The four "a" portions are then combined together and blended in a second operation. Each of the four second blends will now be uniform and equal. Provided the requirements of Table 1.2 are met, and the split following the first blend provides an equal amount to each and every portion for the second blend (and particle size limitations are not violated), the final product will be uniform. The same process can be expanded to much larger samples.

Splitting

Splitting is the process of reducing the sample size while maintaining uniformity. Simply grabbing a sample from the top of a pile or bucket is unlikely to be representative of the whole sample. Random subsampling is difficult to do properly. Each subsample should be much larger than the maximum particle size and the sample should contain at least ten subsamples. Quartering, on the other hand, is a systematic splitting process. It can be performed on both dry and moist materials of virtually any size. Each quartering operation reduces the sample mass by one half. Figure 1.3 provides a schematic of the sequential quartering operation. The material is placed in a pile using reasonable care to maintain uniformity. The pile is split in half and the two portions spread apart. The portions are then split in half in the opposite direction and spread apart. Finally, portions 1

Table 1.2 Minimum required dry sample mass given the largest particle size to maintain uniformity (for 1 percent or 0.1 percent resolution of results).

Largest Particle		Particle Mass	Dry Mass of Sample	
(mm)	(inches)	(Gs = 2.7)	For 1%	For 0.1%
9.5	3/8	1.2 g	120 g	1,200 g
19.1	¾	9.8 g	1,000 g	10 kg
25	1	23 g	2,500 g	25 kg
50	2	186 g	20 kg	200 kg
76	3	625 g	65 kg	650 kg
152	6	5,000 g	500 kg	5,000 kg

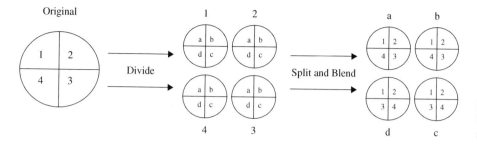

Figure 1.2 Schematic of the blending procedure for large samples.

and 3 (and 2 plus 4) are combined to provide a representative half of the original sample. This method can be repeated over and over to sequentially reduce a sample to the required size. For small samples, the process can be performed on a glass plate using a straight edge. For large samples, use a splitting cloth and shovel.

Another method of splitting a sample is by use of the riffle box. The riffle box also cuts the sample quantity in half for each run through the method. Material is placed in the top of the riffle box and half the material falls to one side of the box on slides, while the second half falls to the opposite side. Containers are supplied with the box to receive the material. Care must be exercised to distribute the material across the top of the box. The sample must be dry or else material will stick to the shoots. The riffle box should only be used with clean, coarse-grained materials. Fines will cause a severe dust problem and will be systematically removed from coarse-grained samples. Figure 1.4 shows the riffle box.

Separating

Separation is the process of dividing the material (usually in two parts) based on specific criteria. For our purposes, the criterion is usually based on particle size, but it could be iron content or specific gravity, as in waste processing, or shape, or hardness. To separate by particle size, a sieve that meets the size criterion is selected, and the sample is passed through the sieve. This yields a coarser fraction and a finer fraction. Sometimes multiple sieves are used in order to isolate a specific size range, such as particles smaller than 25 mm, but greater than 2 mm.

TEST DOCUMENTATION

This is a simple, commonsense topic, but its importance is often overlooked. The only tie between the physical material being tested and the results submitted to a client is the information placed on the data sheets at the time of the test. Data sheets must be filled out accurately and completely with sample and specimen specific information, as well as test station location, initials, and date.

A carefully thought-out data sheet assists with making sure the necessary information is collected and recorded every time. Training on why, where, and when information is required is essential to preventing mistakes. Note that recording superfluous information is costly and can add to confusion. Normally, geotechnical testing is "destructive," meaning that once the specimen is tested, it is generally unsuitable for retesting. It is, however, good practice to archive specimens at least through the completion of a project.

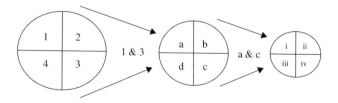

Figure 1.3 Schematic of the sequential quartering procedure.

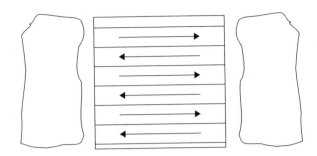

Figure 1.4 Schematic of the riffle box for use with coarse-grained materials.

SPREADSHEETS

Commercially available spreadsheet programs (such as Microsoft's Excel©) can be used to develop a framework for data reduction and results presentation. The ancillary web site for this textbook, www.wiley.com/college/germaine, provides an example electronic data sheet, a raw data set, and an example of what the results should look like for that raw data set. The online component of this textbook allows the instructor to have access to the spreadsheet with the formulas; however, it does not allow the student to have access. The reason for this is simple: if the data reduction is provided as a "canned program" the student simply does not learn how to analyze the data. Providing an example of what the results should look like given raw input data allows the student to write the formulas themselves, with assurance that they have developed them correctly if their results match the example set. This method also allows a certain measure of quality control in that a student can usually spot errors in spreadsheet formulas if the results do not match the example.

Numerous data reduction and results presentation software packages are available. Some provide a convenient tool for processing a constant stream of data in a well thought-out and accurate manner. Others are black boxes that do not explain the assumptions and approximations that underlie the output of the programs. Still others have a good, solid framework, but the format of the output cannot be modified for individual facility needs.

Whether using a commercially available data reduction package or an individualized spreadsheet, the user must have a working knowledge of the analysis and applications to various situations. Stated another way, the user cannot simply approach the software as a black box, but rather must understand the workings of the programs. At the very least, the results should be checked by hand calculations.

In all cases, a reliable quality-control (QC) system must be in place. The QC manual provides some of the most common measures to provide quality control. Many QC techniques involve project-specific knowledge or awareness of the laboratory performing the work, such as how samples flow through the lab, to detect and resolve a problem with the testing.

REPORTING TEST RESULTS

The primary responsibility of the laboratory is to perform the test, make the observations, and properly report the factual information to the requesting agency. The laboratory report must include information about the material tested including the project, a description of the material, and the conditions in which the material was delivered to the laboratory.

The report must also include the test information including the name of the test method and revision number, deviations from the published protocol when applicable, and the method used to process the material before testing. It must include laboratory factual information such as the specific device, the person in charge of the test, and the date of testing. Finally, the report provides the test results after performing the appropriate calculations.

Proper reporting should include tabulated and graphical test results, as well as a statement of procedures. The results must be reported to the appropriate resolution for the individual test and should not include engineering interpretation. Engineering interpretation requires the test measurements to be integrated with the context of the application and is the responsibility of the engineer of record. For example, interpretation of a friction angle from test data requires experience and project-specific application only available to the engineer.

Buried behind the report are specimen size requirements, test limitations, procedural deviations, and rules of significant digits. This level of detail is lost by the time the results are reported to the client. It becomes the professional responsibility of the laboratory to take these issues into account when conducting laboratory testing. The end result of testing is best described as a factual (and hopefully objective) laboratory data report.

A laboratory will usually archive data sheets, electronic files, and a summary of calculation methodology within the laboratory for a certain period of time. This information is usually available for a number of years after completion of the project; however, individual companies have their own policies regarding record retention.

TYPICAL VALUES

Each testing chapter of this text has a section titled "Typical Values." This section is included to provide the reader with a sense of magnitude and range in numerical values expected for each test. Some of these values have been obtained from the literature, while others are from unpublished personal consulting or research records. These values are *not* intended to provide numbers for analysis. Properties of soils can vary significantly through a depth profile, across a site, and among geographical locations as well as with specific testing conditions. The typical values provided should be used as a ballpark comparison with the testing results obtained using the procedures described in the associated chapter.

FURTHER READING AND OTHER REFERENCES

Since this textbook is meant to accompany an undergraduate course, the focus is on presenting the information necessary to perform certain tests, as well as some supporting background information to understand the important factors influencing the results. There are other valuable resources available on the topic of testing.

The three-volume series written by K. H. Head, titled *Manual of Soil Laboratory Testing*, has been published with several revised editions for each volume. The three editions are *Volume 1: Soil Classification and Compaction Tests*, *Volume 2: Permeability, Shear Strength and Compressibility Tests*, and *Volume 3: Effective Stress Tests*. The texts cover most of the same tests discussed in this text, but in much more detail as would be used for a reference by those performing the tests for commercial purposes on a daily basis.

The textbook by T. W. Lambe, *Soil Testing for Engineers*, covers many of the topics of this book. The textbook was intended for use for teaching the subject to students, although numerous engineers carried this reference with them into practice and still have the book on their bookshelves. However, the book was published in 1951, was never updated, and is out of print. Engineering libraries and practicing engineers may have a copy of this valuable resource.

The Naval Facilities Engineering Command (NAVFAC) produces technical documents formerly referred to as "Design Manuals." The three NAVFAC manuals most commonly used in geotechnical work were DM 7.01 (Soil Mechanics), DM 7.02 (Foundations and Earth Structures), and DM 7.3 (Soil Dynamics, Deep Stabilization, and Special Geotechnical Construction). These manuals provide an array of useful information and design procedures, while the sSoil mMechanics volume contains the information relative to geotechnical laboratory testing. The design manuals can be found in numerous places online for free download; however, NAVFAC has revamped their technical document systems. The NAVFAC design documents are now called "Unified Facilities Criteria" or UFC. Refer to NAVFAC's web site and navigate to the "Document Library" for free download of their documents. The geotechnical publications can be found by going to the "Technical" section of the "Document Library," then selecting "Unified Facilities Criteria," "UFC Technical Publications," and finding the list titled "Series 3-200: Civil/Geotechnical/Landscape Architecture."

The U.S. Army Corps of Engineers produces a manual titled *Laboratory Soils Testing,* which can be obtained through their web site at www.usace.army.mil/publications/eng-manuals (Corps of Engineers, 1986). This manual provides a large amount of useful (even if somewhat dated) information on laboratory testing and equipment. The web site has numerous other manuals available for free download as well.

The U.S. Bureau of Reclamation produces a document titled *Earth Manual,* which can be obtained through their web site at http://www.usbr.gov/pmts/writing/earth/earth.pdf (U.S. Bureau of Reclamation, 1998). This manual covers methods of testing, exploration, and construction control.

Numerous soil mechanics textbooks exist. Typically, the textbook used to teach the topic originally is the one used most frequently. For the authors of this text, that book is *Soil Mechanics* by T. W. Lambe and R. V. Whitman. This text is referenced in numerous places in this book.

ASTM International produces standards that include procedures for testing. The ASTM International web site (www.astm.org) allows anyone to browse ASTM standards and view the scope of any standard. The standards can be purchased through ASTM International or accessed in engineering libraries. Engineering schools and companies likely have online access accounts for standards. Individual members pay a rather small annual membership fee and obtain one volume a year, in print, on CD, or online.

The American Association of State Highway and Transportation Officials (AASHTO) produces their own testing and sampling methods and material specifications in the book *Standard Specifications for Transportation and Methods of Sampling and Testing.* Usually, the testing methods are consistent with those produced by ASTM International. The book can be purchased through AASHTO or accessed in engineering libraries.

The Massachusetts Institute of Technology (MIT) has converted many MIT theses to digital form. Although not all theses are available digitally, those that are can be downloaded in .pdf form and viewed by anyone inside or outside of MIT. Only MIT can download the forms that are able to be printed to hardcopy, however. To browse or obtain theses in this way, go to MIT's web site, then navigate to the "Research/Libraries" page, click on "Search Our Collections" and find the entry in the list titled "- theses written by MIT students, electronic" and click on "MIT Theses in DSpace". Alternatively, this can be accessed directly (at least at the time of this writing) at http://dspace.mit.edu/.

Other books and journal articles are referred to as appropriate throughout this text.

REFERENCES

ASTM International. March 2008. *Form and Style for ASTM Standards,* Philadelphia, PA. Web site: http://www.astm.org/COMMIT/Blue_Book.pdf

ASTM International. *Annual Book of Standards,* Volumes 04.08, 04.09, and 14.02. Philadelphia, PA. Web site: www.astm.org

American Association of State Highway and Transportation Officials. *Standard Specifications for Transportation and Methods of Sampling and Testing,* Washington, DC.

Corps of Engineers. 2001. *Geotechnical Investigations,* EM-1110-1-1804, Washington, DC.

Corps of Engineers. 1986. *Laboratory Soils Testing,* EM-1110-2-1906, Washington, DC. Web site: www.usace.army.mil/publications/eng-manuals/em1110-2-1906/entire.pdf.

Durham, Geo. 2008. http://www.durhamgeo.com/testing/soils/ref_soils.html.

Germaine, John T., and Charles C. Ladd. 1988. "State-of-the-Art Paper: Triaxial Testing of Saturated Cohesive Soils." *Advanced Triaxial Testing of Soil and Rock,* ASTM STP 977, Robert T. Donaghe, Ronald C. Chaney, and Marshall L. Silver, Eds., American Society for Testing and Materials, Philadelphia, PA, pp. 421–459.

Head, K. H. *Manual of Soil Laboratory Testing,* John Wiley and Sons, New York. Volume 1: Soil Classification and Compaction Tests; Volume 2: Permeability, Shear Strength and Compressibility Tests; Volume 3: Effective Stress Tests.

Ladd C. C. and D. J. DeGroot, 2003, revised May 9, 2004. "Recommended Practice for Soft Ground Site Characterization: Arthur Casagrande Lecture." Proceedings of the 12th *Panamerican Conference on Soil Mechanics and Geotechnical Engineering,* Cambridge, MA.

Lambe, T. W. 1951. *Soil Testing for Engineers*, John Wiley and Sons, New York.

Lambe, T. W. and R. V. Whitman. 1969. *Soil Mechanics*, John Wiley and Sons, New York.

Lide, David R. (ed.), 2008. *CRC Handbook of Chemistry and Physics*, CRC Press, Boca Raton, FL.

Massachusetts Institute of Technology (MIT). General web site: www.mit.edu; Digital access to theses: dspace.mit.edu.

Mishap Investigation Board. November 10, 1999. *Mars Climate Orbiter, Phase I Report.* ftp://ftp.hq.nasa.gov/pub/pao/reports/1999/MCO_report.pdf.

Naval Facilities Engineering Command (NAVFAC). Web site: www.navfac.navy.mil.

U.S. Bureau of Reclamation, 1998. *Earth Manual: Part 1*, Denver, CO. Web site: www.usbr.gov/pmts/writing/earth/earth.pdf.

Web site: www.mit.edu.

Chapter 2

Phase Relationships

SCOPE AND SUMMARY

This chapter provides information on measuring the phase relationships of a specimen of soil. Phase relationships generally include water content, density, void ratio (or porosity), and saturation. The methods are different for coarse- and fine-grained materials. Procedures are presented for making measurements on fine-grained specimens that can be trimmed into a regular geometry and retain their shape on a bench. Alternative methods are mentioned for measuring the volume of a fine-grained, irregularly shaped, stable specimen. Methods are also presented for coarse-grained materials that will not hold together, and as a result are formed in a rigid container of known volume.

In addition to obtaining the phase relationships, these experiments provide the opportunity to become familiar with general laboratory procedures, equipment, and record keeping.

Water content, density, void ratio, and saturation are four common measurements used to describe the state of geo-materials. They are interrelated, and the preferred measure to characterize a specimen depends on the material and the application. Water content and saturation are most commonly used when dealing with soft soils. Density is most common for compacted materials and rocks. Void ratio is a more universal measure and commonly used in numerical modeling and comparing properties of different soils. This chapter provides specific details for measuring the conditions of sand and clay.

TYPICAL MATERIALS

In general, geo-materials are multiphase particulate systems consisting of solids, liquids, and gases. The liquid phase is usually water, but may also contain dissolved elements or immiscible liquids, such as oil. The gas phase is most often air, but may be methane, natural gas, and so on. Quantifying the relative amounts of solid, liquid, and gas is paramount to characterization of the state of geo-materials.

BACKGROUND

Collectively, the set of equations used to express the mass and volume portions of an element of material are referred to as the phase relationships. Figure 2.1 presents a schematic diagram of an element of material with the volumes portrayed on the left and the masses portrayed on the right. These definitions will be used throughout the text. In general, the gas is assumed to have no mass, which is technically incorrect but acceptable for our purposes. One could expand this representation to more phases such as immiscible fluids, dissolved solids, or the like. Such expansions are common when dealing with transport and resource extraction problems, but add unnecessary complications at this point. This text will focus on the most common situation of soil grains, water, and air. Dissolved salt in pore water is important in many geotechnical applications and will also be discussed in general terms. Explicit corrections of the phase relationships for dissolved salts are presented in Appendix C.

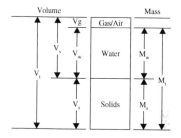

Figure 2.1 Illustration of an element of material with volume and mass definitions.

A water content measurement is routinely performed with almost every geotechnical test. The water content is the most common measure of characterizing the condition of clay. It is important for classification purposes, general description, and describing consistency, as well as being used along with empirical relationships for strength, compressibility, and flow. In some cases, the water content is referred to as the moisture content, but that terminology is not used in this textbook.

Water Content

The water content in geotechnical practice is the ratio of the mass of water to the mass of dry solids, and is usually expressed in percent as shown in Equation 2.1:

$$\omega_C = \frac{M_w}{M_s} \times 100 \tag{2.1}$$

Where:

ω_C = water content (%)
M_w = mass of water (g)
M_s = mass of dry soil (g)

In the standard method, the water content is determined by measuring the mass of the moist specimen, oven drying the specimen to constant mass in a $110 +/- 5°C$ oven, generally overnight, allowing the specimen to cool in a moisture-limited environment (such as a desiccator) or in covered tares, then measuring the mass of the cooled, oven-dry specimen. This method is described in ASTM D2216 Laboratory Determination of Water (Moisture) Content of Soil and Rock by Mass. An alternative ASTM method is titled D4643 Determination of Water (Moisture) Content of Soil by Microwave Oven Heating, which is a quicker method but typically provides less accurate results. Figure 2.2 presents some of the equipment necessary to measure the water content by oven-drying.

Figure 2.2 Desiccator (left); water content tares with covers (right front); desiccant (right rear). Tare covers are not required if the water content specimens are cooled in a desiccator.

Sufficient initial mass must be used to provide a precise and meaningful measurement. Three constraints must be considered when selecting the specimen size: application of significant digits, maximum particle size, and macrofabric of the sample.

The resolution of the scale and the desired number of significant digits sets one minimum size limit for the dry mass. With a scale capable of measuring to 0.01 g, at least a 10 g dry specimen is required to obtain the water content to a resolution of 0.1 percent. Proper accounting of significant digits will automatically take care of this concern.

The second consideration is more specific to geotechnical practice. What is the importance of one grain of material to the water content measurement? Clearly, the largest grain is the most important. Howard (1989) performed a sensitivity analysis to determine the difference in water content caused by removing the largest particle from a test specimen. He simplified the calculation by assuming that the largest particle was completely dry. By removing the dry particle, the measured water content of the remaining material will increase. For coarse-grained material, each grain will be coated by about the same thickness layer of water. This means that the water content for an individual particle should decrease with increasing particle size. One can then generalize Howard's analysis to account for the difference in water content between the finer material and the large particle being removed, yielding Equation 2.2 for the required size of the dry mass:

$$M_d = \frac{10\Delta\omega_C M_p}{E} \tag{2.2}$$

Where:

M_d = dry mass of material required to measure water content (g)
$\Delta\omega_C$ = difference between water content of fine-grained material and that of the largest particle (%)
M_p = mass of the largest particle in the sample (g)
E = acceptable error of measured water content (%)

Equation 2.2 provides a minimum dry specimen mass as a function of the maximum particle size, water content difference, and the desired resolution of the measurement. Equation 2.2 can be normalized by dividing both sides by M_p in order to represent the mass required as a factor times the maximum particle mass. Figure 2.3 presents the results of Equation 2.2 by plotting the error, E, against the normalized dry mass, M_d/M_p, for various values of water content difference.

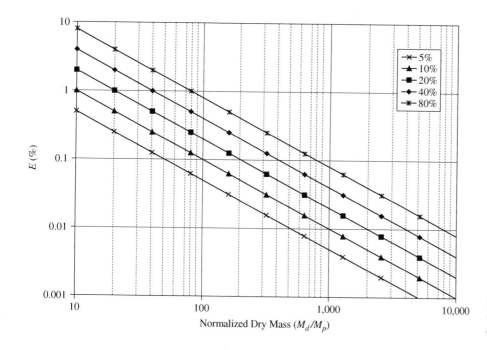

Figure 2.3 Required amount of dry mass to limit error in water content.

While Figure 2.3 is informative, it is difficult to use in practice because the water content is not known in advance of testing. ASTM 2216 simplified the decision making by selecting required minimum dry masses for different maximum particle sizes and for two precision values. As shown in Table 2.1, the normalized dry mass increases as the particle size decreases. This presumably assumes the water content will be higher as the largest particle gets smaller. There is also a 20 g lower limit established for all fine-grained materials.

The third consideration is more difficult to specify because it relates to the engineering application of the results. The variability, or macrofabric, of the sample must be given consideration when selecting the size of the test specimen. Is the goal for the measurement to obtain an average value, or to capture the range in values? For example, when working with a varved clay sample, the various layers may be 1 cm thick. One could choose to make measurements of each individual layer, or collect average measurements over several layers. This decision is important to the application of the results, and must be included as part of the test specification and the report.

The natural water content, ω_N, is a special water content condition and thus has a unique designation. It is the water content when the in situ moisture has been preserved from the time of sampling to the time of performing the test. Preserving the water content can be done in many ways, such as sealing tubes with wax. Further information on this subject is included in Chapter 11, "Background Information for Part II." The

Maximum Particle Size (mm)	Precision ± 0.1%		Precision ± 1%	
	Mass	M_d/M_p	Mass	M_d/M_p
<2	20 g		20 g	
4.75	100 g	600	20 g	120
9.5	500 g	370	50 g	37
75	50 kg	75	5 kg	7.5

Table 2.1 Minimum dry mass for water content specimen to achieve precision of 0.1 or 1 percent.

Source: Adapted from ASTM D2216-05.

distinction between ω_C and ω_N is made because the water content of the specimen can be altered from the natural water content to almost any water content. This can be done intentionally to mimic expected conditions, or it can happen unintentionally, in which case the change may need to be quantified, or the value at the time of setup of a test may need to be determined.

Some other disciplines, such as soil science, define water content as the mass of water divided by the total initial mass of the specimen. In addition, the volumetric water content is used in environmental engineering. Volumetric water content is the volume of water divided by the total volume of the specimen. Awareness of these significant variations in definition can save countless hours of confusion when working with other disciplines, and one should always clearly define the term being reported.

Water contents can exceed 100 percent because the mass of water contained in a specimen can exceed the mass of solids. In peats and vegetable mucks, there can be considerable water trapped by the spongy fibers, and the fibers have a low specific gravity. Diatomaceous materials (Mexico City Clay) have large amounts of water contained in intergranular pores, resulting in very high water contents. Very plastic clays (smectites) have small grains and high surface charge, attracting a relatively large amount of water around each grain. These and other factors can cause measured water contents of 300 percent, and even over 600 percent in some cases.

Salts are often present in the pore fluid when the soil has been deposited in a marine environment, or when the soil is exposed to high rates of evaporation. In extreme cases, such as Salt Lake Clay, the salt content can be high enough to precipitate in the pore space. In most cases, the salt is dissolved in the pore fluid. The salt does not evaporate with the pore fluid when the material is dried, but rather remains as part of the dry mass. If the material contains significant salt in the pore fluid, the phase relationships will be more complicated, and the measured water content will not represent the ratio of the water mass to the grain mass. Methods are described in Chapter 6, "pH and Salinity," for measuring the salt concentration in the pore fluid. Appendix C describes incorporation of salt concentration measurements to correct calculations of physical properties.

Mass Density

Mass density (or simply density) is the mass of soil per unit volume. This parameter is not the weight per unit volume. Unit weight is the mass density multiplied by gravity. Refer to Chapter 1 for discussion on the issue of force versus mass. Density is normally used to describe the state of coarse-grained soils. Mass density is used to calculate the in situ stresses on a soil element. Density is also correlated to many parameters, including strength and the tendency to compress under pressure, as well as assisting in determining other characteristics of soils, as follows later in this chapter. One important use of density is in compaction control.

Two measures of mass density are in common use: total mass density and dry mass density. Total mass density and dry mass density can be determined by direct measurement of the specimen mass in the moist state and dry state, respectively, and the total volume.

The total mass density is given by Equation 2.3:

$$\rho_t = \frac{M_t}{V_t} \tag{2.3}$$

Where:

ρ_t = total mass density (g/cm^3)
M_t = wet mass of soil (g)
V_t = total volume (cm^3)

The dry density is given by Equation 2.4 or Equation 2.5:

$$\rho_d = \frac{M_s}{V_t} \tag{2.4}$$

or

$$\rho_d = \frac{\rho_t}{1 + \frac{\omega_C}{100}} \tag{2.5}$$

Where:

ρ_d = dry mass density (g/cm^3)

Density determination is more difficult than water content because the volume is required. The total volume of a specimen can be obtained in a number of ways. Fine-grained materials will generally hold together in a regular, unsupported shape. In other words, the geometry is stable. This is because the surface tension at the air/water interface around the boundary of the specimen creates negative pore pressure, giving the material strength. The importance of surface tension is easily illustrated by submerging the specimen in water after making the dimensional measurements. Within a few minutes, it will disintegrate (called slaking) as the negative pore pressure dissipates.

If a specimen can sustain its own shape and the shape is regular, the volume can be determined by physically measuring the intact specimen dimensions. Figure 2.4 shows the equipment necessary to trim a fine-grained intact specimen and make the dimensional measurements.

The displacement method can be used to determine the volume of irregularly-shaped specimens. The displacement method involves submerging the intact soil specimen in a fluid and measuring the displaced volume. The fluid must be prevented from intruding into the soil in some way, such as coating the specimen in a thin coat of wax. The method is described more fully in ASTM D4943 Shrinkage Factors of Soils by the Wax Method and is part of Chapter 9, "Atterberg Limits," in this book.

Except in the case of cemented soils, coarse-grained materials generally do not maintain stable geometries. The material must be placed in a container of known volume, and the mass of soil contained in the known volume is obtained in order to calculate the density. The method of placement is a very important consideration because it will determine the density, as well as the mechanical properties of the specimen. In practice, the soil would be placed in the container using a method (compaction, vibration, pluviation) designed to simulate the expected field conditions. Placement of the soil using an

Figure 2.4 Miter box and wire saw used to trim a fine-grained specimen (left). The physical dimensions of the trimmed specimen are then measured using calipers and the specimen volume is calculated; mold of known dimensions and a straight edge to prepare a coarse-grained specimen with a control volume (right).

arbitrary method ignores the fundamental link between density, fabric, and mechanical properties. Laboratory reconstitution methods are discussed in Chapter 11, "Background Information for Part II." Figure 2.4 shows a compaction mold used to prepare a coarse-grained specimen.

Numerous field methods exist for determining the in-place density in the field. The typical methods of direct measurement of in situ density are performed at the ground surface. The direct methods involve excavating a quantity of soil; determining the total mass, dry mass, and water content of the removed soil; and measuring the volume of the excavation. ASTM standards that describe direct methods include the sand cone test (D1556), the rubber balloon method (D2167), and the drive-cylinder method (D2937). ASTM indirect methods for determining the in situ density include the nuclear method (D6938) and time domain reflectometry (TDR) (D6780), among many others. Detailed discussion of these methods is beyond the scope of this textbook; however, ASTM standards and manuals specific to field testing methods are readily available.

Void Ratio

The void ratio quantifies the space available in a soil for flow or compression. The volume of voids includes the volume occupied by both gas and fluid. The volume of solids is the volume occupied by the particles only (i.e., insoluble material). The void ratio is the ratio of the volume of voids to the volume of solids, expressed as Equation 2.6:

$$e = \frac{V_v}{V_s} \tag{2.6}$$

Where:

e = void ratio (dimensionless)
V_v = volume of voids (cm^3)
V_s = volume of solids (cm^3)

The volume of solids is obtained using the relationship given in Equation 2.7:

$$V_s = \frac{M_s}{G_s \rho_w} \tag{2.7}$$

Where:

G_s = specific gravity of soil (dimensionless)
ρ_w = mass density of water (g/cm^3)

The value of specific gravity used in Equation 2.7 is either estimated using experience, or measured as described in Chapter 3, "Specific Gravity." Although the details are beyond the scope of this chapter, note that the specific gravity of soil and the mass density of water are determined at 20°C for this calculation. This point will become clear in the next chapter.

The mass density of water is often assumed to be 1 g/cm^3. In actuality, the value varies with temperature and is about 0.998206 g/cm^3 at room temperature (20°C). The value of the mass density of water used in calculations must obey the rules of significant digits. Refer to Chapter 1, "Background Information for Part I," for more information on the use of significant digits. The mass density of water for various temperatures is available in many locations, including Appendix B of this book.

The mass of solids, M_s, must be limited to the mineral grains. This value is normally obtained by oven-drying the specimen. When working with soils that contain significant salts in the pore fluid, the salt mass must be subtracted from the oven-dry mass before computing the volume of solids. The void ratio must be independent of salt concentration. Appendix C provides the calculations to modify the phase relationships to account of salts.

The volume of voids is calculated as the difference between the total volume and the volume of solids. This is expressed as Equation 2.8:

$$V_v = V_t - V_s \qquad (2.8)$$

Many times, void ratio is a better parameter to use in empirical relationships than density or water content. This is especially true for unsaturated soils. However, as described above, void ratio is more difficult to determine because it requires more information.

The change in void ratio can be used to quantify the volumetric strain in a specimen. That relationship is Equation 2.9:

$$\varepsilon_v = \frac{e - e_0}{1 + e_0} \times 100 \qquad (2.9)$$

Where:
ε_v = volumetric strain (%)
e_0 = initial void ratio (dimensionless)

When subjected to one-dimensional consolidation or compression, the volumetric strain is equal to the axial strain. Therefore, for one-dimensional consolidation, the relationship for axial strain can be expressed in terms of void ratio, as shown in Equation 2.10:

$$\varepsilon_a = \frac{e - e_0}{1 + e_0} \times 100 \qquad (2.10)$$

Where:
ε_a = axial strain (%)

Porosity

Porosity is another parameter used to quantify the volume of voids in an element of soil. Porosity is defined with Equation 2.11 as:

$$n = \frac{V_v}{V_t} \qquad (2.11)$$

Where:
n = porosity (dimensionless)

Notice that porosity expresses the void space in terms of the total volume and has an upper-bound value of 1, unlike void ratio, which is unbounded. In applications where the total volume changes over time, working with porosity is not as useful as working with void ratio. For example, void ratio is a more convenient measure when working with strain-based calculations. However, porosity is very useful in flow problems because porosity directly expresses the available flow space. The relationships between void ratio and porosity can be expressed as Equation 2.12 or Equation 2.13, as follows:

$$n = \frac{e}{1 + e} \qquad (2.12)$$

or

$$e = \frac{n}{1 - n} \qquad (2.13)$$

Saturation

Another useful, and related, parameter is the degree of saturation. Saturation is defined as the volume of water (or fluid) divided by the volume of voids, and is expressed as Equation 2.14:

$$S = \frac{V_w}{V_v} \times 100 \qquad (2.14)$$

Where:

S = degree of saturation (%)

Saturation is used to quantify whether a soil is completely dry, saturated, or somewhere in between. This parameter becomes very important when interpreting soil behavior from consolidation and strength testing. The degree of saturation is a useful measure to evaluate the reasonableness of specimen dimensional measurements and calculated phase relationships. Based on hundreds of measurements on "saturated" clay specimens, the value of saturation can be calculated within $+/-$ 2 percent of the actual value.

The following relationships given in Equation 2.15 and Equation 2.16 are encountered repeatedly when dealing with soils limited to three phases (air, water, and mineral):

$$G_s \omega_C = Se \qquad (2.15)$$

$$\rho_d = \frac{G_s \rho_w}{1 + e} \qquad (2.16)$$

Other important considerations when making phase relationship measurements include:

- Loss of moisture during specimen preparation changes the measured water content that in turn affects the density and void ratio values. Prevent loss of moisture by preparing and measuring specimens in a humid room and make the wet measurement immediately after specimen preparation. Never leave specimens open to the atmosphere for extended periods of time. This is particularly important in climates with low relative humidity.
- Measurement errors in dimensions or masses directly impact the interpreted water content, density, and void ratio. Care must be taken to evaluate measurements as they are obtained to determine whether the values are reasonable. Specimens must have regular geometries (flat surfaces) if making dimensional measurements.
- Time in the oven must be sufficient to remove all the free water in the soil. The time required for a dried soil to reach constant mass is dependent upon soil type and specimen size. Granular soils may achieve constant mass in about 4 hours. Fat clays may require up to 24 hours of oven drying. Usually, constant mass is achieved by placing the soil in a 110 $+/-$ 5°C forced draft oven overnight. When working with soils that are unfamiliar, it is wise to verify that the specimen has reached constant mass. This is accomplished by repeating the oven-drying and massing procedures until there is no change in dry mass. Each time period in the oven must be in excess of one hour, and the specimen must be cooled in a desiccator after each drying cycle.
- Drying temperature is critical to ensure that only free water is removed, rather than both free water and intergranular water. Higher temperatures may remove more intergranular water in a soil, while lower temperatures may not remove all the free water in the soil. In order to obtain reproducible results, the standard oven temperature is controlled to within 5° of 110°C.

- Some ovens have significant variations in temperature across the oven itself. For that reason, the type of oven specified for use is the forced draft oven. A fan prevents "hot spots" in the oven (e.g., closer to the heat source) by circulating the warm air to reach all spaces in the oven uniformly.
- The specimen must be cool to the touch before obtaining the mass. Hot specimens will generate an upward draft that will change the mass read on the scale, similar to blowing air across the scale. Either place the specimen in a desiccator or use a tare cover while cooling completely after oven drying.
- Desiccant is reusable but needs to be regenerated once the product loses its moisture-absorbing capabilities. Regenerate according to the manufacturer's directions. This generally means placing the desiccant in a single layer on a shallow pan in a 210°C oven for about one hour. Desiccant can be purchased in both indicating and nonindicating forms. Indicating desiccant changes color whenno longer effective. Since nonindicating desiccant costs less than half of indicating desiccant, a cost-efficient technique is to mix a small amount of the indicating desiccant with the nonindicating desiccant. The two types can be regenerated together, provided the manufacturer's recommendations for regeneration of the two products is the same.
- Frequently, the value of specific gravity is estimated based on experience with similar materials. An erroneous value impacts the calculation of the volume of solids, which changes the interpreted value of void ratio, porosity, and degree of saturation. The data report should always provide the value of specific gravity used in the calculations, and indicate whether it is assumed or measured.

Values of water content (ω_C), total density (ρ_t), void ratio (e), and saturation (S) from a variety of projects are listed in Table 2.2.

TYPICAL VALUES

No calibration is necessary when using a specimen that can maintain trimmed dimensions. If testing an irregularly shaped soil specimen and the displacement method is used

CALIBRATION

Table 2.2 Typical values of water content, total density, void ratio, and saturation for a selection of soils.

Soil Type	ω_C (%)	ρ_t (g/cm^3)	e	S (%)
Overconsolidated Boston Blue Clay (BBC) (OCR 1.5 to 4)*	31 to 49	1.75 to 1.95	0.9 to 1.4	95 to 100
Slightly Overconsolidated (BBC) Clay (OCR 1.1 to 1.3)*	41 to 51	1.75 to 1.80	1.2 to 1.4	97 to 100
Maine Clay (OCR 1.3 to 4)*	30 to 35	1.9 to 2	0.8 to 1.0	98 to 101
San Francisco Bay Mud (OCR 1.4 to 2.2)**	87 to 98	1.45 to 1.50	2.35 to 2.65	96 to 100
Peat***	250 to 770	1.0 to 1.1	3.9 to 12.4	93 to 101
Smith Bay Arctic Silt (Site W) (OCR 8 to 25)****	29 to 34	1.9 to 2	0.77 to 0.86	95 to 106
Processed Manchester Fine Sand (MFS) (D$_r$ 30% to 87%)*****	23 to 30	1.9 to 2.0	0.62 to 0.81	97 to 99

*Personal database; Assumed G_s = 2.78.

**Personal database; Assumed G_s = 2.70.

***Personal database; Assumed G_s = 1.50.

****After Young, 1986; Assumed G_s = 2.60.

*****After Da Re, 2000; Measured G_s = 2.67.

Equipment Requirements

1. Forced Draft Oven, 110 +/− 5°C
2. Desiccator
3. Scale readable to 0.01 g with a capacity of at least 200 g for determination of water content
4. Scale readable to 0.1 g with a capacity of at least 500 g for determination of trimmed specimen mass
5. Scale readable to 1 g with a capacity of at least 5 kg for determination of cohesionless specimen mass
6. Miter box to trim cohesive soil
7. Equipment to measure dimensions of specimen: calipers readable to 0.02 mm and Pi tape,or wax pot and beaker with water if using displacement method
8. Tares for water contents
9. Mold with regular dimensions and known volume, such as the 4-in. diameter mold used for a Proctor compaction test
10. Thermometer, readable to 0.1°C, for determining the temperature of the water if the water filling method is used to calibrate the mold
11. Utensils, such as wire saw and straight edge, for trimming specimen
12. Wax paper or plastic wrap
13. Small glass or plastic plates
14. Scoop, spatula, and small-diameter (about 5 mm) rod
15. Tongs or gloves to handle hot tares

to determine specimen volume, the mass density of the wax must be known. The reader is referred to ASTM D4943 for further details on determining the mass density of the wax. However, it is unlikely that this method will be used for laboratory instruction.

Calibration of the control volume must be performed for the mold used to form coarse-grained specimens. Use either the direct method of measuring the dimensions and calculating the volume, or use the water filling method.

Dimensional measurement:

1. Measure the height (H_m) of the mold to the nearest 0.02 mm in four locations.
2. Measure the inside diameter (D_m) of the mold to the nearest 0.02 mm in four locations.
3. Calculate the volume of the mold (V_m) to four significant digits using Equation 2.17:

$$V_m = \frac{\pi \cdot D_m^2 H_m}{4} \times \frac{1}{1,000} \qquad (2.17)$$

Where:

V_m = volume of the mold (cm^3)

D_m = inside diameter of the mold (mm)

H_m = height of the mold (mm)

Water filling method:

1. The mold must be watertight.

2. Measure the mass of the empty mold and glass plate to four significant digits (in g), M_m.

3. Fill with equilibrated distilled water, cover with the glass plate to establish the top surface, and dry the excess water. Equilibrated water is water that has been poured into a container, and allowed to sit (usually overnight) to allow the water to come to room temperature and let dissolved air come out of solution.

4. Measure the mass of the mold, water, and plate to four significant digits (in g), M_{wm}.

5. Measure the water temperature to 0.1°C.

6. Calculate the mass of water (M_w) to four significant digits using Equation 2.18:

$$M_w = M_{wm} - M_m \qquad (2.18)$$

Where:

M_w = mass of water (g)

M_{wm} = mass of mold, water and plate (g)

M_m = mass of mold and plate (g)

7. Calculate the volume of the mold (V_m) to four significant digits using Equation 2.19:

$$V_m = \frac{M_w}{\rho_w} \qquad (2.19)$$

Where:

V_m = volume of mold (cm^3)

SPECIMEN PREPARATION

Measurements and calculation of the phase relationships are generally done as part of an engineering or compaction test. As such, this laboratory exercise is somewhat contrived, but necessary to gain laboratory experience. Sample preparation should be done in advance of the laboratory to provide materials with known properties for the testers. It is also desirable to make measurements on three to five "identical" samples.

For cohesive soils, the best and easiest option is to use thin-walled tube samples. Each specimen will require about 5 cm of sample. The material needs to be soft enough to trim with a wire saw. Samples from a depth of less than 50 meters or those with an undrained strength of less than 150 kPa should be soft enough to trim.

For cohesive soils, the less preferred option is to fabricate small block samples using Standard Proctor compaction. Various dry clay powders are commercially available by the bag. The material should be mixed to be slightly wet of optimum, hydrated overnight, and compacted into rigid molds. These can then be treated as block samples for testing.

For coarse-grained soils, concrete sand is readily available. The material should be mixed with 2 to 8 percent water and hydrated overnight. This stock mix can then be used to form the test specimens.

Take precautions to preserve the water content while preparing the specimen.

PROCEDURE

For this laboratory experiment, measurements will be made on trimmed specimens of fine-grained soil and laboratory-prepared coarse-grained specimens to determine the water content, total mass density, dry mass density, void ratio, and degree of saturation.

The water content measurement will be performed in general accordance with ASTM Standard Test Method D2216, while the determinations of mass density and the void ratio will be performed using basic laboratory physical measurements. An estimation of specific gravity will be required. Refer to Chapter 3, "Specific Gravity," for information on the specific gravity of soil.

For intact, fine-grained soils:

Note: When working with medium to soft fine-grained samples, do not pick up the sample with your hands. Finger pressure will increase the degree of disturbance. Rather, cover the surface with a piece of wax paper or parchment and manipulate the specimen with small plastic or glass plates. Always cover the surface with wax paper or parchment to prevent the soil from adhering to the flat rigid surface.

1. Begin with an intact block of soil or section from a tube sample.
2. Use the miter box and wire saw to create one flat surface.
3. Place wax paper on the cut surface and place the cut surface on the plate.
4. Trim one side as the second cut. The flat surface should be 3 to 5 cm long. Cover with wax paper.
5. Rotate the sample 90° on the plate and align using side support of miter box. Trim the second side as the third cut. Transfer the wax paper to the freshly cut surface.
6. Repeat step 5 two more times to create a rectangular section.
7. Place wax paper on one side and rotate the block onto the plastic plate.
8. Align the block along the side support and trim the remaining surface.
9. Measure each of the three dimensions ($L_{1,n}$, $L_{2,n}$, $L_{3,n}$) of the block specimen at four locations (at the center of each edge) to the nearest 0.02 mm. Average the measurements on each side to determine L_1, L_2 and L_3.

Continue with step 10.

For coarse-grained soils:

1. Use scoop to fill mold about one-third with moist soil.
2. Penetrate sample 25 times with a rod to densify soil.
3. Repeat steps 1 and 2 to fill the mold.
4. Strike the surface flat with the straight edge.

Continue with step 10.

For either fine- or coarse-grained soils, continue as outlined below:

10. Measure the mass of a tare (M_c) in grams to four significant digits.
11. Place the specimen in the tare.
12. Measure the moist specimen mass and tare (M_{tc}) in grams to four significant digits.
13. Place the specimen and tare in the drying oven until the specimen reaches constant mass.
14. Remove the specimen and tare from the oven and allow it to cool completely in the desiccator.
15. Measure the dry specimen mass and tare (M_{dc}) in grams to four significant digits.

Calculations

1. Calculate the total volume of the fine-grained specimen to four significant digits using Equation 2.20. Use the volume of the mold (calculated using Equation 2.17 or Equation 2.19) for the volume of the coarse-grained specimen.

$$V_t = L_1 \times L_2 \times L_3 \times \frac{1}{1,000} \qquad (2.20)$$

Where:

V_t = total volume of the specimen (cm³)

L_1 = average of four measurements of the length of side 1 (mm)

L_2 = average of four measurements of the length of side 2 (mm)

L_3 = average of four measurements of the length of side 3 (mm)

For each specimen:

2. Calculate the mass of the moist specimen (M_t) using Equation 2.21:

$$M_t = M_{tc} - M_c \qquad (2.21)$$

Where:

M_{tc} = mass of the moist specimen and tare (g)

M_c = mass of the tare (g)

3. Calculate the mass of the dry specimen (M_s) using Equation 2.22:

$$M_s = M_{dc} - M_c \qquad (2.22)$$

Where:

M_{dc} = mass of the dry specimen and tare (g)

4. Calculate the mass of water (M_w) using Equation 2.23:

$$M_w = M_{tc} - M_{dc} \qquad (2.23)$$

5. Calculate the total mass density (ρ_t) to four significant digits using Equation 2.3.
6. Calculate the water content (ω_c) to the nearest 0.1 percent using Equation 2.1.
7. Calculate the dry mass density (ρ_d) to four significant digits using Equation 2.4 or Equation 2.5.
8. Calculate the volume of solids (V_s) to three significant digits using Equation 2.7.
9. Volume of voids (V_v) to three significant digits using Equation 2.8.
10. Calculate the void ratio (e) to the nearest 0.001 using Equation 2.6.
11. Calculate the porosity (n) to 0.001 using Equation 2.11.
12. Calculate the degree of saturation (S) to the nearest 0.1 percent using Equation 2.14.

Report

Report the water content, total mass density, dry mass density, void ratio, porosity, and degree of saturation for each specimen.

PRECISION

Criteria for judging the acceptability of test results obtained by this test method have not been determined using an Interlaboratory Study (ILS).

However, based on an extensive personal database consisting of hundreds of saturated specimens, a reasonable expectation is that the value of saturation of two properly performed tests on the same soil by the same operator in the same laboratory within a short period of time of each other should not differ by more than 4 percent.

Based on these results, a reasonable expectation of the range in void ratio results performed under the same conditions should be no greater than 0.045. Likewise, the water content should not differ by more than 3.2 percent of the average water content between two tests performed under the same conditions.

DETECTING PROBLEMS WITH RESULTS

If the standard deviation for one set of measurements exceeds the estimates provided above, then evaluate the techniques of the individual performing the test. The possible sources of problems are sloppy placement of soils in water content tares, poor dimensional measurement techniques, an error in dry mass, or insufficient cooling time after oven-drying specimens. If the test results do not fall within the typical ranges, the likely cause of error is systemic, such as an error in the volume of the mold, incorrect temperature measurement, or a miscalculation.

Several methods of isolating the causes of errors are possible. Systematic errors due to equipment deficiencies can be identified by repeating the calibration of the mold. Procedural and technique errors are best identified by performing the test on a coarse material with known phase relationships (such as glass beads).

REFERENCE PROCEDURES

ASTM D2216 Laboratory Determination of Water (Moisture) Content of Soil and Rock by Mass.

REFERENCES

Refer to this textbook's ancillary web site, www.wiley.com/college/germaine, for data sheets, spreadsheets, and example data sets.

Da Re, Gregory. 2000. "Physical Mechanisms Controlling the Pre-failure Stress-Strain Behavior of Frozen Sand," PhD Thesis, Department of Civil and Environmental Engineering, Massachusetts Institute of Technology, Cambridge.

Young, Gretchen. 1986. "The Strength Deformation Properties of Smith Bay Arctic Silts," MS Thesis, Department of Civil Engineering, Massachusetts Institute of Technology, Cambridge.

Chapter 3

Specific Gravity

This chapter provides background and procedures to perform the specific gravity test using the water submersion method. Techniques are presented for both fine-grained soil and coarse-grained soil. It will be much easier to perform the tests on coarse-grained soils for qualifying technicians and instructional laboratories. Fine-grained soils take longer to test and require more care in processing the material.

SCOPE AND SUMMARY

Specific gravity is typically determined on geo-materials ranging from peat to rock. The test is also run on other materials, including glass, cement, and iron ore.

TYPICAL MATERIALS

Specific gravity is defined as the ratio of the mass of a given volume of soil particles to the mass of an equal volume of distilled water (at 4°C). This is given numerically by Equation 3.1:

$$G_{s_{4^\circ}} = \frac{\rho_s}{\rho_w}$$
(3.1)

BACKGROUND

Where:

$G_{s_{4°}}$ = specific gravity at 4°C (dimensionless)

ρ_s = mass density of solids (g/cm³)

ρ_w = mass density of water (g/cm³)

The temperature of 4°C is used as the scientific reference temperature because it corresponds to the highest water density. In soil mechanics, 20°C is typically used as the reference temperature because this is the most common application temperature. For reference, the mass density of water changes by 0.2 percent between 4°C and 20°C, which is about equal to the precision of the test method. On the other hand, the mass density variation of the particles over the same temperature range is insignificant. Throughout the remainder of this text, G_s will be used with the understanding that it is referenced to 20°C. Specific gravity is required to compute the phase relationships in almost all engineering tests in soil mechanics. The equations for phase relationships are found in Chapter 2, "Phase Relationships."

Specific gravity is not useful as a criterion for soil classification because the variation is rather small from mineral to mineral. Typical values of specific gravity are provided for a variety of geo-materials later in this chapter. The table demonstrates the relatively small range of values for several common soil materials. This small range demands a high precision in the test method in order to make it worthwhile to perform the test for a specific application.

Two methods are used to experimentally determine specific gravity. One is the liquid submersion technique and the other is the gas pycnometer technique. The gas pycnometer test method is designated ASTM D5550 Specific Gravity of Soil Solids by Gas Pycnometer. ASTM standard test method ASTM D854 Specific Gravity of Soil Solids by Water Pycnometer uses the submersion technique, which is the subject of this chapter. The water submersion technique is applicable to measuring the specific gravity of heavy (relative to water), nonreactive particles. The submersion method can be used for particles with low specific gravity values or particles that react with water (e.g., gypsum) by replacing water with kerosene or other liquids. The mass density versus temperature relationship for the replacement liquid may have to be determined experimentally or found in other resources.

A key component of determining the specific gravity of a material with the submersion method is to have a precisely controlled volume. Iodine flasks and pycnometers (Figure 3.1), among other volumetrics, are readily available for establishing a controlled volume.

With proper experimental techniques, both volumetrics yield equivalent results, but the iodine flask is preferred because it reduces the subjectivity when setting the control volume.

The volumetric must be calibrated (with a matched plug for the iodine flask) to account for the variations associated with temperature. In general, one must measure the mass and temperature of the volumetric filled with water. While the calibration must be done experimentally, the theoretical equation can be used to better understand the important factors.

Equation 3.2 expresses the mass of the filled volumetric as a function of temperature:

$$M_{B+W_T} = M_B + V_{B_{TC}} \left\{ 1 + (T - T_C)\epsilon_g \right\} (\rho_{w_T} - \rho_{a_T}) \qquad (3.2)$$

Where:

M_{B+W_T} = mass of the volumetric and water at temperature T (g)

M_B = mass of volumetric (g)

$V_{B_{TC}}$ = volume of volumetric at temperature T_C (cm³)

T = temperature of the bottle during individual measurement (°C)

T_C = temperature of the bottle at calibration condition (°C)

Figure 3.1 Iodine flask (left); pycnometer (right).

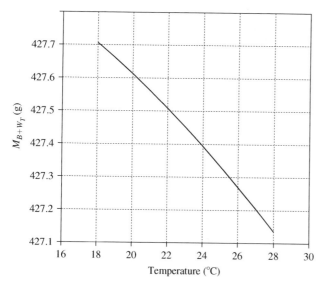

Figure 3.2 Typical calibration curve (mass of volumetric filled with water over a range of temperatures).

ε_g = coefficient of cubical expansion of glass = $0.100 \times 10^{-4}/°C$
ρ_{w_T} = mass density of water at temperature T (g/cm^3)
ρ_{a_T} = mass density of air = 0.0012 g/cm^3

The mass density of water is obtained from standard tables. A table of these values is provided in Appendix B as well as with the flask calibration data sheet.

When the mass of the volumetric and water are calculated and plotted over a range of temperatures, a calibration curve will be developed. A hypothetical calibration curve is presented as Figure 3.2.

Generally, the experimental error in the mass measurement is larger than the changes caused by the cubical expansion of glass and ignoring the displaced mass of air. In addition, the manufacturer's quoted volume of the volumetric is only approximate at ±0.2 mL.

Therefore, it is best to obtain the volume of the volumetric by experimental methods, assuming the term $(T - T_C)\varepsilon_g$ in Equation 3.2 reduces to zero and the mass density of air is zero. Based on these assumptions, the remaining terms in Equation 3.2 can then be rearranged to Equation 3.3:

$$V_{B_{TC}} = \frac{M_{B+W_T} - M_B}{\rho_{w_T}} \tag{3.3}$$

During the calibration, the mass of the volumetric filled with water and the corresponding temperature is measured at least three times at a temperature or a range of temperatures between 18 and 30°C. Take care to ensure the temperature is uniform throughout the control volume, use distilled, equilibrated water, and make sure the volumetric is dry outside and within the neck.

In general, the following measurements are required to perform the submersion technique on a soil: mass of the volumetric; mass of the volumetric and water by calibration; mass of the volumetric, water, and soil (repeated three times); the temperature at each measurement; and the mass of the dry soil.

Computations to determine the specific gravity at the test temperature (G_{S_T}) are performed using Equation 3.4:

$$G_{S_T} = \frac{M_S}{\left(M_{B+W_T} + M_S\right) - M_{B+W+S_T}} \tag{3.4}$$

Where:
M_{B+W+S_T} = mass of the volumetric, water and soil at temperature T, determined experimentally (g)
M_s = mass of dry soil (g)
M_{B+W+S_T} and M_{B+W_T} are at identical volumes and temperatures

A graphical representation of the denominator of this equation is provided as Figure 3.3.

The measured specific gravity must then be corrected to 20°C by accounting for the change in water density and assuming the soil particle density remains constant. This is performed using Equation 3.5:

$$G_s = G_{S_T} \frac{\rho_{w_T}}{\rho_{w_{20}}} \tag{3.5}$$

Other important considerations for the specific gravity test include:

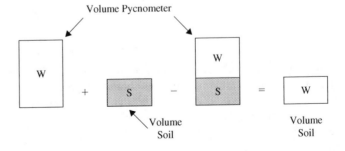

Figure 3.3 Graphical representation of volumes within the specific gravity determination to obtain the displaced water mass.

- Working with differences in large masses presents a challenge in meeting the mass measurement resolution required for this test. Subtraction of the large numbers in the denominator to get the volume of the soil controls the resolution achievable with the test method. For example, if the volumetric mass is 175.52 g, the mass of the volumetric and water is 427.66 g, the mass of dry soil is 38.23 g, and the mass of the bottle, water, and soil is 452.16 g, the maximum number of significant digits is four. If the mass of dry soil is reduced to less than 10 grams or the resolution of the scale is reduced to 0.1 grams, the maximum number of significant digits achievable is three, which is insufficient for the test.

- Temperature variations within the volumetric present the largest source of measurement error. Ensure that the temperature of the volumetric has equilibrated fully. Otherwise, the estimate of the density of the fluid within the volumetric (and thus the mass of the water filled volumetric) will be based erroneously on the point of temperature measurement.

- Calibration of the volumetric must be done very carefully as this value is incorporated into all calculations. An error in the calibration will bias the results but will not be detected in the standard deviation of one set of measurements. The calibration has to be performed only once for a volumetric and cap combination, and that calibration can be used for subsequent measurements. The mass of the empty volumetric and the accuracy of the thermometer are important sources of bias error in the test. The same thermometer should be used for calibration and testing.

- Cleanliness of the neck of the volumetric is crucial. Small amounts of soil retained on the glass can alter the measured mass of solids. However, more significant is the change in seal geometry between the plug and the frosted glass, which subsequently changes the volume of the volumetric to a degree substantial enough that the precision estimates will not be met.

- Cleanliness of the glass affects the angle of contact between the glass and water, which changes the volume of water contained in the meniscus above the reading interface. Care must be taken to obtain an accurate reading of the meniscus if using a pycnometer instead of an iodine flask. Reading at the bottom of the meniscus is standard protocol.

- Water that is equilibrated at room conditions will provide acceptable results. Dissolved air will not alter the test results. Soil must be deaired to remove air trapped between particles and in crevices on the surface of particles. Clays are more difficult to deair than coarse-grained soils.

- Some soils are deposited within salt water environments, such as marine clays, or contain evaporate products. The pore fluid of such soils typically contains salts. Salts will dissolve in the water during the specific gravity test, changing the water density. When the water is evaporated by drying in order to determine the dry mass of solids, the salt is left behind and interpreted as contributing to the mass of solids. When performing a specific gravity test, the presence of salts will lead to erroneously high interpreted values unless accounted for appropriately. For example, a clay soil with a measured water content of approximately 40 percent and a pore fluid salt concentration of typical sea water (35 g/L), the error is on the same order as the precision estimates for within laboratory testing presented in ASTM International Test Method D854 (0.007). Correction of test data to account for the presence of salt in some index tests is presented in Appendix C.

- Drying causes interstitial layers of some clay minerals to collapse. Upon reintroduction of water, the clay particles will not rehydrate to the same size and a different value of specific gravity will result. Therefore, do not dry clay soil prior to determining the specific gravity. As Figure 3.4 indicates, the specific gravity can change dramatically with drying temperature for some soils. When in doubt about whether a soil is affected by drying, use wet soil for the experiment.

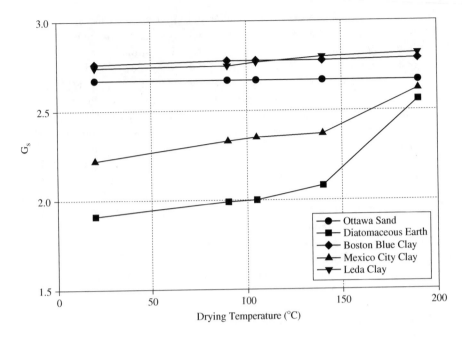

Figure 3.4 Specific gravity versus drying temperature for five soils (Adapted from Lambe, 1949).

Table 3.1 Typical values of specific gravity

Soils in general*	2.65 to 2.85	K-Feldspars**	2.54 to 2.57
Average for clays	2.72	Montmorillonite***	2.35 to 2.7
Average for sands	2.67	Illite***	2.6 to 3.0
Organic clay	~2.0	Kaolinite***	2.6 to 2.68
Peat****	1.0 or less	Biotite**	2.8 to 3.2
Quartz**	2.65	Haematite****	5.2

*Lambe, 1951.
**Lambe and Whitman, 1969.
***Mitchell, 1993.
****Head, 1980.

Source unless indicated otherwise: Personal experience.

TYPICAL VALUES   Typical values of specific gravity are listed in Table 3.1.

Equipment Requirements

1. One iodine flask (250 or 500 ml) for each soil to be tested
2. Digital thermometer (readable to 0.1°C)
3. Cooler (picnic or other insulated enclosure)
4. Water bottle for removing excess water
5. Small water bottle to store thermometer in cooler between readings
6. Scale readable to 0.01 g with a capacity of at least 500 g when using 250 mL iodine flasks and at least 1,000 g when using 500 mL flasks
7. Equipment to deair volumetric: vacuum pump or water aspirator with gage, hot plate, burner, or a combination of a vacuum source and a heat source
8. Evaporation dish with twice the capacity of the volumetric

Each volumetric must be calibrated to obtain the volume and empty mass. These data will be used to compute the mass of the water filled volumetric as a function of temperature.

1. Determine mass of clean, dry volumetric with cap (M_B).

2. Fill with equilibrated, distilled water to above calibration level and place the cap in the resting position on top of the volumetric. The resting position for the cap has the bottom surface completely submerged in the water, taking care that an air bubble is not trapped under the end.

3. Place in cooler for at least 3 hours to temperature equilibrate.

4. Cap off, dry excess water, determine mass (M_{B+W_T}). While capping off the volumetric, be sure the bottom of the cap remains under water to avoid trapping air. Insert the cap into the hole and press firmly with a slight twisting action. The cap will lock into a tight position.

5. Remove cap and measure temperature as quickly as practical. Remove the cap with a twisting action to release connection with the volumetric.

6. Add water to above calibration level and place the cap in the resting position.

7. Repeat steps 3 through 6 at least two more times.

8. Obtain the volume of the volumetric for each measurement using Equation 3.3.

9. Compute the volume of the volumetric (V_B) as the average of at least three measurements.

10. An easy method to evaluate the quality of the data (and the testing procedure) is to compute the standard deviation of the volume. The standard deviation should be less than 0.04 cc. The technique should be adjusted and the process should be repeated until this level of repeatability is achieved.

Use approximately 50 to 100 grams dry mass of soil for silty sands and up to 200 grams dry mass for gravel and coarse sands. Coarse-grained material can be oven dried before the test. This will accelerate the test and make it practical for a laboratory instruction class. Use a smaller mass (30 to 40 grams) for clays because clays are difficult to deair. Do not dry clay prior to determining its specific gravity but rather use Equation 3.6 to determine the approximate wet mass for the test:

$$M_t = M_S\left(1 + \frac{\omega_C}{100}\right) \tag{3.6}$$

Where:
  M_t = wet mass of soil (g)
  ω_C = estimated water content of material (%)

The specific gravity analysis will be performed in general accordance with ASTM Standard Test Method D854.

For fine-grained soils:

1. Obtain the equivalent of 30 to 40 g of dry soil.

2. Mix soil in blender or hand shaker with about 80 g of distilled water.

3. Transfer slurry into volumetric. Do not fill more than half of the volumetric.

4. Use one of the following methods to deair the slurry: vacuum one hour, vacuum and heat 10 minutes, or boil for 3 minutes.

Figure 3.5 Example of a mariet tube setup using a sponge on the end of a water introduction tube attached to a Marriott type bottle. This maintains a clear interface between soil/water slurry, and water containing no soil particles.

For coarse-grained soils:

1. Obtain the equivalent of about 200 g of coarse soil, or if oven-dried, obtain the mass (M_S) to 0.01 g.
2. Transfer all the material into the volumetric.
3. Cover the material with distilled water.
4. Apply vacuum to volumetric for 3 minutes and agitate gently by hand to remove air.

For either fine- or coarse-grained soils, continue as outlined below:

5. Fill volumetric with distilled equilibrated water to above calibration level. Do this slowly when working with clay soils so there is clear water in the top of the volumetric. The use of a sponge (Figure 3.5) or other distribution device attached to the end of the tube on a Marriott type bottle helps maintain a clear interface. Introduce the water by gravity using a small total head.
6. Place volumetric with the plug in the resting position in cooler (Figure 3.6). The minimum equilibration time required depends on the method used to deair the soil and the type of soil tested. The practical limits of a typical instructional laboratory

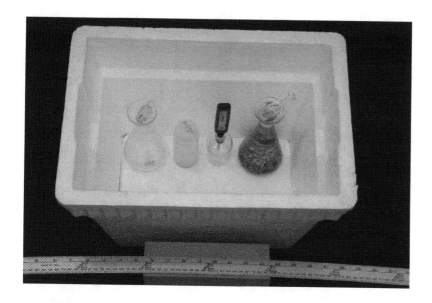

Figure 3.6 Cooler containing iodine flasks equilibrating to room temperature.

Figure 3.7 Stopper at an angle in proper storage position.

time slot may dictate how long the specimens are equilibrated. Deairing methods using heat will typically take longer to equilibrate, while those using vacuum only can be equilibrated relatively quickly. Equilibration time can be reduced when testing coarse-grained soils because the ratio of water to soil in the volumetric is less. The lower bound on equilibration time is 20 minutes for coarse-grained soils using vacuum only for deairing, and about 3 hours for fine-grained soils using heat. It is often convenient for commercial laboratories to equilibrate overnight.

7. Transfer the volumetric to an insulated surface. At this time, the volumetric will still be in the storage position with stopper unplugged and resting partially submerged (Figure 3.7).

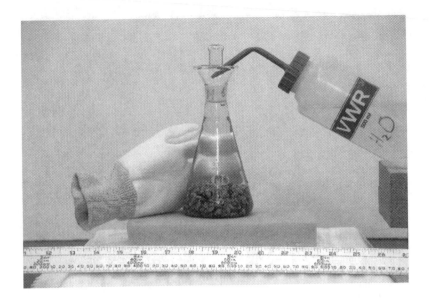

Figure 3.8 Using a squeeze bottle to remove excess water after the stopper has been inserted.

Figure 3.9 Using a lint-free paper towel to remove water droplets from the neck.

8. Cap off volumetric, extract excess water with suction bottle (Figure 3.8), dry rim with strips of paper towel (Figure 3.9), determine mass to 0.01 g (Figure 3.10). This is M_{B+W+S_r} Do this quickly and handle the volumetric with gloves to prevent temperature changes.

9. Remove cap and set aside temporarily. Use a digital thermometer to measure the temperature (T) in the volumetric to 0.1°C (Figure 3.11). It is best to store the thermometer in a small container of water in the cooler to maintain the thermometer at about the same temperature as the water.

Figure 3.10 Determining the mass of the volumetric.

Figure 3.11 Determining the temperature.

10. Add water to above the calibration level and return the plug to the resting position.

11. Repeat steps 6 through 10 at least two more times using about 10 minutes for temperature equilibration between measurements. For class it is informative (but not necessary) to collect data for different temperatures.

12. After the three sets of measurements, obtain the dry mass of soil. Select an evaporating dish and record the mass (M_c) and number. Do not use any metal objects to scrape the soil from the inside of the volumetric. This will scratch the inside surface of the glass. For coarse-grained soil, empty the volumetric into an evaporating dish and rinse the volumetric clean with a squirt bottle. For fine-grained soil, pour off some of the clear water. Cap the volumetric with a rubber stopper and shake vigorously to break up the soil that sticks to the bottom. Pour the slurry into an evaporating dish. Add more water, agitate to loosen the particles, and pour into evaporating dish. Repeat this exercise until all the particles are removed from the volumetric.

13. Oven-dry at 110°C to constant mass and determine the dry mass of the dish and soil (M_{sc}) to 0.01 g.

Calculations

1. Compute the dry mass of soil used in the test using Equation 3.7:

$$M_s = M_{sc} - M_c \qquad (3.7)$$

Where:

M_s = mass of the dry soil (g),
M_{sc} = mass of the dry soil and dish (g)
M_c = mass of the empty dish (g)

2. Compute the specific gravity for each measurement at the measurement temperature using Equation 3.8:

$$G_{S_T} = \frac{M_S}{\left\{ \left(M_B + V_{B_{TC}} \rho_{W_T} + M_S \right) - M_{B+W+S_T} \right\}} \qquad (3.8)$$

3. Correct each value of specific gravity to 20°C using Equation 3.5.

4. Compute the average and standard deviation of the individual measurements to get the final result for the specific gravity.

Report

Report the average value and standard deviation of the specific gravity at 20°C to the nearest 0.001.

PRECISION

Criteria for judging the acceptability of test results obtained by this test method are given as follows as based on the interlaboratory program conducted by the ASTM Reference Soils and Testing Program.

- *Within Laboratory Repeatability:* Expect the standard deviation of your results on the same soil to be on the order of 0.007.
- *Between Laboratory Reproducibility:* Expect the standard deviation of your results compared to others performing the test on the same soil type to be on the order of 0.02.

If the standard deviation for one set of measurements exceeds 0.005, then evaluate the techniques of the individual performing the test. The possible sources of problems are insufficient temperature equilibration, poor cleaning technique, poor control when setting volume, or sloppy handling of the volumetric. If duplicate measurements exceed the within laboratory repeatability provided above, sources of experimental error are likely related to insufficient deairing, errors in the dry mass, or volumetric calibrations. If the test results do not fall within the typical ranges or exceed the reproducibility limit, the likely cause of error is systemic, such as an error in the mass of the volumetric, error in the temperature measurement, or equipment out of calibration.

Several methods of isolating the causes of errors are possible. Systematic errors due to equipment deficiencies can be identified by performing the volumetric calibration over a range of temperatures. Procedural and technique errors are best identified by performing the test on a coarse material with a known specific gravity (such as glass beads).

DETECTING PROBLEMS WITH RESULTS

ASTM D854 Specific Gravity of Soil Solids by Water Pycnometer.

REFERENCE PROCEDURES

REFERENCES

Refer to this textbook's ancillary web site, www.wiley.com/college/germaine, for data sheets, spreadsheets, and example data sets.

Head, K. H. 1980. *Manual of Soil Laboratory Testing,* Volume 1: Soil Classification and Compaction Tests, Pentech Press, London.

Lambe, T. W. 1949. "How Dry is a Dry Soil?," *Proceedings of the Highway Research Board.*

Lambe, T. W. 1951. *Soil Testing for Engineers,* John Wiley and Sons, New York.

Lambe, T. W., and R. V. Whitman. 1969. *Soil Mechanics,* John Wiley and Sons, New York.

Mitchell, J. K. 1993. *Fundamentals of Soil Behavior,* John Wiley and Sons, New York.

Chapter 4

Maximum Density, Minimum Density

SCOPE AND SUMMARY

This chapter will detail the procedures to perform the maximum and minimum index density tests on dry, coarse-grained material. The funnel method of deposition will be used to prepare a specimen at the loosest state. This is one of several methods to get this index density. The slow deposition method will be used to prepare a specimen at the densest state. Once again, this is only one of several methods.

TYPICAL MATERIALS

Maximum and minimum density indices are useful for evaluating the relative density of coarse-grained materials. Soils with a maximum particle size of up to 75 mm (3 in.) may be tested, provided the equipment is appropriately sized.

BACKGROUND

The mechanical properties of coarse-grained materials are a strong function of density. The stable range in densities for a specific material varies considerably depending on the size distribution and shape of the particles. For this reason, relative density (D_R) is commonly used to characterize granular material rather than absolute density. Relative density scales a given density state between its loosest and densest states, expressed as Equation 4.1:

$$D_R = \frac{\rho_{max}}{\rho_d} \cdot \frac{\rho_d - \rho_{min}}{\rho_{max} - \rho_{min}} \times 100 \qquad (4.1)$$

Where:

D_R = relative density (%)
ρ_d = dry mass density (g/cm³)
ρ_{min} = minimum mass density (g/cm³)
ρ_{max} = maximum mass density (g/cm³)

Relative density can also be expressed in terms of void ratio, as presented in Equation 4.2:

$$D_R = \frac{e_{max} - e}{e_{max} - e_{min}} \times 100 \qquad (4.2)$$

Where:

e = void ratio (dimensionless)
e_{max} = maximum void ratio (dimensionless)
e_{min} = minimum void ratio (dimensionless)

Material properties are similar at equal relative densities. Relative density then provides a powerful means to compare different materials.

The minimum mass density corresponds to the maximum void ratio. Likewise, the maximum mass density corresponds to the minimum void ratio. Working in terms of void ratio is preferable because it removes particle density from the representation, making it easier to make comparisons between materials at different states.

The numerical values of the loosest and densest states are not absolute numbers. They are considered index densities and will vary depending on the method of preparation. Several methods exist to measure both the maximum and minimum index densities. Unfortunately, there is no common correction from one to the next. This section describes the common methods and provides some insight about the materials to which they best apply.

Loosest State (Maximum Void Ratio, e_{max})

The minimum density (maximum void ratio) is the loosest state at which the material can be sustained in a dry condition. For a collection of uniformly sized spheres, such as glass beads, arranged in a cubic packed configuration the void ratio would be 0.92, regardless of the sphere diameter. This provides a convenient reference when evaluating maximum void ratio data. In general, granular materials will have e_{max} values greater than 0.92.

A very important consideration when measuring the maximum void ratio is preventing electrostatic charge. When working with fine sands, the repulsive force can be significant enough to cause large overestimates of the maximum and minimum void ratio. This can be a serious problem in dry atmospheres. Additionally, moisture on the particles can cause the particles to densify more and lead to an underestimate of e_{max} and e_{min}.

The key to achieving very loose conditions is to deposit the particles with the maximum possible hindrance. Forcing each particle to stay in the exact orientation and location that it is in at the moment it contacts a surface will create the loosest possible condition. This is ideally achieved by supporting each particle with neighboring particles at the instant of deposition.

Fast deposition is an effective method to achieve the loosest state. This concept is employed in the inverted container (ASTM D4254 Minimum Index Density and Unit Weight of Soils and Calculation of Relative Density, Method C) and the shaken container methods. It is also the basis for the trapdoor method. These methods are aptly described by their names. The inverted container method consists of filling a container with a specified mass of soil, covering the container, inverting the container, then reverting to the original position. The volume is then determined. The shaken container

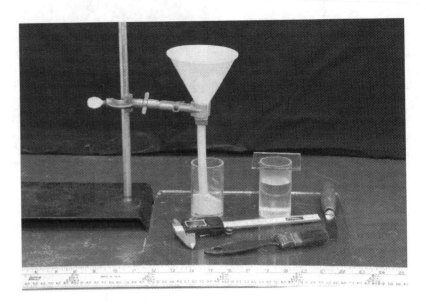

Figure 4.1 Equipment setup used for the funnel method to determine minimum index density.

involves filling a container with a specified volume of soil, shaking the container, then determining the resulting volume. The trapdoor method is performed by placing a mass of soil in a container outfitted with a trapdoor, opening the trapdoor into another container, and determining the resulting volume.

The funnel method (ASTM D4254 Method A) and the double tube method (ASTM D4254 Method B) also achieve high levels of hindrance but allow the material to flow in mass to the resting position. The funnel method uses a funnel and tube to deposit soil into a mold, while keeping the fall height small. The double tube method employs a tube filled with soil placed in the middle of the mold, then the tube is removed, allowing the soil to fill the mold. Finally, deposition through water attempts to slow the terminal velocity in combination with high rates of deposition. All these methods achieve very loose states. The measure of success for each method depends on the grain size and the distribution of particle sizes.

The procedures provided in this chapter for measuring the minimum index density are generally consistent with ASTM D4254 Method C, the funnel method. An example equipment setup using this method is presented as Figure 4.1.

Densest State (Minimum Void Ratio, e_min)

The maximum density (minimum void ratio) is the densest packing of the particles achievable without crushing the contacts. For uniform spherical particles, this condition would be analogous to cubic closest packing. The void ratio for this condition is equal to 0.35, independent of particle size. Natural granular materials are typically looser than this reference value.

Hindrance is again the key factor in achieving this limiting condition. For the dense condition, hindrance should be as small as possible. In other words, each particle should have the opportunity to move around after contacting the surface to snuggle into place with its neighbors as tightly as possible.

This is the concept behind the pepper shaker method, where the particles are rained from a container very slowly to achieve the dense condition. By depositing one particle at a time, the particle has time to rotate and translate into the nearest depression. Depositing from a distance adds a bit of impact energy to tighten the base layer even more.

The vibrating table method (ASTM D4253 Maximum Index Density and Unit Weight of Soils Using a Vibratory Table) is based on the same concept, but attempts to adjust all the particles at once using vertical vibration from a vibratory table, while imparting a surcharge on the specimen to lock the particles in place. The vibration simply provides many opportunities for the particles to adjust. The specimen can either be moist or dry when using Method D4253.

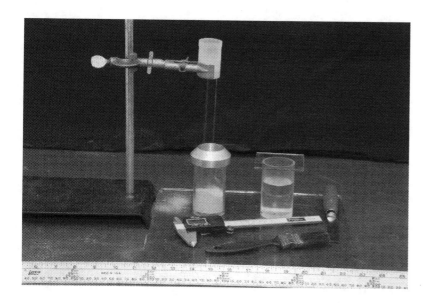

Figure 4.2 Equipment setup used for the pepper shaker method to determine maximum index density.

Other methods are also available, such as a wet tamping method (Head, 1980) for sands. This method is an approach that locks the particles in place using shearing action as the submerged surface is tamped with a mass attached to a vibrating hammer.

Many educational laboratories do not have vibrating tables at their disposal. Therefore, even though D4253 is the ASTM test method used to measure the maximum index density, in this text the concepts of preparing a dense specimen are provided using the pepper shaker method. While the pepper shaker method may not provide the same values of maximum density as the ASTM method, the results will generally be comparable. An example equipment setup using this method is presented in Figure 4.2.

TYPICAL VALUES

Typical values of the maximum void ratio, minimum void ratio, maximum mass density, and minimum mass density of selected soils are presented in Table 4.1.

CALIBRATION

The volume of the mold can be determined either through direct measurement of the dimensions of the mold, or through the water filling method. The water filling method is presented below.

1. Measure the mass of the water tight mold (M_m) to 1 gram (or four significant digits).
2. Fill the mold with equilibrated, distilled water to the calibration level. Measure the mass of the mold and water (M_{wm}) to 1 gram.
3. Calculate the mass of water (M_w) to 1 gram using Equation 4.3:

Table 4.1 Typical values of maximum void ratio, minimum void ratio, maximum mass density, and minimum mass density of selected soils.

Soil Type	e_{max}	ρ_{min}(g/cm³)	e_{min}	ρ_{max}(g/cm³)
Processed Manchester Fine Sand*	0.909	1.408	0.580	1.701
2010 Industrial Quartz**	0.955	1.355	0.640	1.616
Ticino Sand***	0.930	1.380	0.570	1.700

*After Andersen, 1991; values determined using D4253 and D4254.

**After Sinfield, 1997; e_{min}/ρ_{max} determined using vibrating table and surcharge, methods for determining e_{max}/ρ_{min} unknown.

***As appearing in Larson, 1992 (from Franco, 1989); methods unknown.

Equipment Requirements

1. Forced draft oven capable of maintaining a temperature of 110 +/- 5°C throughout the oven.

2. Mold of appropriate diameter according to maximum particle size: For soils with 100% passing the 4.75 mm (No. 4) sieve, use a mold with a nominal capacity of 100 cm³; for soils with 100% passing the 9.5 mm (3/8 in.) sieve, use a mold with a nominal capacity of 1,000 cm³; for soils with 100% passing the 19 mm (3/4 in.) sieve, use a mold with a nominal capacity of 2,830 cm³ (0.1 ft³); for soils with larger particle sizes up to 75 mm (3 in.) in diameter, use a mold with a nominal capacity of 14,200 cm³ (0.5 ft³).*

3. Scale with a capacity according to mold size. When using the 100 cm³ mold, use a scale with a capacity of at least 600 grams and readable to 0.01 grams; when using the 1,000 cm³ mold, use a scale with a capacity of at least 4 kilograms and readable to 0.1 grams; when using the 2,830 cm³ mold, use a scale with a capacity of at least 15 kg and readable to 1 g. When using the 14,200 cm³ mold, use a scale with a capacity of at least 40 kg and readable to 5 g.*

4. Funnel with a capacity of about twice the volume of the mold. The funnel must have a spout attached. The spout must have a diameter of about 13 mm (0.5 in.) when the soil has a maximum particle size of 4.75 mm (No. 4 sieve) or smaller, or about 25 mm (1 in.) when the soil has a maximum particle size of 9.5 mm (3/8 in). For soils with larger maximum particle sizes, a scoop or shovel will be used in place of a funnel and spout.*

5. Straight edge

6. Container with holes in the base. The holes should be about twice the maximum particle size.

---

*A poorly-graded sand (SP), such as Ticino sand, is well suited for laboratory instructional purposes. Since this type of soil has a maximum particle size of 4.75 mm (No. 4 sieve), the 100 cm³ mold can be used, along with a 600 g capacity scale (readable to 0.01 g) and a funnel with a 13 mm spout.

$$M_w = M_{wm} - M_m \qquad (4.3)$$

Where:

M_w = mass of water (g)
M_{wm} = mass of water and mold (g)
M_m = mass of mold (g)

4. Calculate the volume of the mold (V_m) to 1 cm³ using Equation 4.4:

$$V_m = \frac{M_w}{\rho_w} \qquad (4.4)$$

Where:

V_m = volume of mold (cm³)
ρ_w = mass density of water (g/cm³) (refer to Appendix B for the mass density of water versus temperature)

Oven-dry at least 12 kg of poorly-graded sand. The sand can be reused for both the maximum and minimum density tests. After drying, remove the sand from the oven and allow it to cool in a covered container, preventing the soil from absorbing moisture from the air.

The minimum density analyses will be performed in general accordance with ASTM Standard Test Method D4254. Since a large number of laboratories do not have the vibrating table necessary to measure the maximum index density according to ASTM D4253, the procedures presented below use the shaker method.

1. Measure the mass of the mold (M_m) to 1 gram.
2. Using the funnel and extension tube to fill the density cup with sand. Be sure to keep the tube full of sand during the deposition process. Allow the sand to slide out the end of the tube and deposit in a spiral motion.
3. Overfill the mold to about 13 mm (0.5 in.) to 25 mm (1 in.) above the upper edge of the mold.
4. Trim the sand level with the rim of the mold, taking care not to densify the soil.
5. Measure the mass of the mold and the dry sand (M_{sm}) to 1 g.
6. Tap the side of the cup and notice how much the sand densifies.
7. Empty the mold and repeat steps 2 through 5 several times to determine a consistent number.

1. Measure the mass of the mold (M_m) to 1 gram.
2. Place the extension collar on the mold.
3. Slowly deposit the sand in the mold from a height of about 25 cm (10 in.). Use the container with holes in the base to control the rate of filling. It is easiest to tap the side of this container to get a uniform deposition rate.
4. Fill the sand to the top of the collar.
5. Rotate and remove collar, lifting vertically.
6. Scrape the sand level with the rim of the mold using a straight edge.
7. Measure the mass of the mold and the dry sand (M_{sm}) to 1 g.
8. Empty the mold and repeat steps 2 through 7 several times to determine a consistent number.

The void ratio is calculated for each measurement as described in Chapter 2, "Phase Relationships." That information is repeated here for convenience.
    The mass of solids is calculated using Equation 4.5:

$$M_s = M_T - M_m \qquad (4.5)$$

Where:
    M_s = mass of solids (g)
    M_{sm} = mass of mold and soil (g)

    Calculate the mass density using Equation 4.6. Use the subscript "min" when the measurements have been made using the minimum density method. Use the subscript "max" when the measurements have been made using the maximum density method.

$$\rho = \frac{M_s}{V} \qquad (4.6)$$

Obtain the volume of solids using Equation 4.7:

$$V_s = \frac{M_s}{G_s \rho_w}$$ (4.7)

Where:
V_s = volume of solids (cm^3)
G_s = specific gravity of soil (dimensionless)

The volume of voids is calculated as the difference between the total volume as measured during the calibration process, and the volume of solids. This is expressed as Equation 4.8:

$$V_v = V - V_s$$ (4.8)

Where:
V_v = volume of voids (cm^3)

Calculate the void ratio using Equation 4.9. Use the subscript "max" when the measurements have been made using the minimum density method. Use the subscript "min" when the measurements have been made using the maximum density method.

$$e = \frac{V_v}{V_s}$$ (4.9)

Report

Report the maximum density and minimum density to 0.001 g/cm^3 and the maximum and minimum void ratios to 0.001.

PRECISION

Criteria for judging the acceptability of test results obtained by this test method are given as follows as based on the ILS conducted by the ASTM Reference Soils and Testing Program using ASTM D4253 and D4254.

- *Within Laboratory Repeatability:* Expect the standard deviation of your results on the same poorly graded sand (SP) to be on the order of 0.5 lbf/ft^3 (0.008 g/cm^3) for the minimum density test and on the order of 0.6 lbf/ft^3 (0.01 g/cm^3) for the maximum density test.
- *Between Laboratory Reproducibility:* Expect the standard deviation of your results compared to others performing the test on the same poorly graded sand (SP) to be on the order of 2.5 lbf/ft^3 (0.04 g/cm^3) for the minimum density test and on the order of 1.0 lbf/ft^3 (0.016 g/cm^3) for the maximum density test.

DETECTING PROBLEMS WITH RESULTS

If the standard deviation for one set of measurements exceeds the repeatability estimates above, then evaluate the techniques of the individual performing the test. The possible sources of problems are poor trimming technique, insufficient control of the fall height, or sloppy handling of the mold with soil. If duplicate measurements exceed the within laboratory repeatability provided above, sources of experimental error are likely related to errors in the dry mass or (in the case of the minimum density test) densifying the soils during trimming or handling. If the test results do not fall within the typical ranges or exceed the reproducibility limit, the likely cause of error is systemic, such as an error in the mass or volume of the mold, or moisture in the soil.

Several methods of isolating the causes of errors are possible. Systematic errors due to equipment deficiencies can be identified by performing the mold calibration again. If moisture is detected in the soil, dry the soil and perform the testing in a low humidity room. Procedural and technique errors are best identified by performing the test on uniform glass beads and comparing the results to the estimates provided in the background section.

ASTM D4254 Minimum Index Density and Unit Weight of Soils and Calculation of Relative Density

In-house laboratory procedure to measure the maximum index density in place of ASTM D4253 Maximum Index Density and Unit Weight of Soils Using a Vibratory Table

REFERENCE PROCEDURES

Refer to this textbook's ancillary web site, www.wiley.com/college/germaine, for data sheets, spreadsheets, and example data sets.

Andersen, G. R. 1991. "Physical Mechanisms Controlling the Strength and Deformation Behavior of Frozen Sand," ScD Thesis, Department of Civil and Environmental Engineering, Massachusetts Institute of Technology, Cambridge.

Franco, C. 1989. "Caratteristiche Sforzi-Deformazioni-Resistenza Delle Sabbie," PhD thesis, Politecnico di Torino, Torino, Italy.

Head, K. H. 1980. *Manual of Soil Laboratory Testing: Volume 1: Soil Classification and Compaction Tests,* Pentech Press, London.

Larson, D. 1992. "A Laboratory Investigation of Load Transfer in Reinforced Soil," PhD thesis, Department of Civil and Environmental Engineering, Massachusetts Institute of Technology, Cambridge.

Sinfield, J. V. 1997. "Fluorescence of Contaminants in Soil and Groundwater Using a Time-Resolved Microchip Laser System," ScD thesis, Department of Civil and Environmental Engineering, Massachusetts Institute of Technology, Cambridge.

REFERENCES

Chapter 5

Calcite Equivalent

SCOPE AND SUMMARY

This chapter provides background and detailed procedures to perform the calcite equivalent test using the hydrochloric acid digestion method and a pressure vessel. The test is performed on oven-dried material that is ground to a fine powder passing the 0.425 mm (No. 40) sieve. It will be much easier to perform this test on finer-grained soils for qualifying technicians and instructional laboratories.

TYPICAL MATERIALS

Calcite equivalent is typically determined on both coarse- and fine-grained materials. The test is most common when looking for evidence of calcium cementation or trying to quantify the existence of carbonate-based minerals. The test may be performed on rock provided that the material is processed into a fine powder.

BACKGROUND

Calcium carbonate ($CaCO_3$) is reacted with hydrochloric acid (HCl) to form calcium chloride ($CaCl_2$) plus carbon dioxide gas (CO_2) and water (H_2O). Equation 5.1 provides the quantitative balance:

$$CaCO_3 + 2HCl \rightarrow CaCl_2 + CO_2 + H_2O \qquad (5.1)$$

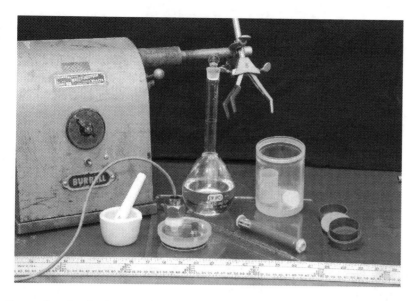

Figure 5.1 Typical experimental setup for Calcite Equivalent Test.

One molecule of calcium carbonate reacts with two molecules of hydrochloric acid to form one molecule each of calcium chloride, carbon dioxide, and water. The calcium chloride is a salt and the carbon dioxide is a gas.

This equation, combined with the ideal gas law, can be used to estimate how much pressure will be generated in the reaction vessel of a given volume as a function of the calcium carbonate mass. When 1 gram of calcium carbonate is completely reacted with hydrochloric acid in a sealed 0.5 L container, approximately 55 kPa is generated. Larger amounts of calcium carbonate will generate larger pressures if fully reacted. Therefore, this point must be kept in mind when choosing the pressure capacity of the vessel and the initial mass of the components to be reacted.

Rather than work with a theoretical relationship between mass and pressure, it is common practice to develop a calibration equation using known amounts of reagent-grade calcium carbonate powder. This calibration equation is then used to compute the equivalent amount of calcite that would result in the same pressure when digesting an unknown mineral. Since the mineral calcite is pure calcium carbonate, the measurement is referred to as the "calcite equivalent". This acknowledges the fact that each carbonate-based mineral has a different atomic structure resulting in a different gas to mass ratio. The calcite equivalent test provides a quick index that can be augmented with detailed mineralogy if more precise information is required.

The material is ground into a powder that passes a 0.425 mm (No. 40) sieve because the grain size controls the surface area exposed to the acid as well as the depth of acid penetration. The two characteristics, in turn, control the rate of reaction. The test is preformed by reacting the oven-dry powder with one normal (1N)[1] hydrochloric acid in a closed pressure vessel and measuring the resulting pressure. A typical setup for the calcite equivalent test is shown in Figure 5.1.

Calcite equivalent is defined as the percentage (by mass) of a given dry material that is composed of digestible carbonate-based minerals. This is given numerically by Equation 5.2:

$$CE = \frac{M_{CaCO_3}}{M_s} \times 100 \qquad (5.2)$$

Where:

CE = calcite equivalent (%)
M_{CaCO_3} = mass of calcite equivalent (g)
M_s = mass of dry soil (g)

---

[1] A 1N solution has exactly one gram equivalent weight per liter of solution.

The device is calibrated using reagent-grade calcium carbonate. Other carbonate species will react with HCl; however, the mass to gas ratio will be different. This method is essentially the same as ASTM D4373 Rapid Determination of Carbonate Content of Soils.

Other important considerations include:

- The calibration curve depends on the volume of the free space inside the vessel as well as the compliance of the measuring device. Therefore, each vessel must be calibrated along with internal components as a unique set.
- The mass of the specimen also occupies space in the pressure vessel. The calibration is performed with about 1 g of material. Appreciable variations in the initial dry mass will change the free space in the vessel and result in testing errors.
- Likewise, the volume of dilute hydrochloric acid occupies space in the pressure vessel. The 20 mL of 1N HCl solution is more than enough to digest 1 g of calcium carbonate. The volume should be kept constant in all tests to maintain consistent free space in the vessel.
- The size of the specimen is very small. This introduces considerable uncertainty especially for coarse-grained soil in the test result. Triplicate tests should be considered to evaluate consistency in material handling and processing.
- Pressure vessel seals need to be maintained properly and checked regularly. A seal leak will create a distinctive peak in the pressure versus time relationship and yield low calcite equivalent results. Any reduction in pressure during the test is due to a leak. The seals should be cleaned after each test to prevent the undigested mineral grains from scratching the seal surface.
- Agitation of the mixture will prevent stagnation in the reaction and maintain a uniform acid concentration. A simple wrist action shaker set for moderate oscillation works very well. Agitation will alter the reaction rate.
- Reaction time is an important variable when testing some slow-reacting minerals. Detailed pressure versus time data should be collected when testing unfamiliar materials. Some reactions can take several hours.
- Grain size will affect the rate of reaction as well as the final calcite content. When testing slow-reacting minerals, it is advisable to decrease the grain size to passing the 0.150 mm (No. 100) sieve to accelerate the process and attain a complete reaction.
- Either a pressure gage or a pressure transducer can be used to measure the pressure versus time during the test. Although not required, transducer readings versus time can be recorded with a data acquisition system. Pressure gages must be read by hand for the duration of the test. In addition, pressure gages have significantly lower resolution and larger hysteresis[2] than pressure transducers, making it more difficult to detect pressure leaks.
- The mineral type has two effects on the test. Since each mineral has a different atomic composition, the amount of carbon dioxide generated per gram of solid will vary. As seen in Table 5.1, this variation can be either larger or smaller than the calcite equivalent and the variation can be large. The second mineral effect is the rate of digestion. When calibrating with calcium carbonate powder, the reaction will be almost instantaneous. However, as seen in Figure 5.2, the rate of reaction is mineral specific and can take a considerable amount of time. This rate of reaction behavior can provide an indication of the mineral type.

TYPICAL VALUES　　　　Typical values of calcite equivalent in soils are listed in Table 5.2.

---

[2] Hysteresis refers to the difference in readings when the measured property is increasing in value versus decreasing.

Mineral	Calcite Equivalent (%)	Mineral	Calcite Equivalent (%)	
Calcite	100.0	Siderite	86.4	**Table 5.1** Calcite equivalent of pure minerals.
Dolomite	108.6	Strontianite	67.8	
Magnesite	118.7	Witherite	50.7	
Rhodochrosite	87.1	Cerrusite	37.5	

Source: After Martin, 1996.

1. Reagent-grade hydrochloric acid
2. Reagent-grade calcium carbonate

1. Prepare a stock solution of 1N HCl by diluting 80 mL of concentrated HCl in 1 L of distilled water. Start with about 500 mL of water, then add the acid and finally add water to the calibration mark. Store in a polyethylene container.
2. Clean and dry the pressure vessel and lubricate the seal.
3. Insert the tipping rod into the vessel.
4. Place the aluminum tare on the scale and zero.
5. Add the approximate amount of reagent calcium carbonate and record mass (M_{cc}) to 0.001 g. For the calibration curve, perform about 5 different masses between 0 and 1 g.
6. Pour the calcium carbonate into the vessel on one side of the rod. Be sure to transfer all the material. (It is good practice to check the mass of the empty tare.)
7. Place the 25 mL plastic container on the scale and zero.
8. Fill with about 20 +/− 0.2 g of the stock HCl solution. This assumes that 1 g HCl has a volume of 1 mL, and is sufficient for most purposes.
9. Place this container in the vessel on the other side of the rod.
10. Close the container cover and record the zero pressure reading (P_I).
11. If using a data acquisition system, start taking readings every second.

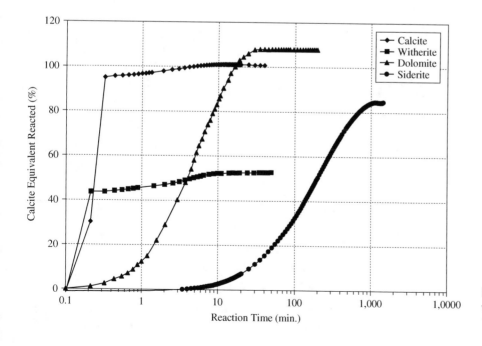

Figure 5.2 Rate of reaction of various minerals. After Martin, 1996.

Table 5.2 Typical values of calcite equivalent in soils.

Soil Type	Calcite Equivalent (%)	Soil Type	Calcite Equivalent (%)
Pierre Shale*	1 to 64	Eastern Atlantic Deep-sea Calcareous Clay**	40 to 90
Rio Grande Valley Alluvium (Depth: 0.3 to 1.2 m)***	3 to 9	Rio Grande Valley Alluvium (Depth: 2.2 to 3.5 m)***	5 to 32
Rio Grande Valley Alluvium (Depth: 6.2 to 7.5 m)***	15 to 40	Rio Grande Valley Alluvium (Depth: 10.2 to 12.7 m)***	15
Rio Grande Valley Alluvium (Depth: 15.8 to 19 m)***	12 to 30	New Mexico and West Texas "cap rock" (Depth: 0 to 3+ m)***	60 to 90

*McKown and Ladd, 1982.
**Demars, 1982.
***Beckwith and Hansen, 1982.

Equipment Requirements

1. One reactor pressure vessel with about 500 mL capacity and able to contain 200 kPa pressure
2. Plastic container to hold acid inside pressure vessel, approximately 25 mL
3. Pressure transducer (or pressure gage) with 200 kPa capacity plus power supply and voltmeter
4. Plastic trip bar that fits in base of pressure vessel
5. Silicone seal lubricant
6. Polyethylene container with a capacity of greater than 1 liter to hold stock acid solution
7. Mechanical (preferably wrist action) shaker
8. 0.425 mm (No. 40) sieve
9. Mortar and pestle
10. Glass plate, approximately 20 cm by 20 cm
11. Metal straight edge
12. Small aluminum tare
13. Long tweezers (>10 cm in length)
14. Timer readable to the second
15. Scale readable to 0.001 g with a capacity of at least 100 g
16. Safety gloves, goggles, apron, and other appropriate safety gear
17. Reagents
    (a) Reagent-grade hydrochloric acid
    (b) Reagent-grade calcium carbonate

12. Close the vent valve and record the sealed pressure reading (P_S).
13. Start the timer and tip the vessel to spill the acid on the powder.
14. Gently shake the vessel for 1 minute.
15. Record the 1-minute pressure reading (P_F).
16. Allow the vessel to sit for another minute and record the pressure reading.

17. If the second reading is lower than the first, then the vessel is leaking and the test is invalid.

18. Open the vent valve, remove the cap, and clean the vessel.

19. Repeat the process starting from step 2 with a different mass of calcium carbonate until sufficient data are obtained to define the calibration curve. Five points should be adequate.

20. Construct a plot of mass (M_{cc}) of calcium carbonate versus pressure (P_F-P_I) (kPa). Perform a linear regression analysis to obtain the slope (CF_{cc}) of these data. The line should pass through the origin. Use the slope of this line as the calibration curve for the vessel. The units for the slope are mass per unit pressure.

SPECIMEN PREPARATION

Select a representative amount of material for the evaluation. One should generally start with a mass that is the greater of either 100 times the largest particle or about 5 g. This material should be oven-dried. Using the mortar and pestle, reduce the particle size such that all the material passes the 0.425 mm (No. 40) sieve. Pile the material on the clean glass plate. Use the straight edge to quarter the pile twice in order to obtain a test specimen of about 1 g.

PROCEDURE

The calcite equivalent analysis will be performed in general accordance with ASTM Standard Test Method D4373.

1. Clean and dry the pressure vessel and lubricate the seal.

2. Insert the tipping rod into the vessel.

3. Place the aluminum tare on the scale and zero.

4. Add the test specimen to the tare and record mass (M_t) to 0.001 g. The test specimen should be about 1 g unless the calcite equivalent is below 10 percent, in which case the mass should be increased to 2 g.

5. Pour the test specimen into the vessel on one side of the tipping rod. Be sure to transfer all the material. (It is good practice to check the mass of the empty tare.)

6. Place the 25 mL plastic container on the scale and zero.

7. Fill with about 20 +/− 0.2 g of the stock HCl solution. This assumes that 1 g HCl has a volume of 1 mL, and is sufficient accuracy for most purposes.

8. Place this container in the vessel on the other side of the rod.

9. Close the container cover and record the zero pressure reading (P_I).

10. If using a data acquisition system then start taking readings every 5 seconds.

11. Close the vent valve and record the sealed pressure reading (P_S).

12. Start the timer and tip the vessel to spill the acid on the powder.

13. Lock the container in the shaker and start shaking at a moderate rate. Be sure the fluid does not splash into the transducer housing.

14. If recording by hand, take a reading every minute.

15. Continue shaking for at least 10 minutes or until the pressure is stable.

16. Record the final pressure reading (P_F).

17. Any reduction in pressure indicates a leak and invalidates the test.

18. Stop the shaker and remove the vessel.

19. Open the vent valve and remove the cover.

20. Discard the contents and clean the vessel.

Calculations

Note: If a transducer is used to measure the vessel pressure then it will be necessary to record both the output of the transducer as well as the input voltage used to power the transducer. In addition, a calibration equation will be used to convert electric output to pressure. Refer to the introductory chapter of Part II for more specific information on the application of transducers.

1. Compute the equivalent mass of calcium carbonate digested based on the measured pressure using Equation 5.3:

$$M_D = (P_F - P_I) \times CF_{CC} \qquad (5.3)$$

Where:
  M_D = equivalent mass of calcium carbonate digested (g)
  P_F = final pressure (kPa)
  P_I = initial pressure (kPa)
  CF_{cc} = slope of the calibration equation for the vessel (g/kPa)

2. Compute the Calcite Equivalent using Equation 5.4:

$$CE = \frac{M_D}{M_S} \times 100 \qquad (5.4)$$

Where:
  M_s = mass of dry soil

Report

Report the calcite equivalent content to the nearest 0.1 percent. If transducer readings are recorded over time, present the rate of reaction curve.

PRECISION

Criteria for judging the acceptability of test results obtained by this test method have not been determined by ASTM. However, it is reasonable to expect that two tests performed properly on the same material will produce a calcite equivalent value within about 2 percent.

DETECTING PROBLEMS WITH RESULTS

If the standard deviation for one set of measurements exceeds 1 percent, individual techniques should be evaluated. The largest source of scatter is poor material handling techniques. Other potential sources of problems are insufficient reaction time, poor massing technique, or a leak in the device. If the results do not fall within the typical ranges, the likely cause of error is systemic, such as an error in the mass of the solids or an incorrect calibration.

Troubleshoot problems using combinations of nonreactive quartz sand and reagent calcium carbonate powder. Be sure to include large grains that will require grinding prior to testing. Verification experiments should have a longer duration than normal tests in order to check for leaks.

REFERENCE PROCEDURES

ASTM D4373 Rapid Determination of Carbonate Content of Soils.

REFERENCES

Refer to this textbook's ancillary web site, www.wiley.com/college/germaine, for data sheets, spreadsheets, and example data sets.

Beckwith, G. H., and L. A. Hansen. "Calcareous Soils of the Southwestern United States." *Geotechnical Properties, Behavior, and Performance of Calcareous Soils*, ASTM STP 777, K. R. Demars and R. C. Chaney, Eds., American Society for Testing and Materials, Philadelphia, PA.

Demars, K. R. "Unique Engineering Properties and Compression Behavior of Deep-Sea Calcareous Sediments." *Geotechnical Properties, Behavior, and Performance of Calcareous Soils*, ASTM STP 777, K. R. Demars and R. C. Chaney, Eds., American Society for Testing and Materials, Philadelphia, PA.

McKown, A., and C. C. Ladd. 1982. "Effects of Cementation on the Compressibility of Pierre Shale." *Geotechnical Properties, Behavior, and Performance of Calcareous Soils*, ASTM STP 777, K. R. Demars and R. C. Chaney, Eds., American Society for Testing and Materials, Philadelphia, PA.

Martin, R. T. 1996. Written communication.

Chapter **6**

pH and Salinity

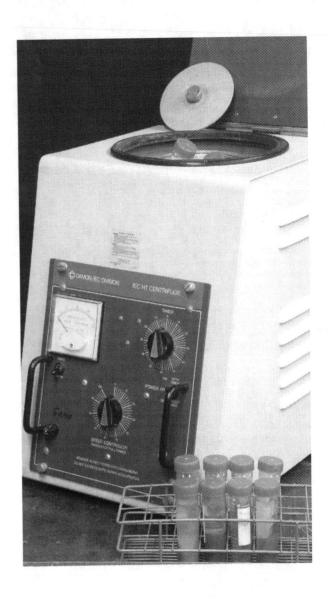

SCOPE AND SUMMARY

This chapter presents background and procedures for the measurement of both the pH and the salinity of the pore fluid in soils. The measurements are applicable to the pore fluid of all types of soils but the procedures in this chapter are restricted to fine-grained material and peat. Salinity and pH are commonly used to characterize the pore fluid. There are several methods available to make such measurements. The chapter provides rapid, simple methods that are approximate, yet sufficiently adequate for most geo-engineering applications.

TYPICAL MATERIALS

Salinity and pH are two common measures used to characterize pore fluids of all geo-materials. Pore fluids are commonly extracted in the laboratory when working with fine-grained and peat materials. The samples do not need to be intact, but the chemical composition must be preserved during transport. When working with free-draining materials such as sands, it will be necessary to extract the pore fluids while in the field.

BACKGROUND

The chemical composition of the pore fluid is important to the behavior of fine-grained soils, the durability of structures in contact with soils, the evaluation of contaminant transport, and the use of ground water. The salinity and pH of the pore fluid have a significant impact on the mechanical properties of fine-grained soils. Both influence the thickness of the double layer around clay particles and consequently the interaction between these particles. In general, the finer particles are more sensitive to pore fluid chemistry than larger particles. The conditions of the pore fluid during deposition (e.g., a marine environment) will control the initial fabric or arrangement of particles.

Changes in the pore fluid chemistry subsequent to deposition can have a significant influence on the mechanical behavior. One classic example is the high degree of strength sensitivity observed when salt is leached from clay deposits. Pore fluid chemistry can enhance the rate of corrosion of materials in contact with soil as well as systems used to extract groundwater. Measurements of pore fluid chemistry are used extensively to evaluate the extent of contamination into deposits, infiltration of salt water into aquifers, and monitoring the security of landfill liners.

Although pH and salinity measurements are applied in very different ways in engineering practice, both parameters contribute to soil characterization and provide information about the chemical properties of the pore fluid. The procedures used to make the measurements are very similar. Both measurements require processing the sample to extract pore fluid in the laboratory and provide an introduction to important analytic techniques. Figure 6.1 shows the typical equipment used to extract pore fluid from a soil specimen.

The measurement of pH and salinity have been selected for this chapter because they are common in the geotechnical arena, provide useful information for engineering practice, and illustrate methods used to test pore fluid chemistry. Therefore, these two measurements are grouped together in this one chapter. Evaluation of pH and salinity are only two of a host of chemical assessment measurements. Many of these measurements are made using similar procedures and ion-specific probes. Others are measured by performing analytical chemical assays on the extracted pore fluid.

pH

pH is a dimensionless representation of the concentration of free hydrogen ions in an aqueous solution. It is calculated as the negative of the logarithm (base ten) of the concentration of hydrogen ions. In this context, concentration is the number of moles

Figure 6.1 Equipment used to extract pore fluid from a soil specimen.

(of hydrogen atoms) per liter of water. pH commonly ranges in value from 0 for a very strong acid to 14 for a very strong base, although these are not absolute limits. For each whole-number decrease in pH there is a 10-fold increase in the number of hydrogen ions in the solution. Even pure water contains electronically charged free ions due to the dissociation of water molecules. One liter of pure water contains 10^{-7} moles (equal to grams for hydrogen because the atomic weight is 1) of positively charged hydrogen ions (H+) and an equal number of negatively charged hydroxyl ions (OH−). This liquid is neutral with a pH equal to 7 and has a balanced ionic charge. It is important to note that water equilibrated with air at standard conditions does not have a pH of 7. The carbon dioxide in the air will dissolve (very quickly) in the water, turning the water into a weak acid having a pH of about 5.5. Since pH is represented using a logarithmic scale, this reaction creates more than a 30-fold increase in the hydrogen ions.

When working with soils, pH is measured either by contacting the surface of the soil with a pH sensitive device or by submersion of the device in the pore fluid. The surface measurement can result in lower pH values because the soil particles in contact with the probe have surface charge and can bias the measurement. The surface contact measurement is most common when conducting corrosion analysis of facilities installed in the ground because the method more closely mimics the field situation. Pore fluid measurements generally require addition of fluid to the sample to yield enough particle-free fluid for the measurement. Measurements are made on each sample using a distilled water solution and a dilute salt solution of 0.01 M calcium chloride ($CaCl_2$). The addition of a salt solution is used to reduce the effect of natural salts in the soil on the measurement.

Equipment is available to measure pH using either a hydrogen specific probe or disposable indicating paper. pH paper is made with a variety of color-indicating organic compounds. These dyes are designed to provide a specific color in the presence of a known hydrogen concentration. The paper is available in different ranges and sensitivities. In general, papers are limited to pH increments of 0.1.

The ion-specific probe provides a more accurate measurement over a wider range of values but must be maintained properly and calibrated routinely. pH probes are used in combination with a very high inductance voltmeter and generally sold as a unit called a pH meter. The probe is an electrode consisting of a very thin glass bulb that is plated to be sensitive to hydrogen ions. The inside of the bulb is filled with a reference solution, typically calomel. An electrical contact is submerged in the reference solution. An electrical potential (differential voltage) is created across the glass, which is a function of the hydrogen ion concentration. A second electrical contact is required to measure the voltage across the glass bulb. This contact is submerged in a salt solution inside the probe that is in contact with the fluid outside the probe. The salt solution connection to the outside liquid is achieved through a hole in the glass filled with a permeable membrane. The salt solution slowly leaks from the probe. The probe must be stored in a salt solution and the reference solution needs to be replaced on a regular basis. Always refer to the manufacturer's instructions for proper maintenance.

Three ASTM standards are available to measure pH for geotechnical purposes: D4972 pH of Soils, D2976 pH of Peat Materials, and G51 Measuring pH of Soil for Use in Corrosion Testing. The method used in G51 is a surface contact measurement and normally performed in the field. The procedures contained in D4972 and D2976 are essentially the same method of measurement with a few noncritical differences in the discussions. This chapter makes use of D4972 Method A (pH Meter) to measure the pH. Figure 6.2 shows the equipment necessary to measure pH using the pH meter.

Important considerations for the measurement of pH include:

- Frequent calibration verification of the pH probe is critical to consistent measurements. Buffer solutions are available in a variety of pH values, with 4, 7, and 10 being the most common. Erratic readings or incorrect readings in the buffer solutions require immediate action. Most meter systems provide the ability to

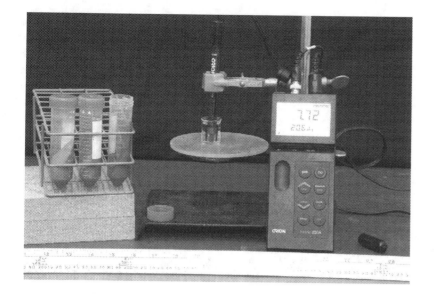

Figure 6.2 Equipment used to measure the pH of pore fluid: centrifuges tubes with soil and pore fluid separated (left); probe positioned in a 10 mL beaker (center); pH meter (right).

recalibrate based on buffer solutions. While this is a great feature, it still requires proper stocking of the buffer solutions.

- Proper care of the pH probe is absolutely necessary. The reference salt solution needs to be replenished on a regular basis and the probe should be cleaned routinely and always stored in the salt solution. There are many products on the market to help with proper maintenance of the probe.
- While it would seem that distilled water, with a pH in the middle of the scale, would provide a convenient calibration point, this is far from the truth. Distilled water has low ionic strength, which results in very high electrical conductivity and makes the task of measuring pH extremely difficult.
- The test methods require addition of distilled water to the soil in order to provide enough fluid for submersion of the pH probe. Diluting the pore fluid will change the pH. The magnitude of the error increases as the soil pH deviates from 7 and as the amount of dilution increases. If more accurate values are required, then the measurements should be corrected for the dilution.

Salinity

Salinity is the amount of salt that is dissolved in the pore fluid. This is sometimes referred to as the Total Dissolved Solids (TDS). Salinity is expressed in terms of grams per liter. Seawater is generally at a salinity of about 35 grams of salt per liter of seawater (g/L) but it varies somewhat with geographic location and season. Sea salt has a complex chemical composition. It is mostly sodium chloride (about 85 percent), with lesser amounts of sulfate, magnesium, calcium and potassium in decreasing concentrations.

When soils are deposited in a marine environment, the salt remains in the interstitial pore fluid. Geologic processes can alter the salinity of the pore fluid after deposition. Reduction of the void ratio by compression expels salt with the pore fluid and generally does not change the salinity. Evaporation will increase the salinity, as evidenced in salt flats and by the clays in the Great Salt Lake region of Utah, where the pore fluid salinity can be up to 300 g/L.

Leaching is a common process that leads to a reduction in the salinity. Artesian pressure below sedimentary clay layers produces upward flow of fresh water, which displaces the saline pore fluid. Diffusion of salt through the network of pore space to create a balance in concentration can also be important in sedimentary deposits. Diffusion can either increase or decrease local salt concentrations.

The presence of salt in the pore space has various repercussions. Salt can be important for project-level concerns, which are beyond the scope of this text. In such situations, measurement of salinity may be required simply to evaluate the in situ conditions. More to the point for measuring the properties of soils is the fact that salinity influences mechanical behavior, the interpretation of some test measurements, and can damage equipment.

Salinity in the pore fluid causes an error in many common index measurements. The magnitude of this error increases with increasing salinity. This is the case for water content, void ratio, dry mass, specific gravity, and particle size analysis using sedimentation. All of these test results can be explicitly corrected provided the salinity is known. Appendix C provides a more detailed discussion relative to corrections as well as equations for the phase relationships that account for salinity.

While not a matter for data correction, the influence of salinity on mechanical behavior is also important for understanding the measurements. Salinity influences everything from the liquid limit to the undrained strength of cohesive soils. Having a measurement of the salinity provides the information to properly compare measurements on different samples and consider changes that may occur if salinity changes over time. Salinity, similar to pH, alters the electrolyte concentration in the pore fluid. The ions alter the thickness of the double layer and ultimately the physical interaction of the particles. Much like pH, salinity changes must be substantial to be important. Soil behavior changes would be expected as salinity changes by a factor of 2. Leached or quick behavior requires salinity in the single digits, but not all soils with low levels of salinity will be leached or quick.

There are at least three methods in common use to measure the salinity of fluid. They each make use of very different principles of measurement and require different levels of effort and equipment. All three methods require separation of the pore fluid from the soil.

The most straightforward approach is to make gravimetric measurements. This method requires extraction of a known volume of pore fluid, oven-drying the fluid to remove the water, and then determining the mass of the evaporate. This method measures the TDS and makes no distinction between fine soil particles and salts that were in solution. A high-precision analytic balance is required to obtain reasonable resolution. The method works fine for high salt solutions but quickly loses precision as the salinity decreases.

The second method takes advantage of the fact that the angle of refraction changes with salt content. One such method is described in ASTM D4542 Pore Water Extraction and Determination of the Soluble Salt Content of Soils by Refractometer. The refraction method makes use of a readily available and fairly inexpensive tool, but has a limited salinity range and poor resolution for low concentrations.

The third method is based on measurement of electrical conductivity. The electrical conductivity of distilled water is extremely low because distilled water does not contain free ions, which are the mobile atoms or molecules capable of transporting electrons. The addition of ions to the water provides charge carriers and increases the electrical conductivity. For a given ion (or salt), the change in conductivity is proportional to the change in concentration. The method has the distinctive advantage that equipment is available to measure conductivity over many orders of magnitude. This advantage provides comparable precision (represented as the coefficient of variation) for all ranges of interest. The electrical conductivity method is used in this text. Figure 6.3 shows the equipment necessary to measure the electrical conductivity of the extracted pore fluid.

The method suffers from the fact that it is indirect. All free ions in solution will contribute to the electrical conductivity, but each with its own specific contribution. Each ion has a different equivalent ionic conductance and molecular weight. The measured conductivity will be the sum of the contribution of each ion. This complication is handled by establishing a calibration curve for the salt of interest. Figure 6.4 presents a typical calibration curve for sea salt that has been normalized to the conductivity at 12.5 g/L. Normalizing the curve provides a generic relationship that can be used with different devices and different reference salts.

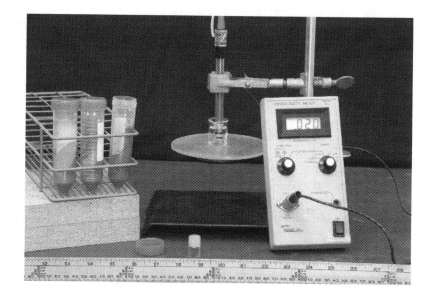

Figure 6.3 Equipment used to measure the electrical conductivity of pore fluid: electrical conductivity probe connected to a handheld meter.

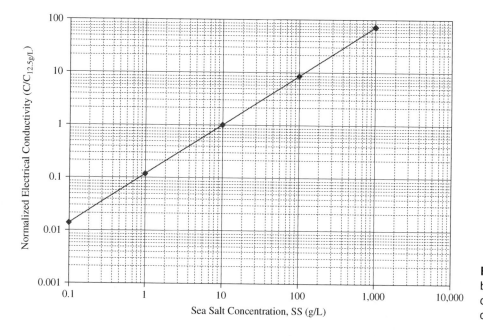

Figure 6.4 Relationship between normalized electrical conductivity and sea salt concentration.

Important considerations for the measurement of salinity include:

- The centrifuge is used to accelerate the rate at which particles will fall out of suspension. The smaller clay particles have the most potential to stay in solution and are also the particles with the most surface charge. These small particles will function as ions and alter the measurement. The supernatant should be clear when making the measurements.
- Cleaning the probe is essential for reproducible results. A small drop of distilled water or saline water from a prior measurement will immediately change the concentration of the small volume of the supernatant. The probe must be meticulously washed and dried in between each measurement.
- Purity of distilled water is important to the measurement since contamination in the water will increase the conductivity. The error is more important as the

salinity of the pore fluid decreases. The fact that the measurement is made on a dilution makes this consideration more important.

- The calibration curve expresses the grams of soluble salt per liter, SS, with seawater salt as the reference salt. When the dissolved salts contributing to the electrical conductivity differ markedly from seawater composition, the absolute value of the grams of salt per liter may be inaccurate. However, even small relative changes in salt concentration are accurately measured as long as the salt composition at the location under investigation does not change markedly. The plot of C/C_o versus SS gives a unique curve whether seawater or reagent-grade sodium chloride is used as the salt.

- The test should not be performed on dried material. Drying will transport salts to the boundaries of the specimen, making it difficult to obtain representative samples. This is especially true of oven-dried samples.

TYPICAL VALUES

Typical values of pH and salinity of selected soils are presented in Table 6.1.

Table 6.1 Typical values of pH and salinity of selected soils.

Soil Type	pH	Salinity (g/L)
Boston Blue Clay*	—	8 to 31
Mexico City Clay**	7.2 to 9.1	1 to 6

*Personal database.
**After Fernández, 1994.

Equipment Requirements

pH

1. Potentiometer equipped with glass-calomel electrode system. Follow the manufacturer's instructions for proper operation of the pH meter. A silver/silver chloride electrode system or similar is also acceptable.

Reagents

1. Water: distilled, deionized water
2. Acid potassium phthalate buffer solution (0.05 M) or commercial pH 4.0 buffer solution
3. Phosphate buffer solution (0.025 M) or commercial pH 7.0 buffer solution
4. Commercial pH probe storage solution
5. Calcium chloride stock solution ($CaCl_2*2H_2O$) (1.0 M). Dissolve 147 g of $CaCl_2*2H_2O$ in water and increase the volume to one liter
6. Calcium chloride solution (0.01 M). Take 10 mL of stock 1.0 M solution (item 5) and add water to increase the volume to 1 L to create a 0.01 M solution.

Salinity

1. Electrical conductivity meter using an alternating current (1000 Hz works well) Wheatstone bridge design. Many commercial units are available.

2. A conductivity probe with a cell constant between 0.4 and 1.0. The design of the cell should be such that air bubbles are not trapped in the cell during filling, it is easy to rinse between readings, and it requires a small volume of solution (about 5 ml).

3. A bench-top analytic centrifuge is desirable to accelerate the separation of particles from the liquid.

Reagents

1. Water: distilled, deionized water

2. Reference sea salt solution; dissolve 12.5 g of sea salt in distilled deionized water and add water to make 1.000 L at room temperature.

General

1. Electronic thermometer (if not integrated in the pH meter)
2. Equipment for water content measurements
3. Equipment to split sample
4. Two 200 ml beakers
5. 100 ml pipette
6. Several 50 ml centrifuge tubes
7. Several 10 ml glass beakers
8. Wash bottle
9. Glass mixing beads (about 5 mm diameter)

CALIBRATION

pH

Calibrate the pH meter at a pH of 4 and 7 using the buffer solutions. Adjustment of the pH meter should follow the manufacturer's directions. At the completion of calibration, the probe should read the proper pH value in each buffer. The probe should be repaired or replaced if the readings are not repeatable in the buffer solutions.

Salinity

The measurement system must be calibrated to obtain the relationship between salt concentration and electrical conductivity. The relationship will depend on the geometry of the probe (area of electrodes and distance between plates characterized by the cell constant), the particular salt, and the conductivity meter. The conductivity meter provides a measure of conductance rather than conductivity. Conductance is the inverse of resistance and measured in Seimens (S). Conductivity is conductance per meter. It is always necessary to obtain a specific calibration for a particular probe and conductivity meter combination. It is sufficiently accurate to account for variations in salt type by determining a normalized calibration curve and then adjusting to the salt of interest. The range of calibration should match the range in expected values.

Prepare the starting calibration point by dissolving 20.00 g of salt in distilled water and adjusting the volume to 200 cc. This provides a calibration point at 100 g/L, which is higher than most natural soils. Remove about 10 cc with a pipette and transfer to a clear volumetric. Measure the electrical conductivity using the following procedure.

1. Measure the distilled water and confirm that the conductance is below 1×10^{-3} mS.

2. Carefully dry off the electrodes and surrounding surfaces. Any excess water will dilute the next measurement.

3. Measure the conductance of the salt solution (C).

4. Rinse the probe with distilled water.

5. Dry off the electrode and surrounding surfaces.

6. Measure the conductance of the distilled water to confirm that the probe is clean.

7. Dry off the electrode and surrounding surfaces.

8. Measure the conductance of the salt solution (C) and verify the value is the same as obtained previously. If not, repeat the measurement, taking care not to systematically dilute the measurement.

9. Reduce the salt concentration by half for the next calibration point. Do this by using a clean pipette to remove 100 cc from the previous solution and transferring to a clear 200 cc volumetric. Add distilled water to increase the volume to 200 cc and mix thoroughly.

10. Measure the conductance of the reduced salt concentration solution using steps 1 through 8.

11. Reduce the salt concentration by half once again, and repeat the measurements. The calibration should be performed for at least 5 concentrations, such as: 100 g/L, 50 g/L, 25 g/L, 12.5 g/L, and 6.25 g/L. The range and number of points can be adjusted depending on the range in expected values.

12. Construct a calibration graph with the normalized conductance ($C/C_{@12.5\,g/L}$) on the y-axis versus the salinity on the x-axis, as shown on Figure 6.4. Plot both axes on log scales.

SPECIMEN PREPARATION

Obtain a sample of about 80 g equivalent dry-mass material passing the No. 10 sieve. If the material contains a significant amount of plus No. 10 material, it should be removed manually or the material should be worked through the sieve while wet. Using a spatula, blend the wet material on a glass plate. Transfer one-half of the material to a bowl and air-dry. This portion will be used for the pH measurement. Drying may take several days, depending on the plasticity. Use one-quarter to measure the water content, ω_N. The final quarter will be used for the conductivity measurement.

PROCEDURE

For the pH Measurement

The pH analysis will be performed in general accordance with ASTM Standard Test Method D4972.

1. Obtain approximately 10 g of air-dried soil. Place the soil into a centrifuge tube and add approximately 10 mL of water.

2. Obtain approximately 10 g of air-dried soil. Place the soil into a centrifuge tube and add approximately 10 mL of the 0.01 M $CaCl_2$ solution.

3. Add a few small (about 5 mm) glass beads to each centrifuge tube and shake rigorously for 30 seconds.

4. Place tubes in centrifuge (be sure to balance the load) and run at about 5000 rpm for 10–20 minutes.

5. The mixture should be at approximately room temperature (15 to 25°C) at the time of pH measurement.

6. Decant the supernatant liquid from each tube into the 10 ml glass beaker.

7. Insert the probe into the clear liquid of each beaker and read the pH to 0.01 on the meter and the temperature to 0.1°C.

8. Rinse the probe with distilled water between measurements.

For the Salinity Measurement

The salinity analysis is not currently documented in any standard test method. The following procedure is based on an in-house protocol (Martin, 1982). Measurement of salinity is very analytical and care during mass measurements is essential for consistent results.

1. Record the mass of a centrifuge tube with cap (M_c) to 0.01 g.
2. Add the wet soil to the centrifuge tube. Use the equivalent of about 15 g dry mass.
3. Record the mass of centrifuge tube, cap, and soil (M_{tc}) to 0.01 g.
4. Add approximately 15 g of distilled water to the centrifuge tube.
5. Record the mass of centrifuge tube, cap, water, and soil (M_{twc}) to 0.01 g.
6. Add a few small (about 5 mm) glass beads to each centrifuge tube and shake rigorously for 30 seconds.
7. Allow 20–30 minutes for equilibration and repeat 30 seconds of shaking.
8. Place tubes in centrifuge (be sure to balance the load) and run at about 5000 rpm for 10–20 minutes.
9. Decant the supernatant liquid from each tube into the 10 ml glass beaker.
10. Clean the probe with distilled water and dry.
11. Measure the conductance of the reference salt solution (C_o).
12. Clean the probe with distilled water and dry.
13. Measure the conductance of distilled water (C_{dw}).
14. Dry the probe.
15. Measure the conductance of the clear supernatant liquid (C) from the sample.
16. Clean and store the probe.

Calculations

pH

No calculations are required for the pH measurement. The meter will provide direct readings of pH and usually temperature.

Salinity

1. Convert the conductance of the liquid supernatant to a normalized reading (NC) using Equation 6.1:

$$NC = \frac{C}{C_o} \qquad (6.1)$$

Where:
NC = normalized reading of the liquid supernatant (dimensionless)
C = conductance in the liquid suspension (S)
C_o = conductance in the reference 12.5 g/L solution (S)

2. Use the normalized conductance to obtain the salinity (SS) of the liquid supernatant using the calibration curve.

3. Calculate the mass of water in the wet soil added to the centrifuge tube (M_w) using Equation 6.2:

$$M_w = (M_{tc} - M_c)\left(\frac{\omega_N}{100}\right) \Big/ \left(1 + \frac{\omega_N}{100}\right) \qquad (6.2)$$

Where:
M_w = mass of water (g)
M_{tc} = mass of centrifuge tube, cap, and moist soil (g)
M_c = mass of centrifuge tube and cap (g)
ω_N = natural water content of soil (%)

4. Calculate the water content of the soil in the centrifuge tube (ω_c) using Equation 6.3:

$$\omega_C = \frac{(M_w + M_{twc} - M_{tc})}{(M_{tc} - M_c - M_w)} \times 100 \qquad (6.3)$$

Where:

ω_c = water content of soil specimen (%)

M_{twc} = mass of centrifuge tube, cap, moist soil, and added water (g)

5. Calculate the salinity of the test specimen using Equation 6.4:

$$RSS = SS \times \frac{\omega_C}{\omega_N}$$

(6.4)

Where:

RSS = salinity of the test specimen (g/L)

SS = salinity of the liquid supernatant (g/L)

Report

Report the average value and standard deviation (if multiple specimens are processed) of the pH in each of the solutions (distilled water and calcium chloride) to 0.01, temperature of the pH solutions at the time of the measurement to 0.1°C, salinity of each of the distilled water and the liquid supernatant in g/L to the nearest two significant digits, and water content to three significant digits.

PRECISION

Criteria for judging the acceptability of test results obtained by ASTM D4972 Method A (pH meter) are not published in the current version of the standard. However, results of a limited study performed by one agency (National Technical Center of the United States Department of Agriculture) using 174 replicates for the water mixture and 32 replicates for the calcium chloride mixture were published in a previous version of the standard. The resulting Within Laboratory Repeatability is given below.

- *Within Laboratory Repeatability:* Expect the standard deviation of your results on the same soil to be on the order of 0.031 (pH units) for the water mixture and 0.139 (pH units) for the calcium chloride mixture.

Criteria for judging the acceptability of salinity results have not been determined. However, based on the resolution of the measurements, expect the standard deviation of your results to be on the order of 2 percent of the average value determined for the same soil.

DETECTING PROBLEMS WITH RESULTS

If the standard deviation for one set of pH measurements exceeds the estimates provided above, then evaluate the techniques of the individual performing the test. The most likely source of the problem is contamination, either of the buffer solutions or the supernatants. If the test results do not fall within the typical ranges, the likely cause of error is systemic, such as an error in the calibration of the probe, or the plating of the glass bulb on the probe is insufficient.

Several methods of isolating the causes of errors are possible. Systematic errors due to equipment deficiencies can be identified by repeating the probe calibration procedure after replacing any expired buffer solutions if they have expired. Replate or replace the glass bulb on the probe if necessary. Procedural and technique errors are best identified by performing the test on a material with a known pH such as drinking water, which typically will be in the range of 6.8 to 7.2.

Salinity levels above 35 g/L indicate the very unusual situation of the pore fluid having salinity greater than that of seawater. If a measurement indicates salinity outside the typical values for the site conditions, evaluate the equipment. Visually inspect the probe to make sure the geometry of the cell has not changed and check the reading in a standard solution.

If the range of a set of salinity measurements on the same soil exceeds the estimates provided above, evaluate the testing method. Likely sources of error are poor sample homogenization practices, sloppy methods of obtaining a matching water content specimen, poor probe cleaning technique, and errors in the various masses. Also verify that small soil particles are not in the supernatant. If any of these issues are identified, repeat the experiment, taking care to completely homogenize the specimen and matching water content, obtain the correct masses, practice better cleanliness techniques, and/or allow a longer time for soil solids to fall out of solution.

REFERENCE PROCEDURES

ASTM D4972 pH of Soils.

In-house procedure to determine salinity using a conductivity probe (Martin, 1982).

REFERENCES

Refer to this textbook's ancillary web site, www.wiley.com/college/germaine, for data sheets, spreadsheets, and example data sets.

Fernández, S. C.1994. "Characterization of the Engineering Properties of Mexico City Clay," MS Thesis, Department of Civil and Environmental Engineering, Massachusetts Institute of Technology, Cambridge.

Martin, R. T. 1982. Written communication.

Chapter 7

Organic Content

SCOPE AND SUMMARY

This chapter provides background and procedures to perform the organic content test by the loss on ignition method. The measurement is performed by measuring the reduction in mass of an oven-dried specimen when it is baked at very high temperature to burn off the organic matter.

TYPICAL MATERIALS

Organic content by loss on ignition is typically determined on sandy soils as well as organic clay, organic silt and mucks for geotechnical purposes. The method can also be used on peats for evaluation of use as fuel.

BACKGROUND

Organic content can be determined by loss on ignition (LOI) or chemical oxidation methods. This chapter deals specifically with the loss on ignition method, which is more straightforward and is typically used in geotechnical practice. The results will be different for the two methods and the chemical oxidation (usually by hydrogen peroxide) more correctly measures organic matter. The loss on ignition will generally yield higher values because the elevated temperature will also drive off water contained in the structure of several clay minerals.

Figure 7.1 Porcelain crucibles and muffle furnace used to conduct the organic content test by loss on ignition. Samples in crucibles are potting soil (left) and clay (right).

Figure 7.2 Soil specimens after ignition at 440°C: potting soil (left); clay (right).

The organic content test by loss on ignition measures loss of mass by ignition when a specimen previously dried in a 105°C oven (as used to determine water content for geotechnical testing purposes) is then placed in a porcelain crucible that is then placed in a furnace at a much higher temperature. Figure 7.1 shows porcelain crucibles and the muffle furnace used for the test.

The ash content is determined as the mass of soil remaining after ignition to the mass of oven-dried soil, expressed as a percentage. The organic content is the difference between 100 percent and the ash content. Figure 7.2 shows samples after ignition.

Two ignition temperatures are used in the reference ASTM standard titled D2974 Standard Test Methods for Moisture, Ash, and Organic Matter of Peat and Other Organic Soils. The lower temperature is 440°C and is typically used for general geotechnical purposes. The higher temperature (750°C) is used for evaluation of peat materials for use as fuel. When testing fine-grained soils, the furnace temperature causes removal of water associated with the structure of the mineral and therefore should be used as a relative indicator only for these soils.

The organic content is used as criteria for acceptability of materials for construction, such as granular materials for road bases and structural fill. Organic materials degrade over time and therefore cause production of gases and result in settlement. Usually, construction specifications will have acceptance criteria of the material being "inorganic" or will give a maximum organic content. The soil science community uses minimum organic content criteria for whether a material can be used as a growing medium. The organic content also assists in classifying peat in ASTM D4427 Standard Classification of Peat Samples by Laboratory Testing.

It is important to note that the Unified Soil Classification System (USCS) does not use the organic content test for determining whether a soil is described as organic. Rather, the USCS uses criteria based on the ratio of the liquid limit of a soil after oven-drying to the liquid limit before oven-drying.

TYPICAL VALUES

Typical values of organic content, as measured by loss on ignition in percent (%), are listed in Table 7.1.

CALIBRATION

Calibration is not required for this test. However, each porcelain crucible should be marked with permanent paint and baked to at least the test temperature before it is used to perform any soil testing.

SPECIMEN PREPARATION

The organic content test is performed on a relatively small amount of material. Therefore, it is very important to obtain representative material. Select an appropriate sized specimen (according to the maximum particle size) and perform a moisture content test according to ASTM D2216. Subsample approximately 10 grams of the oven-dried water content specimen for the organic content test. Large grains of sand and gravel will not contribute to the LOI but will be important for the application of the result. If the subsample intentionally avoids large particles, this must be noted on the data sheet.

PROCEDURE

The organic content analysis will be performed in general accordance with ASTM Standard Test Method D2974.

1. Determine the mass of a porcelain crucible (M_c) to 0.01 g.
2. Obtain approximately 10 g of the oven-dried moisture content specimen.
3. Add the oven-dried soil to the crucible and determine the mass (M_{sc}) to 0.01 g.
4. Place the crucible and contents in a muffle furnace. The crucible must be uncovered. Gradually bring the muffle furnace up to 440°C. Hold the furnace at this temperature until there is no change in mass. This usually takes about 4 to 5 hours of holding at the maximum temperature, but must be confirmed.
5. Remove the crucible from the furnace, cover with aluminum foil, and place in a desiccator to cool.

Table 7.1 Typical values of organic content.

Soil Type	Organic Content (%)	Soil Type	Organic Content (%)
Peats	75–100*	Loam Borrow	4–20***
Organic Matter	75–100**	Peat Borrow	25 (minimum)***
Highly Organic Soils	30–75**	Road Base Material	By specification. Usually less than 5.
Organic Silt/Sand	5–30**	Structural Fill	By specification. Usually "Inorganic" or less than 2.

* ASTM D4427.
** Naval Facilities Engineering Command, 1986, *Design Manual: 7.01.*
*** Massachusetts Highway Department, 1988.

6. Uncover the crucible, obtain the mass of the crucible and ash to 0.01 g.

7. Place the crucible back in the muffle furnace for a period of at least one hour, and repeat steps 5 and 6 until there is no change in mass from the previous measurement. Record the final mass of the crucible and ash (M_{ac}) to 0.01 g.

Calculations

Compute the ash content using Equation 7.1:

$$AC = \frac{M_{ac} - M_c}{M_{sc} - M_c} \cdot 100 \qquad (7.1)$$

Where:

AC = ash content of soil (%)
M_{ac} = mass of crucible and ash (g)
M_c = mass of crucible (g)
M_{sc} = mass of crucible and soil (g)

Compute the organic content using Equation 7.2:

$$OC = 100 - AC \qquad (7.2)$$

Where:

OC = organic content of soil (%)

Report

Report the organic content and ash content to the nearest 0.1% along with the temperature of the muffle furnace. Report the average and standard deviation as well, if multiple measurements were made on representative specimens.

PRECISION

Criteria for judging the acceptability of test results obtained by this test method have not been determined. However, based on a brief study of performing twelve tests on a potting soil specimens with an organic content of approximately 21 percent, two organic content tests performed properly by a single operator in the same laboratory in the same time period should not differ by more than about 4 percent.

DETECTING PROBLEMS WITH RESULTS

If the range of one set of measurements exceeds 4 percent, individual techniques should be evaluated. The likely problems are: incomplete ignition of organics, poor massing technique, insufficient initial specimen mass, insufficient cooling time, or heterogeneity of specimens.

REFERENCE PROCEDURES

ASTM D2974 Moisture, Ash, and Organic Matter of Peat and Other Organic Soils.

REFERENCES

Refer to this textbook's ancillary web site, www.wiley.com/college/germaine, for data sheets, spreadsheets, and example data sets.

Massachusetts Highway Department. 1988. *Standard Specifications for Highways and Bridges.*

Naval Facilities Engineering Command. 1986. *Soil Mechanics: Design Manual 7.01.*

Chapter 8

Grain Size Analysis

SCOPE AND SUMMARY

This chapter provides background information and procedures to perform a grain size analysis on a wide range of soils. The procedures differ depending on the size of the particles and the resolution of the reported results. Manual measurements are used for very large particles. Mechanical sieving is used for the coarse-grained portion and sedimentation is used for the fine-grained portion of the material. Detailed procedures are provided for both the sieve and sedimentation analyses. The grain size analysis is part of the Unified Soil Classification System, which is presented in Chapter 10, "Soil Classification and Description."

TYPICAL MATERIALS

Grain size analyses are performed on essentially all geotechnical particulate materials ranging from clay to boulders. Similar techniques are used in many other professions and industries to characterize manufactured products such as coffee, corn flakes, pharmaceutical powders, and others, and to control industrial processes.

BACKGROUND

Particulate materials are made up of a range and distribution of particle sizes. In most geotechnical applications, this distribution is a continuum of varying sizes over the

Figure 8.1 Photograph of soil sample as a whole, then as broken down into its parts.

Table 8.1 USCS grain size boundaries.

| Soil Component | Sieve Size | | Grain Size (mm) | | Separation Technique |
	Lower Bound	Upper Bound	Lower Bound	Upper Bound	
Boulders	12 in.	—	300	—	Manual Measurement
Cobbles	3 in.	12 in.	75	300	
Gravel					
Coarse	0.75 in.	3 in.	19	75	
Fine	#4	0.75 in.	4.75	19	
Sand					Mechanical Sieving
Coarse	#10	#4	2	4.75	
Medium	#40	#10	0.425	2	
Fine	#200	#40	0.075	0.425	
Fines	—	#200	—	0.075	Sedimentation

represented range for a given specimen. As shown in Figure 8.1, overall a soil sample may appear to be homogeneous while actually consisting of particles ranging from sand to silt sizes.

Grain size analysis refers to discerning the percentage of particles (by dry mass) within a specified particle size range across all the sizes represented for the sample. The distribution of particle sizes is used to distinguish the maximum particle size and the major portion of the particle sizes, as well as to characterize the soil, such as by the Unified Soil Classification System (USCS). The USCS uses the grain sizes shown in Table 8.1 for distinguishing between soil types.

The result of this analysis is a grain size distribution (GSD), which also may be referred to as a particle size distribution (PSD). In geotechnical engineering, a typical GSD has the percent finer (N) plotted on the y-axis and the particle size (using a log scale) on the x-axis. The percent finer refers to the percentage of the dry soil mass that is composed of smaller-diameter grains than a given particle size. The percent finer results are presented on a natural scale from 0 to 100 percent, with the 100 percent value at the top of the y-axis. The particle size results are typically plotted in millimeters, with the largest value at the left-most end of the scale and smaller values as the x-axis advances to the right. The x-axis is presented on a log scale in order to adequately represent the vast range of particle sizes. An example grain size distribution is shown as Figure 8.2, using the Unified Soil Classification System (USCS) definitions of particle sizes. The figure also shows the opening size of the sieves included in a standard set.

A GSD is useful for visualizing the breakdown of particle sizes. The plot can also be used to estimate the size of material for which a certain percentage is finer (D_X). Grain size analyses produce information used in the design of filters, as well as for characterizing the shape of the curve.

The D_{10}, D_{30}, and D_{60} values are used to calculate two dimensionless parameters, namely C_u and C_c, which characterize the position and shape of the grain size distribution curves. The D_{10} value is also referred to as the "effective diameter" in some engineering applications. The coefficient of uniformity (C_u) characterizes the spread in sizes within the major portion of the material and is given by Equation 8.1:

$$C_u = \frac{D_{60}}{D_{10}} \tag{8.1}$$

Where:

C_u = coefficient of uniformity (dimensionless)
D_{60} = particle diameter corresponding to 60 percent finer by dry mass on the grain size distribution curve (mm)
D_{10} = particle diameter corresponding to 10 percent finer by dry mass on the grain size distribution curve (mm)

The coefficient of curvature (C_c) characterizes the shape of the central portion of the curve and is given by Equation 8.2:

$$C_c = \frac{D_{30}^2}{D_{60}D_{10}} \tag{8.2}$$

Where:

C_c = coefficient of curvature (dimensionless)
D_{30} = particle diameter corresponding to 30 percent finer by dry mass on the grain size distribution curve (mm)

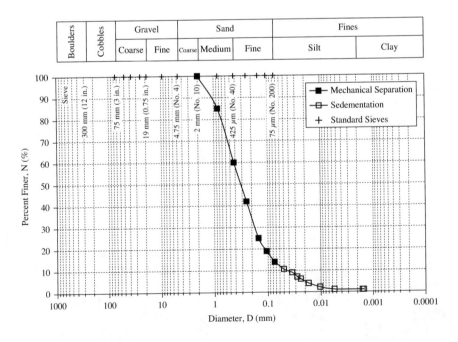

Figure 8.2 Grain size distribution using the USCS definitions of particle sizes.

The values of C_u and C_c are used within the USCS classification system to characterize coarse-grained soils (gravels and sands) as poorly graded or well graded, as will be discussed in Chapter 10.

While it may seem trivial, the task of correctly measuring a GSD can be very difficult. In the most general situations, geotechnical engineers work with individual particles with dimensions as large as a few meters to as small as a fraction of a micron. This is 7 orders of magnitude in size difference, 14 orders of magnitude in surface area difference, and 21 orders of magnitude in mass difference for a single particle! That being the case, different methods of determining grain size must be used for various size ranges. In fact, the diameters of very large particles are literally measured by hand, one particle at a time. Mid-sized particle diameters are measured by mechanical separation or sieving. Fine-grained particle diameters are determined by settling velocity. Table 8.1 lists the appropriate separation technique based on particle size. Each method has different criteria for establishing the particle diameter and at times it can be a challenge to integrate results together at the boundaries.

Flexibility is the key to performing high-quality particle size analyses while being timely and efficient. The most appropriate method and the exact steps will vary depending on the material and the project needs. However, there are five essential considerations when performing the analysis: (1) specimen size, (2) alteration, (3) cementation, (4) pretreatment, and (5) sieve set selection.

In addition to the usual rules of significant digits, the specimen size controls the reportable resolution of the test result. This is a simple matter of considering the impact of adding or removing one particle from the test specimen. The largest-diameter particle has the most impact. To report the results to 1 percent, the initial specimen dry mass must be at least 100 times the mass of the largest particle. For 0.1 percent resolution, the initial specimen mass must be 1000 times the largest particle mass. While this may seem to be a basic concept, the impact on the profession has been profound due to the increase in minimum specimen sizes. Table 8.2 provides values for the minimum required mass of dry material in a specimen as a function of maximum particle size. Values are included for a 1 percent and a 0.1 percent reporting resolution and for the most commonly encountered maximum sizes. The calculated limits differ slightly from those presented in ASTM D6913 Particle-Size Distribution (Gradation) of Soils using Sieve Analysis, where the values are adjusted for various practical considerations. Both sets of values are included in Table 8.2 for comparison.

Alteration is a term used to characterize changes in the GSD caused during physical processing of the material. Alteration can be the formation of new particles (creating aggregates from smaller particles, such as during drying) or breakage of grains to form

Largest Particle		Particle Mass	Minimum Dry Mass of Specimen			
			Constant Factor		By ASTM D6913	
(mm)	(inches)	(G_s = 2.7)	For 1%	For 0.1%	For 1%	For 0.1%
152	6	5000 g	500 kg	5000 kg	NA	NA
76	3	625 g	65 kg	650 kg	70 kg	NA
50	2	186 g	20 kg	200 kg	25 kg	NA
25	1	23 g	2,500 g	25 kg	3,000 g	NA
19	3/4	9.8 g	1,000 g	10 kg	1,200 g	NA
9.5	3/8	1.2 g	120 g	1200 g	165 g	NA
4.75	#4	0.15 g	15 g	150 g	75 g	200 g
2.00	#10	0.011 g	1.1 g	11 g	50 g	100 g
0.425	#40	0.0001 g	0.01 g	0.1 g	50 g	75 g

Table 8.2 Required dry mass of sample as a function of maximum particle size and reporting resolution.

smaller particles (such as due to too much mechanical energy imparted on the soil during sieving). In general, the fine fraction of a soil should not be dried prior to testing. Drying often causes irreversible alteration of the GSD. Even air-drying changes the GSD of some soils. On the other hand, sieving must be done on dry material; therefore, the fine fraction is removed prior to drying and sieving. Very large particles can be air-dried to save oven space but oven-drying is necessary for sands and gravels to remove all the water. Use of an oven also accelerates the testing process.

Cementation or bonding of grains forms larger aggregates and can cause errors in the GSD. When working with coarse sand, cementation is easy to identify. However, when the grain size becomes small, bonding is less obvious and breaking down the cementation can be very difficult. If cementation or bonding is identified in the material, the decision to debond the particles must be made according to the engineering application of the grain size analysis. For example, if residual soils are being compacted in the field and the process of placement will break the material into a finer, cohesive matrix, the grain size distribution of interest may be that following a laboratory compaction test. If the decision is made to debond the material, the debonding method must be selected to simulate field conditions.

Pretreatment of the material prior to measuring the GSD will alter the results due to removal of material as well as debonding aggregates. In most situations, it is appropriate to perform the particle size analysis on the "as-received" material. However, in certain situations, removal of organics, carbonates, iron oxides, or evaporates may be warranted. Methods are available to treat for each separately. If all treatments are applied, the final analysis provides the GSD of only the mineral constituents. This is common practice in soil science but seldom desired for geotechnical practice.

Sieve set selection must be chosen to obtain a reasonably detailed representation of the GSD, without creating unreasonable effort. A typical curve would be represented by about seven measurements. The sieves should be selected to match the range of sizes present in the specimen and ideally would provide data points evenly spaced across the GSD when plotted on a log scale. A second consideration in sieve set selection is sieve overloading. Overloading occurs when there is a large amount of soil in one size increment, such that the particles do not have sufficient opportunity to pass through the sieve opening. If a large fraction of the specimen is represented by a very narrow size range (called gap grading), then more sieves will be required in that size range to spread out the individual sieve load. Overloading of sieves must be prevented either by changing the sieve set to distribute the load more evenly, by sieving the specimen in portions, or by using a sieve with a larger surface area.

In the most general case, a sample would contain a wide range of particle sizes from boulders to clays. For such a wide range in sizes, the grain size analysis will involve a combination of blending, separating, and splitting the material in order to control the mass used at various stages of the analysis.

Manual and Mechanical Analyses

The largest particles (i.e., greater than 150 mm [6 in.] in diameter) are usually separated from the rest of the material and then massed and measured by hand. This is referred to as the "manual" method. The particles are handled individually while measuring the length, width, and height. For nonspherical particles, the particle diameter is taken as the intermediate (second largest) dimension, which is conceptually equivalent to the smallest square hole the particle would fit through. The diameter and mass are recorded for each particle. Particles are then grouped within a size increment. Usually, the increments are spaced such that the maximum diameters increase by a multiple of approximately two.

The GSD for the coarse fraction of soil consisting of particle diameters 150 mm and smaller is analyzed using sieves, referred to as the "mechanical" method. Sieves are shown in the photograph in Figure 8.3. Since sieves have square openings, the size criterion is based on round or platelike soil particles fitting through a square hole.

Figure 8.3 Photograph of sieves.

A series of sieves is set up with the largest openings in the top sieve, gradually reducing in size as the sieves get closer to the bottom. The choice of sieve openings controls the fineness of the GSD. A cover is placed on top and a solid pan placed on the bottom. This set is referred to as a rack. A soil specimen is placed in the top sieve of the rack, and the rack is shaken to force the particles down through the sieves until each particle rests on a sieve that it cannot pass through. The dry mass retained on each sieve size is then used to compute the percent passing. This process takes a continuous distribution of sizes and lumps them into discrete size ranges.

Sieve sizes in the United States were originally developed with designations of either nominal opening size or with a number designation. Sieves with nominal openings smaller than 3/8 in. (9.5 mm) are listed as the number of openings per an inch. Currently, the preferential designation is to refer to the sieves by their opening size in millimeters. ASTM E11 Standard Specification for Wire Cloth and Sieves for Testing Purposes lists the designation and pertinent information for the standard sieves in the United States. Many "standard" sieve sets are used in practice and vary by organization and application. For example, the Tyler series was established so that, in most cases, the opening sizes change by a factor of the square root of two as the size becomes successively larger. This sieve set is used in the concrete industry to determine the Fineness Modulus. Three common standard sieve sets are presented in Table 8.3. Many other opening sizes are commercially available.

Generally, sieves are used for the portion of soil greater than 75 μm (No. 200 sieve) in particle diameter. Sieves are available down to the 20 μm sieve, but at that scale the force of gravity on particles is small and electrostatic forces are high. This combination of factors makes it impractical to use these small sieve sizes for quantitative particle separation. The specimen fraction that is finer than the 75 μm (No. 200) sieve is analyzed using nonmechanical means. Sedimentation is the common method used to determine the particle size distribution for this fraction.

Sedimentation

Sedimentation methods are used for determining grain size distributions of fine-grained soils. Sedimentation describes the process of particles falling through a fluid, and is used to separate the particles by size in space and time. The concept is that the smaller particles take longer to drop out of solution. Sedimentation is then combined with another method that measures the quantity of particles corresponding to the specific size.

Determination of the particle size during sedimentation is based on Stoke's Law. Stoke's Law provides a governing equation based on the hydrodynamics of a single

Table 8.3 Three common standard sieve sets.

ASTM D6913		Tyler Series		U.S. Bureau of Reclamation	
Sieve Designation		Sieve Designation		Sieve Designation	
Alternative	Standard	Alternative	Standard	Alternative	Standard
Lid.	—	Lid.	—	Lid.	—
				4 in.	100 mm
3 in.	75 mm	3 in.	75 mm		
2 in.	50 mm	2 in.	50 mm	2 in.	50 mm
1–1/2 in.	37.5 mm				
		1.06 in.	26.5 mm		
1 in.	25.0 mm			1 in.	25.0 mm
3/4 in.	19.0 mm			3/4 in.	19.0 mm
		0.742 in.	18.85 mm		
		0.525 in.	13.33 mm		
				1/2 in.	12.5 mm
3/8 in.	9.5 mm			3/8 in.	9.5 mm
		0.371 in.	9.423 mm		
		No. 3	6.35 mm	No. 3	6.35 mm
No. 4	4.75 mm	No. 4	4.75 mm	No. 4	4.75 mm
		No. 6	3.35 mm	No. 6	3.35 mm
		No. 8	2.36 mm	No. 8	2.36 mm
		No. 9	1.981 mm		
No. 10	2.00 mm	No. 10	2.00 mm	No. 10	2.00 mm
				No. 12	1.7 mm
		No. 14	1.4 mm		
				No. 16	1.18 mm
No. 20	850 μm	No. 20	850 μm	No. 20	850 μm
		No. 28	589 μm		
				No. 30	600 μm
		No. 35	500 μm		
No. 40	425 μm			No. 40	425 μm
		No. 48	295 μm		
				No. 50	300 μm
No. 60	250 μm	No. 60	250 μm	No. 60	250 μm
		No. 65	208 μm		
				No. 70	212 μm
No. 100	150 μm	No. 100	150 μm	No. 100	150 μm
No. 140	106 μm			No. 140	106 μm
		No. 150	104 μm		
No. 200	75 μm	No. 200	75 μm	No. 200	75 μm
		No. 270	53 μm	No. 270	53 μm
		No. 400	38 μm	No. 400	38 μm
Pan	—	Pan	—	Pan	—

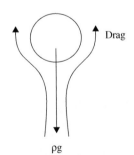

Figure 8.4 Forces acting on a spherical particle falling through a fluid.

spherical particle falling in a stationary fluid. The particle is driven downward by gravitational forces and retarded by drag forces. Figure 8.4 illustrates the concept.

A particle will accelerate downward in the fluid until the drag forces equal the gravitational force. Once these forces are in equilibrium, the particle will achieve terminal velocity (v). For a smooth particle, the terminal velocity can be expressed as a function of the density difference between the particle and fluid, the particle diameter, and the fluid viscosity, as shown in Equation 8.3. The terminal velocity is proportional to the diameter of the particle squared. This simple principle is used to separate particles by size in both time and position within a container of liquid.

$$v = \frac{(\rho_s - \rho_f)gD^2}{18\mu} \qquad (8.3)$$

Where:

v = terminal velocity (cm/s)
ρ_s = mass density of solids (g/cm^3)
ρ_f = mass density of fluid (g/cm^3)
g = acceleration due to gravity (cm/s^2)
D = diameter of the particle (mm)
μ = viscosity of fluid (mPa-s)

There are important assumptions implicit in sedimentation analysis. The first assumption is that Stoke's Law is valid, which means that the particles are spherical and smooth, there is no interference between particles, no side wall effects (no difference in currents between middle and sides of container), all particles have the same density, and flow is laminar. The laminar flow restriction sets an upper size limit of about 0.1 mm. At the other extreme, small particles will exchange momentum with water molecules. This Brownian motion effect sets a lower limit for sedimentation measurements at about 0.0002 mm, although time limitations make it impractical to use the method for particles much smaller than 0.001 mm.

If startup effects are ignored, then the velocity can be taken as the distance of travel divided by the travel time, or H/t. Assuming the fluid is water and the temperature remains constant during the entire experiment, substitutions can be made and Equation 8.3 rearranged to solve for the largest particle diameter that remains in suspension as a function of time and position. Put another way, at a given time, if the fluid properties and the distance of fall are known, the diameter of the largest particle in suspension at that location can be determined. This yields Equation 8.4:

$$D = \sqrt{\frac{18\mu}{\rho_w g(G_s - 1)} \cdot \frac{H}{t}} \qquad (8.4)$$

Where:

H = distance particle falls (cm)
ρ_w = mass density of water (g/cm^3)
G_s = specific gravity (dimensionless)
t = time for fall (s)

Refer to Appendix B for the mass density and viscosity of water as a function of temperature.

At time equal to zero in a sedimentation test, the fluid will have the same concentration of particles at every location. Each particle size falls at a different terminal velocity, but all particles of a particular size fall at the same rate independent of starting position. As a result, at a later time during the sedimentation process, the distribution of particle sizes is different at every depth within the cylinder. Any particle of size D or larger will have fallen below the point of measurement in the suspension. This is shown schematically in Figure 8.5.

The model provides a determination of particle size and must be combined with a method to measure the mass of particles at a given location and time. Numerous

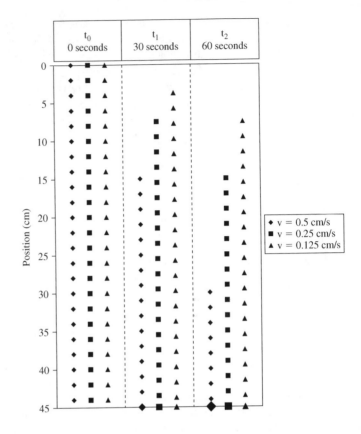

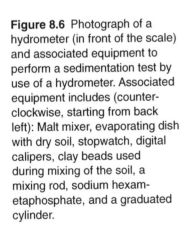

Figure 8.5 Schematic of the positions of different particle sizes, and therefore different settling velocities, at different times in the sedimentation cylinder.

Figure 8.6 Photograph of a hydrometer (in front of the scale) and associated equipment to perform a sedimentation test by use of a hydrometer. Associated equipment includes (counterclockwise, starting from back left): Malt mixer, evaporating dish with dry soil, stopwatch, digital calipers, clay beads used during mixing of the soil, a mixing rod, sodium hexametaphosphate, and a graduated cylinder.

techniques have been used to quantify the amount of material remaining in suspension during sedimentation. These techniques include suspension density by hydrometer, suspension density by force difference, suspension extraction by the pipette, change in intensity by radiation (gamma rays), and change in scattering of light. The most common techniques in geotechnical practice make use of either the hydrometer or the pipette.

This text uses the hydrometer method, which is also the basis of ASTM D422 Particle-Size Analysis of Soils. Figure 8.6 shows a picture of a hydrometer, along with some of the equipment necessary to perform a sedimentation test using a hydrometer.

The hydrometer is a buoyancy-based device that balances the weight (mass times gravity) of the device with the weight of displaced fluid.

Figure 8.7 shows the important geometric features of a typical hydrometer. In brief, a known quantity of soil is mixed with water into a slurry. A hydrometer is placed in the solution to determine the specific gravity of the fluid at a series of times. The fraction of particles remaining in suspension is calculated using the specific gravity of the particles.

The hydrometer integrates the fluid density everywhere it is submerged. The expanded bulb section makes the hydrometer most sensitive to depth range h. The center of buoyancy (c_b) of this section is taken as the reading location that is equal to the distance H from the free surface. Using the typical length of the expanded section and Equation 8.4, the range in particle size within the measurement zone is between a factor of 1.4 and 2.0, depending on the depth of submersion of the hydrometer. As soil particles settle below the hydrometer and the specific gravity of the suspension decreases, the hydrometer moves deeper into the suspension. This creates a dependence between the reading, r, and the depth, H, from the free surface to the center of buoyancy. The stem creates a change in buoyant weight as the hydrometer moves up and down, adjusting to fluid density, and provides a mechanism to observe different readings of fluid density. The sensitivity of the hydrometer is controlled by the stem diameter.

ASTM D422 specifies a type 151H hydrometer, which measures the specific gravity of the suspension, or a type 152H, which measures the density of solids in the suspension. The 151H hydrometer has a maximum capacity of 1.038, while the 152H hydrometer has a capacity of 60 g/L. This text is based on the type 151H hydrometer.

The scale on the hydrometer is calibrated to provide values of specific gravity of the suspension referenced to distilled water at 20°C. When using a hydrometer in geotechnical applications, it is necessary to separately account for the specific gravity of the fluid. The fluid density changes with temperature and the soluble salt concentration, both from the dispersant and salt in the soil. The "zero" value of the hydrometer can be shifted by taking a reading in water with the dispersant and at the same temperature as the sedimentation test. This ignores any salt in the soil. Equation 8.5 provides an expression that quantifies the mass of material contained within the average measurement depth zone, h, normalized to the dry mass of soil added to the cylinder. This quantity is also the percent finer material at reading m (N_m).

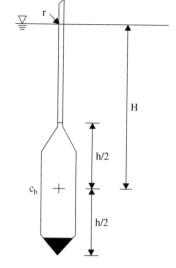

Figure 8.7 Schematic diagram of specific gravity hydrometer along with dimensional definitions.

$$N_m = \frac{G_s}{G_s - 1}\left(\frac{V}{M_D}\right)\rho_C(r_m - r_{w,m}) \times 100 \qquad (8.5)$$

Where:

N_m = percent finer material at reading m (%)
V = volume of suspension (cm³)
M_D = dry soil mass of hydrometer specimen (g)
ρ_c = mass density of water at the calibration temperature (g/cm³)
r_m = hydrometer reading in suspension at time, t, and temperature, T (dimensionless)
$r_{w,m}$ = hydrometer reading in water with dispersant at the same temperature as for r_m (dimensionless)
m = reading number

Note that the calibration temperature mentioned above for the definition of ρ_C is the temperature at the time the hydrometer was calibrated (i.e., the scale set) by the manufacturer. The calibration temperature will be noted in the calibration documentation for the hydrometer and is most likely 20°C.

Practical Considerations

There is not a single standard test method for grain size analysis that provides a methodology applicable to the entire range of particle sizes. The new ASTM test method for sieve analysis is ASTM D6913 Particle-Size Distribution (Gradation) of Soils using

Sieve Analysis. The sieve analysis was formerly combined with the hydrometer analysis in D422 Particle-Size Analysis of Soils. However, the interlaboratory study conducted by the ASTM Reference Soils and Testing Program described in Chapter 1, "Background Information for Part I," showed that D422 did not provide sufficiently reproducible results. Therefore, D6913 was written to rectify some of the problems encountered when running sieve analyses as part of the program. A new draft standard addressing the hydrometer portion of the GSD is being balloted within ASTM but is not available as of the date of this writing. The combination of the two standards will eventually replace D422. The procedures described in this chapter are generally consistent with D6913 for sieve analysis as well as D422 for hydrometer analysis, although the results of the important findings from the interlaboratory study are included as well.

Given the wide range in materials encountered in practice, that three particle size dependent methods may be required to measure the GSD, and the dramatic difference in effort associated with each method, it is convenient to define four basic protocols. The following paragraphs describe each basic protocol, along with the particular conditions for application.

- *Simple Sieve:* applicable for samples with a maximum particle size less than 19 mm (3/4 in.) and when no analysis is required for the distribution of fines. The basic process is to disaggregate the material, remove the fines, oven-dry the material, and then perform the sieve analysis. A companion specimen is used to obtain a water content and compute the initial dry mass so the percent of fines can be calculated.
- *Simple Sedimentation:* applicable for fine-grained soils when the amount of material greater than the 75 μm (No. 200 sieve) will be quantified, but the GSD for this portion will not be determined. The specimen is processed while wet and split into two portions. One portion is used to determine the water content and the other portion is used as the specimen for hydrometer analysis. At the end of the test, the hydrometer specimen is oven-dried for confirmation of the dry mass calculated from the companion water content.
- *Combined Sieve/Sedimentation:* applicable for samples with a maximum particle size less than 19 mm when the GSD of fines is required. Moist material is processed on a fairly coarse sieve, such as the 2 mm (No. 10) sieve. The coarser fraction is oven-dried and sieved. The finer fraction is split into at least three portions while moist. One portion is used to measure the water content, while sedimentation analysis is performed on the second portion and an intermediate sieve analysis is performed on the third portion. The third portion is washed over the 75 μm (No. 200) sieve, wasting the finer fraction. The portion retained on the 75 μm sieve is oven-dried and sieved. The results are numerically combined.
- *Composite Sieve:* applicable for samples with a maximum particle size greater than 19 mm particles, which will require separation and analysis in segments. The procedure may include manual measurement and a Combined Sieve/Sedimentation. The moist material is separated on Designated Sieve A. The coarse fraction is air-dried and processed. The finer fraction is split to reduce the mass. Depending on the distribution of particles, the finer fraction may require separation on a smaller Designated Sieve B. The intermediate portion is dried and sieved. Again, the mass of the finer portion may need to be reduced by splitting. Depending on the characteristics of the material, either a Combined Sieve/Sedimentation or a Simple Sieve is performed on the finer portion. The results are numerically combined.

Considering the complexity and the limitations of both the sieve analysis and the sedimentation analysis, it is encouraging that precision estimates are generally in the range of 2 percent. Reporting test results to 1 percent is common practice and should be encouraged as the standard. Increasing the resolution to 0.1 percent requires much larger specimens, and is only realistic for limited circumstances with a maximum particle size of 19 mm (0.75 in.) or less.

The simple sieve analysis is only applicable to materials with a maximum particle size of 19 mm or smaller and when the GSD for fines in not required. The material should be processed at the natural water content. The soil must not be oven-dried prior to removal of the fine fraction.

A combination of the blending and splitting procedures is used to obtain a representative specimen from the sample. The specimen should be at least twice the required minimum dry mass listed in Table 8.2 based on the estimated maximum particle size. In all cases, this would not be more than 2 kg for a reporting resolution of 1 percent.

The specimen is split in two, where one-half of the specimen is used to determine the water content, while the wet mass is determined on the other half, which will be sieved. This material is thoroughly mixed with and covered by a solution of distilled water and 5 g of sodium hexametaphosphate, and soaked for at least 16 hours. This mixture is then worked by hand or with a rubber spatula to break up any clods and dislodge any fines from the coarse particles.

The fine-grained fraction is separated from the coarse-grained fraction by washing over a reinforced 75 μm (No. 200) sieve covered by a 425 μm (No. 40) sieve. No more than 200 g of the material is transferred at a time to the 425 μm sieve. The material is washed with a water jet on the 425 μm sieve until clean, then the material retained on the upper sieve is transferred to a drying bowl. The material can be worked by hand if needed. The material remaining on the 75 μm sieve is washed. Because the surface of the 75 μm sieve must not be rubbed, only a water jet can be used to work the particles across the surface of the sieve. The material passing the 75 μm sieve is discarded. The retained material is transferred to a drying bowl. The rest of the material is then washed using the same process. As a final step, the 75 μm sieve is reverse washed into the drying bowl to verify that both sieves are empty. The coarse material is oven-dried and used for the sieve analysis.

The most appropriate sieve sizes are selected based on either the standard laboratory set or the perceived distribution of grains. All the material should pass through the largest sieve size, which is taken as the maximum particle size. Before sieving, each sieve is checked for damage and any stuck particles are removed from the mesh. Particles must not be forced through the openings. The appropriate sieve brush must be used to clean each sieve. The mass of each empty sieve is measured and the sieve rack is assembled with the pan on bottom.

Once dried, the soil is placed in the top sieve and covered. The rack is placed in the sieve shaker and shaken for at least 10 minutes. Too much shaking will cause particle breakdown, and therefore total time in a sieve shaker must be limited to 20 minutes.

The efficiency of the shaking action is based upon the shaking method and the condition of the shaker. Pure vibration (small up and down motion) does not effectively allow particles to move around, change orientation, and attempt to pass through holes. Vibration with tapping is better at giving particles a chance to find the appropriate size hole to fall through. The best shaking action is the gyratory motion.

Each sieve should be checked to verify the effectiveness of the shaking action. After mechanical sieving, remove each sieve from the rack, place it over a piece of white paper, and manually tap the side to make sure soil doesn't fall through. If particles appear on the paper, add the particles to the next smaller sieve, return the rack to the shaker, and repeat shaking for a second 10-minute cycle. If particles still appear on the paper after the second shaking cycle, the shaker is ineffective and needs service.

When sieving is complete, the sieves are separated and the mass of each sieve plus the retained soil is measured. The mass of material retained on each sieve is calculated. The mass of dry soil retained on each sieve (or sum of each sieve size) is given the designation, $M_{d,i}$, where i designates the sieve and starts from the largest opening (top) and increases down to the pan.

The dry mass of the entire sieve specimen is computed from the total mass of the sieve specimen and the water content of the companion specimen as Equation 8.6:

$$M_D = \frac{M_T}{\left(1 + \dfrac{\omega_c}{100}\right)} \tag{8.6}$$

Where:
M_D = dry mass of specimen (g)
M_T = total mass of specimen (g)
ω_C = water content of companion specimen (%)

The percent finer than each sieve size is then obtained from the sum of the masses retained on the larger sieves, using Equation 8.7:

$$N_m = \frac{M_D - \displaystyle\sum_{i=1}^{m} M_{d,i}}{M_D} \times 100 \tag{8.7}$$

Where:
$M_{d,i}$ = dry mass of soil retained on sieve i (g)

Important considerations for sieve analyses are:

- In order to obtain accurate masses of soil retained on each sieve, easily check for overloading, and allow for sieving in portions, masses should be recorded one sieve at a time. Determine the mass of the clean sieve (tare) before sieving and then mass the sieve plus soil after shaking. Subtract the two measurements to obtain the mass of dry soil retained on that sieve. This is also good practice because shaking can continue after massing if it is discovered that further sieving is necessary.
- The sieve analysis relies on each particle having the opportunity to fall through a hole at the optimal orientation. For this to happen, the sieve rack must be shaken sufficiently and the particle must have access to the opening. The number of layers of particles in a sieve provides a rational measure to judge whether an individual particle has a reasonable chance of having access to a hole. The mass of a single layer (M_{SL}) of spherical particles on a given sieve can be approximated as Equation 8.8:

$$M_{SL} = 0.52 \frac{\pi D_s^2}{4} D_p G_s \rho_w \tag{8.8}$$

Where:
M_{SL} = the mass of a single layer of spherical particles on a sieve (g)
0.52 = the volume ratio of a sphere in a cube (dimensionless)
D_s = the diameter of the sieve (cm)
D_p = the diameter of the particle or sieve opening (cm)

For example, a uniform layer of particles on a 200 mm (8 in) diameter sieve with openings of 4.75 mm (No. 4) will have a mass of approximately 210 grams per layer. Judgment is required to select how many layers provide a reasonable limit. Three layers of 4.75 mm diameter material will have a mass of about 630 grams, which is too much material on that sieve size. On the other hand, one layer on the 75 μm (No. 200) sieve will have a mass of 3.3 grams, resulting in three layers having a total mass of 10 grams. This is too little mass when considering the desired resolution of results and the effort that would be required to achieve complete sieving of this fine fraction. To account for the practicalities of sieving, ASTM uses a gradually increasing number of layers as the sieve size decreases. These limits are provided in Table 8.4.

When performing a sieve analysis, check the mass on each sieve compared to the criteria provided in Table 8.4. If the mass on any sieve exceeds the overloading

Sieve Opening Size			
(mm)	(inches)	No. of Layers	Maximum Mass Retained on a Sieve (g) [200 mm (8 in.) Diameter Sieve]
19	3/4	1	900 g
9.5	3/8	1.25	550 g
4.75	#4	1.5	325 g
2.00	#10	2	180 g
0.85	#20	3	115 g
0.425	#40	4	75 g
0.25	#60	5	60 g
0.15	#100	6	40 g
0.106	#140	6	30 g
0.075	#200	6	20 g

Table 8.4 Overloading limits for various sieve sizes, assuming $G_s = 2.7$.

Source: After ASTM D6913-04, Table 3. Copyright ASTM INTERNATIONAL. Reprinted with permission.

criteria, then the material should be sieved through the entire rack in portions small enough to stay below the limit on every sieve, use larger-diameter sieves, or change the sieve set distribution to reduce the load on overloaded sieves.

- The selection of sieves included in the rack is important for appropriate characterization of the particle size distribution, as well as preventing overloading. If the material is poorly graded, a uniform spacing of sieve sizes will not capture the characteristics of the curve. A typical shaker holds a rack consisting of the pan and six sieves, which are generally adequate to define the curve. It is usually possible to use a standard set of sieves, but when too much material is retained on a sieve, it is necessary to employ one of the techniques described above to prevent sieve overloading.
- Clean each sieve with the appropriate-stiffness brush before each use. Sieve brushes are manufactured for various sieve opening sizes. The brushes help remove particles remaining in the mesh. When using the brushes, it is important to remove particles while not damaging the sieve or changing the opening size of the mesh. Heavy wire brushes can be used for coarse sieves. Fine wire brushes can be used for sieves with openings equal to or larger than 850 μm (No. 20 sieve). Finer sieves must be cleaned with nonmetallic bristle brushes, such as horsehair or plastic. When cleaning sieves sized 150 μm (No. 100 sieve) or finer, brushes with soft, small-diameter bristles must be used.
- Sieves must be maintained in good condition. Inspect sieves for tearing, damage, and trapped particles before each use. Also check for proper interlocking. Connection damage will lead to loss of fines. Finally, follow the scheduled inspection program and written documentation for the laboratory.
- Calibration of sieves is a matter of establishing that the opening sizes are correct. Out of calibration sieves will have openings that are too large. Pressing particles through the sieve mesh while cleaning or using stiff brushes on fine sieves will damage sieves. Calibration is a difficult and expensive task, with the effort increasing as the sieve opening size decreases. Calibration methods include checking each hole with calipers or a gage rod, inspection with optical scanners, and sieving with calibration particles.
- The standard shaking time is 10 minutes. This should be adequate for most soils with an effective shaker. Smaller sieve-opening sizes and large loads create concern. Always verify shaker effectiveness manually, which is the

reference method in ASTM for performing a sieve analysis. A quick check can be performed by tamping each sieve over white paper and verifying that no particles fall onto the paper.

- Since the material is first washed on the 75 μm (No. 200) sieve, fines remaining in the pan after sieving are a cause for concern. In theory, the pan should be completely empty. While this does not alter the basic calculation, it does indicate that either the wet processing is not effective, the sieving action is creating fines, or the 75 μm sieve is out of calibration.
- Processing the material in the wet condition should eliminate problems with particle bonding. When working the material in the soaking bath, carefully check for clumps and aggregates. If the material has significant clumping, work the material manually with a rubber spatula. Confirm that the washed material contains individual grains using a hand magnifying glass.

The Simple Sedimentation Analysis

The simple sedimentation analysis is based on the hydrometer method and is only relevant for the fine-grained fraction of soil. The material larger than the 75 μm sieve will be quantified, but the GSD will not be measured on this coarse-grained fraction. In general, results should be reported to 1 percent. The type 151H hydrometer has a capacity of about 50 g of soil in suspension. The initial specimen mass should be adjusted to account for the coarse fraction while having the mass of the fine fraction as close as practical to the capacity.

The dry mass of material is needed in order to complete the hydrometer calculations. The most accurate procedure is to transfer all the material from step to step throughout the test, and then measure the dry mass as the last step. This requires more careful handling of the material and a large drying oven, but yields the most accurate results. It is the procedure presented in this chapter. The second option is to measure the initial water content on a companion specimen and compute the dry mass from the total mass used for the hydrometer test. This avoids the need to dry the fine-grained material prior to sedimentation or evaporate a large amount of water after the test. This procedure is faster and less demanding, but also less accurate.

The standard procedure for the sedimentation test is to pretreat the soil in order to deflocculate the clay particles. Sodium hexametaphosphate is used as the dispersing agent at a concentration of 5 g/L in the final slurry volume. The distilled water is added to the soil and dispersant to a volume of about 400 mL, covering the soil. The dispersant neutralizes the surface charge on particles, which is a primary cause of floc formation. It is best to add a measured quantity of dry dispersant to the solution rather than work with volumes of a stock solution, since solutions of sodium hexametaphosphate hydrolyze within one week. Allow the soil, water, and dispersant mixture to soak for at least 16 hours prior to mixing.

The soil and dispersant must be thoroughly mixed to break up existing flocs and cemented particles. Several techniques are effective for mixing fine-grained soils, including air-jet dispersion, ultrasonification, mechanical reciprocating shaking, and high-speed malt mixer blending. The time for mixing will depend on the apparatus used and the plasticity of the soil, and ranges from one minute to an hour.

The deflocculated mixture is transferred to the cylinder and filled to 1000 mL with distilled water. Temperature is important to both the particle terminal velocity and the hydrometer measurement. The test should not be started until the temperature of the mixture is equilibrated with room temperature, or with the bath temperature if using a temperature-controlled tank.

The suspension is mixed using an agitation rod or by inverting the cylinder several times by hand. A rubber stopper can be used to cap the cylinder during inversions. Be sure sediment does not remain on the bottom of the cylinder. Once mixed, the cylinder is placed in the upright position and a timer is started at the same instant. A hydrometer is quickly inserted into the suspension and readings taken with time for the first

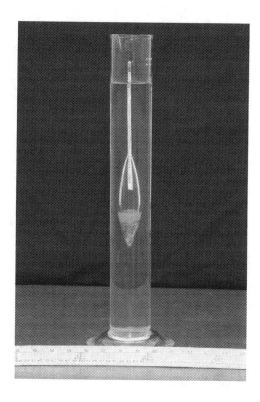

Figure 8.8 A hydrometer placed in a suspension within a graduated cylinder.

2 minutes. The suspension is remixed, and then the timer is restarted. The hydrometer is inserted just in time to take a 2-minute reading. The hydrometer is removed and stored in a water-filled cylinder at the same temperature. The hydrometer is inserted just in time to take the 4-minute reading. This process is repeated for readings roughly doubling in time interval for up to 4 days. It is always possible to remix and restart the sedimentation process to provide convenient reading times. A picture of a hydrometer placed in a suspension within a graduated cylinder is shown as Figure 8.8.

At the end of the sedimentation process, the suspension is mixed to dislodge the particles from the bottom, then the mixture is poured into a large evaporating dish, using a water jet to thoroughly rinse the inside of the cylinder. The mixture is oven-dried, taking several days, and the final dry soil mass is measured. This mass will include the 5.00 g of dispersant, that must be subtracted to get the total dry mass used in Equation 8.5.

Important considerations for hydrometer analysis include:

- The 75 μm particles and larger will fall out of solution within the first 30 seconds of the test. This makes it very important to set up quickly after mixing and obtain stable hydrometer readings at early times. It is also the reason for leaving the hydrometer in the suspension for the first 2 minutes and taking readings as necessary.
- Insertion and removal of the hydrometer must be done carefully to prevent remixing the upper portion of the suspension. To place the hydrometer, hold it by the stem using two fingertips. Position the hydrometer over the center of container holding the suspension. Slowly lower the hydrometer into the suspension until the bulb is covered, then continue until a small upward force can be felt. Release the stem and make sure it slowly rises in the solution by about one division. Lowering the hydrometer a small distance below the anticipated reading allows the hydrometer stem to be wet prior to reading. Let the hydrometer bob this slight amount until it comes to rest. Take the reading at the appropriate time. With practice, it should be possible to insert and read the hydrometer in 10 seconds.
- The hydrometer should be removed from the solution after the 2-minute reading. With some soils, the particles will collect on the glass of the hydrometer. This

increases the mass of the hydrometer and alters the readings. For times greater than 2 minutes, the hydrometer is inserted for the reading and then immediately removed from the suspension. An immersion correction is applied to these readings to account for the upward movement of the surface of the suspension. The immersion correction is one-half the volume of the hydrometer bulb divided by the cross-sectional area of the cylinder. The method to obtain the correction will be discussed in the calibration section.

- The hydrometer test can optimistically yield results to a resolution of 1 percent. This requires careful control of the experiment. Setting a target resolution of 0.5 percent and using Equation 8.5 to calculate the required tolerances yields a reading resolution of +/- 0.2 of a division. In addition, the volume of the suspension needs to be controlled to 5 mL.

- The hydrometer is designed to be read at the water surface. Since the suspension is not transparent, the reading must be taken at the top of the meniscus. Provided the reading in the dispersant is also read at the top of the meniscus, the zero offset will correct for this difference when determining the percent passing of the material. The determination of the diameter at a particular reading is based on the fall height, which must be corrected for the meniscus rise.

- Many soils have salt in the pore fluid. The salt will be in solution during the hydrometer measurements, causing a shift in the zero reading of the hydrometer. In addition, the salt will cause an error in the dry mass. If the soil contains a significant amount of salt, this needs to be measured and accounted for in the calculations. Chapter 6, "pH and Salinity," describes how to measure salt concentration, while Appendix C indicates how to use this measurement to correct the calculations of physical properties.

- The method used to mix the slurry can be important. While the high-energy malt mixer is very effective, it can generate fines by breaking down coarse-grained materials. A less aggressive mixing option should be used when working with slurries containing high proportions of coarse-grained particles.

- The practical size limits of the hydrometer measurement are from about 0.1 mm to about 0.001 mm. The upper end of the range is controlled by the time required to start the experiment, while the lower end is controlled by test duration. It requires nearly 30 days to reach the 0.0002 mm size, at which point Brownian motion effects invalidate the method.

- Chemical cementing agents can bond clay particles together, causing a significant shift in the GSD toward larger-size fractions. Treatment of the material to remove cementing agents is possible. Organics can be digested with hydrogen peroxide (and moderate heat), which is standard practice in the British method. Carbonates can be removed with HCL washing, which is routine for the Soil Conservation Service. Iron oxides are a strong cementing agent found in many surface and residual deposits and can be removed using a Dithionite-Citrate-Bicarbonate (DCB) Extraction. All these treatments add significant time and effort to the sedimentation test.

- The shape and density of the grains can be important to the results. The sedimentation process assumes the shape is spherical and the density equal for all sizes. Clearly, the fine silt- and clay-sized particles are more likely to be plate-shaped and have larger mineral densities than the larger particles.

The Combined Sieve/ Sedimentation Analysis

The combined sieve and sedimentation analysis is applicable for samples with a maximum particle size less than 19 mm and when there is also a need to characterize the GSD of fines. This would generally apply when the fines exceed 10 percent. Given the challenges of processing the material during a combined analysis, the results should be reported to 1 percent resolution.

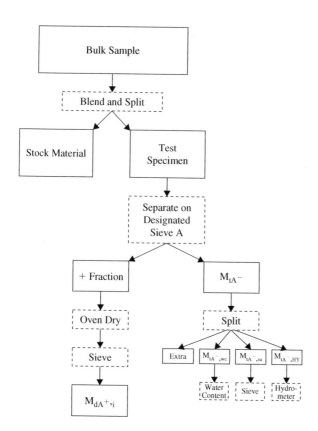

The following labels appear in the flow chart:

Bulk Sample → Blend and Split → Stock Material / Test Specimen → Separate on Designated Sieve A → + Fraction / M_{tA}^{-}

+ Fraction → Oven Dry → Sieve → $M_{dA}^{+},_{i}$

M_{tA}^{-} → Split → Extra / $M_{tA}^{-},_{wc}$ / $M_{tA}^{-},_{sa}$ / $M_{tA}^{-},_{HY}$

$M_{tA}^{-},_{wc}$ → Water Content
$M_{tA}^{-},_{sa}$ → Sieve
$M_{tA}^{-},_{HY}$ → Hydro-meter

Figure 8.9 Flow chart for the sequence of steps in a Combined Sieve/Sedimentation analysis.

The Combined Sieve/Sedimentation analysis involves a fairly complex decision process. As a result, Figure 8.9 presents a schematic flow chart of the sequence of steps involved in performing a combined analysis.

In most cases, the material will arrive at the laboratory in the moist condition. The analysis should be performed without drying the fine fraction to a water content below this condition. The maximum particle size and the water content are estimated. The maximum particle size determines the quantity of dry material required for the test specimen according to Table 8.2. Again, it is best to use approximately twice the required mass. Thoroughly blend and split the moist sample to acquire the moist test specimen. The necessary total mass is the target dry mass times 1 plus the water content. This specimen will have a maximum dry mass of just over 2 kg. It is better to have too much material than too little. Record the moist mass of the specimen.

The next step is to separate the moist material into a coarser and a finer portion using what is termed the Designated Sieve. ASTM D422 suggests the 2 mm (No. 10) sieve as the Designated Sieve. However, any sieve from the 9.5 mm (3/8 in.) to the 0.425 mm (No. 40) sieve can be used as long as it is indicated on the data forms. This is a matter of balancing the amount of effort necessary to work the material through the sieve with the quantity of material passing the sieve. Ultimately, the goal is to have an optimal amount of fines for the hydrometer test, but not so many coarse particles passing the Designated Sieve that processing the hydrometer specimen becomes difficult. In general, the more coarse material in the sample, the smaller the Designated Sieve. The material must be worked through the Designated Sieve. A rubber spatula is most helpful during this process, but care should be given so that the spatula does not excessively damage the sieve. Separation sieves used for processing are usually not those used in the sieve shaker, so extreme care does not need to be taken while working the soil through the separation sieve. It is important to separate the material completely and inspect the coarse fraction to be sure clumps are destroyed and passed through the sieve. Be sure to dislodge fine particles from the larger particles.

Transfer the coarser fraction to a bowl and oven-dry. Perform a simple sieve on the oven-dried material of the coarser fraction. Determine the mass of the finer moist fraction and once again blend this specimen. Split the finer fraction into portions depending on the mass and the size of the Designated Sieve. Each portion must at least meet the minimum size requirement for the maximum particle size, taken as the opening size of the Designated Sieve. For example, if separating on the 4.75 mm (No. 4) sieve, the required minimum dry mass of each portion would be 75 g per ASTM D6913 for a reporting resolution of 1 percent. Use one portion to measure the water content. This will be used to compute the dry mass of the finer fraction. Measure the moist mass of the second portion and proceed to perform a simple sieve analysis, which will give the GSD of the intermediate-sized particles. Measure the moist mass of the third portion and proceed to perform a simple sedimentation analysis, which will give the GSD of the fine-grained portion. Discard the fourth portion or, if it appears that more fines are required for the sedimentation test, combine this with the sedimentation portion, record the combined moist mass, and perform a simple sedimentation test.

The grain size distribution curve is obtained by numerically combining the results from the various stages of the laboratory procedure. The first task is to determine the dry mass of the material using Equation 8.9:

$$M_D = \sum_{i=1}^{n} M_{dA^+,i} + M_{tA^-}\left(\frac{1}{1 + \omega_c/100}\right) \tag{8.9}$$

Where:

M_D = the dry mass of the test specimen (g)

n = number of sieves used in the coarser fraction sieve analysis plus the pan, with the largest diameter corresponding to sieve number 1

$M_{dA^+,i}$ = the dry mass retained on each individual sieve and the pan during the sieve analysis of the portion of material retained on Designated Sieve A (g)

M_{tA^-} = the total mass of the finer fraction passing the Designated Sieve A (g)

ω_c = the water content measured on the finer fraction passing the Designated Sieve A (%)

The measurements collected on the oven-dried coarser fraction are used to compute the percent passing for each sieve m (N_m) using Equation 8.10:

$$N_m = \frac{M_D - \sum_{i=1}^{m} M_{dA^+,i}}{M_D} \times 100 \tag{8.10}$$

Where:

N_m = the percent passing sieve number m (%)

m = the sieve number in the calculation sequence, with the largest diameter corresponding to sieve number 1

The dry mass of material that is coarser than the Designated Sieve A (M_{DA^+}) is obtained by summing the mass retained on each of the sieves used in the coarser sieve analysis. This summation does not include the material that is collected in the pan. The material in the pan should be distributed within the finer fraction but is assumed to be small, and hence is ignored for this calculation. Equation 8.11 provides the calculation for the dry mass retained on the Designated Sieve A.

$$M_{DA^+} = \sum_{i=1}^{m} M_{dA^+,i} \tag{8.11}$$

Where:

M_{DA^+} = the dry mass of the material that is coarser than the Designated Sieve
A (g)

The calculations can be continued to the next smaller set of diameters. These data are collected using the sieve analysis on the portion passing the Designated Sieve A. The percent passing for this size range extending down to the 75 μm sieve is presented in Equation 8.12:

$$N_m = \frac{M_D - M_{DA^+} - \dfrac{M_{tA^-}}{M_{tA^-,sa}}\left(\sum_{i=1}^{m} M_{dA^-,i}\right)}{M_D} \times 100$$

(8.12)

Where:

$M_{tA^-,sa}$ = the total mass of the portion passing the Designated Sieve A and being used for the finer fraction sieve analysis (g)

$M_{dA^-,i}$ = the dry mass retained on each individual sieve during the sieve analysis of the finer fraction (g)

The calculation finally proceeds to the fines fraction. The measurements are obtained using the hydrometer analysis on a portion of the material passing the Designated Sieve A. The calculation for the percent finer of the total mass can then be obtained for each hydrometer reading using Equation 8.13:

$$N_m = \frac{M_D - M_{DA^+}}{M_D} \cdot N_{HY,m}$$

(8.13)

Where:

$N_{HY,m}$ = the percent finer for hydrometer reading m as calculated using Equation 8.5 (%) The dry mass used in Equation 8.5 is the dry mass of the hydrometer specimen. Use the dry mass of material collected at the end of the test (preferred method) or calculate the dry mass by dividing the moist mass of the hydrometer specimen ($M_{tA^-,HY}$) by one plus the water content of the material passing Designated Sieve A.

Important considerations for combined analyses:

- The various considerations for the simple sieve and the sedimentation test also apply to the Combined Analysis.
- The overlap measurement for the 75 μm size for both the sieve and the sedimentation is often a source of difficulty. In addition to the different criteria used to define the particle size, the material is processed differently for the two methods. The different size criteria should only cause a minor difference, so when the discrepancy is unsatisfactory the material handling steps should be scrutinized carefully.
- Separation on the Designated Sieve A and subsequent blending and splitting of the finer material is key to consistency and efficiency. Problems with the separation will result in an appreciable quantity of material in the pan when performing the sieve analysis on the coarser fraction. Poor technique in the blending and splitting of the finer fraction will result in testing errors that will likely go undetected.
- There is no mass balance calculation in the Combined Analysis because the starting dry mass is unknown. This eliminates the possibility of checking that all the individual measurements add up to the starting mass.

The composite sieve analysis is required for all samples containing particles larger than 19 mm. Once the maximum particle size exceeds 19 mm, the specimen size becomes too large to process the material as a single unit. This requires separation and analysis

The Composite Sieve Analysis

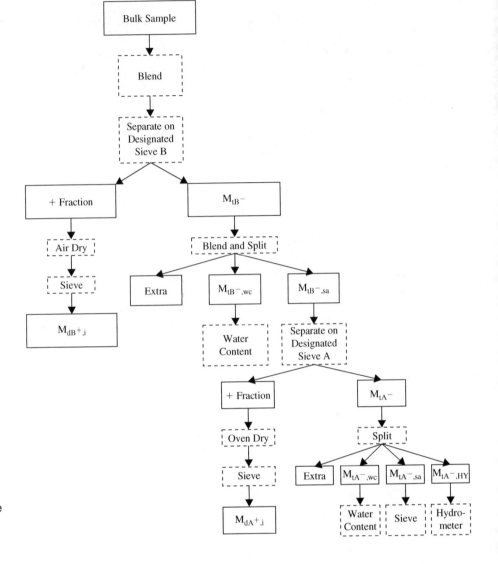

Figure 8.10. Flow chart for the sequence of steps in a Composite Sieve analysis.

in segments. Generally, the composite sieve analysis will be used for large maximum particle sizes, and results will be reported to 1 percent. The exception is for materials having a smaller maximum size, in which case results can be reported to 0.1 percent resolution. Depending on the maximum particle size, the procedure may require manual measurement and a Combined Sieve/Sedimentation. ASTM D6913 is based on a composite sieve analysis without specification of the sedimentation analysis.

Like the Combined Sieve/Sedimentation analysis, the Composite Analysis involves multiple steps and decisions based on the outcome of each step. As an aid, Figure 8.10 presents a schematic flow chart of the sequence of steps involved in performing a composite analysis with two separations.

The first step is to estimate the maximum particle size and determine the required mass of the specimen for the analysis. Transfer this material to a clean surface or a quartering cloth. Blend the material and split as necessary to obtain a representative test specimen. Depending on the sizes involved, this can be a considerable challenge.

The next task is to separate the material on the Designated Sieve B. This separation is intended to cull off most of the mass, making the subsequent processes as easy as possible. A good candidate sieve would be the 19 mm (3/4 in.) sieve since the minus 19 mm material might be used for other testing. Remove the very large particles by hand, brush off the fines, and set these particles off to the side. Be sure to clean finer material off the coarser material with a brush and inspect for thoroughness. Separate the material using the Designated Sieve B.

Air-dry the coarser Designated B fraction and determine the GSD. This material can be air-dried because the hygroscopic water content will be too small to impact the results. Oven-drying is also acceptable, but few laboratories will have sufficient oven capacity. Any particles larger than the available sieves should be measured by hand or sorted into groups with a gage plate. A gage plate can be as simple as a square opening cut in a piece of plywood. Gage plates corresponding to 75 mm (3 in.), 100 mm (4 in.), and 152 mm (6 in.) square openings will be helpful. Use standard sieves to separate the remaining material. Determine the dry mass of each size fraction, as well as the mass of any material finer than the Designated Sieve B, which was carried along with the larger particles. This finer fraction represents a testing error and should be less then 0.2 percent of the maximum particle mass. If it is larger, the processing procedures should be modified.

Determine the moist mass of the fraction finer than Designated Sieve B. Blend this material and split by quartering as many times as required to produce one finer portion that is about 4 times the minimum required specimen size for the Designated Sieve B. The rest of the moist finer fraction should be stored for other testing needs. Split the finer portion in two portions. Use one portion for a water content determination. Measure the moist mass of the second portion and use this material with the procedures outlined for the Combined Sieve/Sedimentation analysis, although the sedimentation test may not be required.

The grain size distribution curve is obtained by numerically combining the results from the various stages of the laboratory procedures. The equations are similar to those used in the combined sieve/sedimentation analysis. The first separation is made on the Designated Sieve B and the second separation is made on the Designated Sieve A. This maintains consistency in the terminology when performing the combined analysis and provides better integration of the two stages of the analysis. The dry mass of the test specimen is computed using Equation 8.14:

$$M_D = \sum_{i=1}^{n} M_{dB^+,i} + M_{tB^-} \left(\frac{1}{1 + \omega_{c,B^-}/100} \right) \tag{8.14}$$

Where:

M_D = the dry mass of the test specimen (g)

n = number of groupings and sieves used in the coarser fraction sieve analysis plus the pan, with the largest diameter corresponding to number 1

$M_{dB^+,i}$ = the dry mass in each size fraction, retained on each individual sieve and the pan during the sieve analysis of the portion of material coarser than Designated Sieve B (g)

M_{tB^-} = the total mass of the finer fraction passing Designated Sieve B (g)

ω_{c,B^-} = the water content measured on the finer fraction passing Designated Sieve B (%)

The measurements collected on the air-dried fraction that is coarser than the Designated Sieve B are used to compute the percent passing for each portion or sieve m (N_m) using Equation 8.15:

$$N_m = \frac{M_D - \sum_{i=1}^{m} M_{dB^+,i}}{M_D} \times 100 \tag{8.15}$$

Where:

N_m = the percent passing grouping or sieve number m (%)

m = the grouping or sieve number in the calculation sequence, with the largest diameter corresponding to number 1

The dry mass of all material that is coarser than the Designated Sieve B ($M_{DB}+$) is obtained by summing the mass measured by hand and retained on each of the sieves used in the coarser fraction analysis. This summation does not include the material that is collected in the pan during the sieve analysis. The material in the pan should be distributed within the finer fraction but is assumed to be small, and hence is ignored for this calculation. Equation 8.16 provides the calculation for the dry mass retained on the Designated Sieve B.

$$M_{DB^+} = \sum_{i=1}^{m} M_{dB^+,i} \qquad (8.16)$$

Where:

$M_{DB}+$ = the dry mass of the material that is coarser than Designated Sieve B (g)

The next size fraction to consider is the material that passes the Designated Sieve B, but is retained on Designated Sieve A. These data are collected by performing a sieve analysis on the oven-dried material retained on Designated Sieve A. The mass must be adjusted for the reduction in mass of the material passing Designated Sieve B. The percent passing for this size range is presented in Equation 8.17:

$$N_m = \frac{M_D - M_{DB^+} - \dfrac{M_{tB^-}}{M_{tB^-,sa}}\left(\displaystyle\sum_{i=1}^{m} M_{dA^+,i}\right)}{M_D} \times 100 \qquad (8.17)$$

Where:

M_{tB^-} = the total mass of the portion passing Designated Sieve B (g)

$M_{tB^-,sa}$ = the total mass of the portion passing Designated Sieve B and being used for the finer fraction sieve analysis (g)

$M_{dA^+,i}$ = the dry mass retained on each individual sieve during the sieve analysis of the oven-dried material that is retained on Designated Sieve A (g)

m = the sieve number in the calculation sequence, with the largest diameter corresponding to number 1

The dry mass of material that is between Designated Sieve B and Designated Sieve A ($M_{DA}+$) is obtained by summing the mass retained on the sieves during analysis of the coarser A fraction and adjusting for the mass reduction of the material that passed Designated Sieve B. This summation does not include the material that is collected in the pan during the sieve analysis. The material in the pan should be distributed within the finer fraction but is assumed to be small, and hence is ignored for this calculation. Equation 8.18 provides the calculation for the dry mass retained on Designated Sieve A.

$$M_{DA^+} = \frac{M_{tB^-}}{M_{tB^-,sa}} \sum_{i=1}^{m} M_{dA^+,i} \qquad (8.18)$$

Where:

$M_{DA}+$ = the dry mass of the material that is coarser than the Designated Sieve A (g)

The next size fraction is the material that passes Designated Sieve A and is coarser than the 75 μm sieve. These data are collected using the sieve analysis on the oven-dried material passing Designated Sieve A. The mass must be adjusted for two mass reductions. The first accounts for the split on Designated Sieve B, and the second for the split on Designated Sieve A. The percent passing for this size range is presented in Equation 8.19:

$$N_m = \frac{M_D - M_{DB^+} - M_{DA^+} - \left(\dfrac{M_{tB^-}}{M_{tB^-,sa}}\right)\left(\dfrac{M_{tA^-}}{M_{tA^-,sa}}\right)\left(\displaystyle\sum_{i=1}^{m} M_{dA^-,i}\right)}{M_D} \times 100 \qquad (8.19)$$

Where:

M_{tA^-} = the total mass of the portion passing the Designated Sieve A (g)

$M_{tA^-,sa}$ = the total mass of the portion passing the Designated Sieve A and being used for the finer fraction sieve analysis (g)

$M_{dA^-,i}$ = the dry mass retained on each individual sieve during the sieve analysis of the oven-dried material passing the Designated Sieve A (g)

m = the sieve number in the calculation sequence, with the largest diameter corresponding to number 1

The calculation finally proceeds to the fines fraction. The measurements are obtained using the hydrometer analysis on a portion of the material passing the Designated Sieve A. The calculation for the percent finer of the total mass can then be obtained for each hydrometer reading using Equation 8.20:

$$N_m = \frac{M_D - M_{DB^+} - M_{DA^+}}{M_D} \cdot N_{HY,m} \qquad (8.20)$$

Where:

$N_{HY,m}$ = the percent finer for hydrometer reading m as calculated using Equation 8.5 (%) The dry mass used in Equation 8.5 is the dry mass of the hydrometer specimen. Use the dry mass of material collected at the end of the test (preferred method) or calculate the dry mass by dividing the moist mass of the hydrometer specimen ($M_{tA^-,HY}$) by one plus the water content of the material passing Designated Sieve A.

Important considerations for composite sieve analyses include:

- The various considerations for the simple sieve, the sedimentation test, and the Combined Analysis also apply to the Composite Analysis.
- Obtaining a representative specimen from a very large sample or from a unit in the field is a significant challenge. Very large amounts of material are being processed and it is easy to bias the blending and sampling operation.
- It is wise to work with more than the minimum required material for each step in the process. This will reduce some of the potential for sampling bias.
- Of the multiple handling stages, the quartering step to split the finer fraction of Designated Sieve B has the most potential to bias the results.
- Material collected in the pan during the sieving analysis of the coarser B and coarser A fractions represent errors in the test. Each quantity should be evaluated as a ratio of the processed mass and considered a potential shift in the finer fraction of the curve. Ratios on the order of 0.002 (0.2 percent) would be reasonable to accept.
- While the results of the Composite Analysis are normally reported to 1 percent of the total specimen dry mass, the distributions of each of the finer fractions will be measured to a much higher resolution. These more precise distributions are applicable to the portion being measured. Take, for example, a sample having a maximum particle size of 300 mm and 3 percent passing the 75 μm sieve. The sedimentation analysis would measure the distribution of the fine fraction to 1 percent resolution, resulting in several measurements of percent passing. However, when the fine fraction distribution is converted to a percentage of the entire specimen, it would result in a maximum of three increments over the entire range of the hydrometer analysis (1 percent, 2 percent, and 3 percent).

TYPICAL VALUES

Typical values of various parameters related to grain size analysis are listed in Table 8.5.

Table 8.5 Typical values of various parameters related to grain size analysis.

Soil Type	D_{60} (mm)	D_{30} (mm)	D_{10} (mm)	C_u	C_c	<75 μm (%)
Boston Blue Clay*	0.0015–0.003	N/A	N/A	N/A	N/A	70–90
Smith Bay Arctic Silt (Site W)**	0.002–0.004	N/A	N/A	N/A	N/A	88–100
Frederick Sand***	0.85	0.4	0.24	3.5	0.78	2
Ticino Sand****	0.54	0.43	0.35	1.5	0.98	0
Processed Manchester Fine Sand (MFS)*****	0.16	0.11	0.077	2.08	0.982	9
2010 Industrial Quartz*****	0.60	0.43	0.29	2.07	1.06	<1.5

*Personal database.
**After Young, 1986; Values estimated using grain size distribution figure.
***After ASTM D6913; values estimated.
****After Larson, 1992; except for C_u, values estimated using grain size distribution figure.
*****After Da Re, 2000; values of <75 μm estimated from grain size distribution figure.

Equipment Requirements

Sieve Analysis

1. Sieve set, typically 200 mm (8 in.) diameter
2. Scale readable to 0.01 g with a capacity of at least 1000 g
3. Brushes for cleaning sieves: Coarse wire and fine wire for sieves equal to or larger than 850 μm (No. 20); fine nonmetallic for sieves sized 150 μm (No. 100) and smaller; coarse nonmetallic can be used for the intermediate sieves.
4. Sieve shaker
5. Fines washing station
6. Drying oven (110 +/− 5°C)
7. Equipment to process soil: splitting cloth, glass plate, scoop, straight edge

Sedimentation Analysis

1. Hydrometer, Type 151H. (Type 152H can be used instead, if preferred, with modifications to the procedures and calculations.)
2. Digital Thermometer (readable to 0.1°C)
3. Calipers (readable to 0.01 mm)
4. 1000 mL (nominal capacity) cylinder
5. Constant temperature water bath or enclosure (+/−1°C)
6. Agitation rod or large rubber stopper
7. Scale readable to 0.01 g with a capacity of at least 2000 g
8. Drying oven (110 +/− 5°C)
9. Evaporating dish with a capacity of greater than 1200 mL
10. Large-mouth 250 mL plastic container with screw lid

When first receiving a new sieve, the user must check that the sieve is marked with the ASTM Designation E11, the name of the manufacturer, and a permanent marking (such as an etching) of the serial number for the sieve. This demonstrates that the sieve opening size has been properly established. Then, prior to first use, visually check the sieve for general conditions. If deviations exist, return the sieve to the manufacturer for replacement.

For sieves in routine use, perform the inspection about every six months. Make sure that excess soil particles have been removed, using an ultrasonic bath to aid this process if necessary. In addition, for the No. 100 sieve and finer, check the sieve visually during each sieving operation. The most convenient time to do this is likely during removal of the retained material on the sieve. If defects are observed, the sieve should be removed from service and the soil in that sieve and underlying sieves must be resieved.

Other tools are available for verification of sieves. Calibrated sands can be purchased from the National Institute of Standards and Technology (NIST) and the AASHTO Materials Reference Laboratory (AMRL). The opening sizes can be checked (usually 30 openings are checked) using calipers or other measuring devices for the larger opening sieves and optical scanners or gage rods for the smaller sieve openings. ASTM E11 provides guidance for verifying that sieve openings are within the specified tolerance. Various geotechnical equipment manufacturing companies will also provide verification services for a fee.

CALIBRATION Sieves

The hydrometer must be calibrated prior to testing to obtain information for three factors: the meniscus rise, the effective depth for any particular reading, and the changes in fluid density with temperature and dispersing agent.

Hydrometer

1. Meniscus rise: Hydrometers are designed to be read at the fluid surface, as illustrated by the dashed line in Figure 8.11. The fluid is wetting to the glass so the suspension will always climb up the hydrometer. Since the sedimentation test suspension is opaque, reading at the fluid surface is impossible at times, so it is common practice to make the reading at the top of the meniscus. The hydrometer reading must always be taken at the top of the meniscus, even when it is possible to see through the fluid. Reading at the top of the meniscus affects both the distance the particles have fallen and the calculation of the quantity of particles in the suspension. An explicit meniscus correction (C_{mr}) must be determined and applied to the distance of fall. Record the meniscus correction as the increment in reading between the free water surface (the proper reading) and reading at the top of the meniscus. Make sure the hydrometer is clean prior to inserting into a volumetric cylinder containing old water. Confirm that the water rises on the glass; if not, the glass is dirty (such as with oil film) and must be cleaned prior to use. The meniscus correction is always a positive number. The meniscus affect on the calculation of quantity of particles in suspension is discussed as part of item number 3.

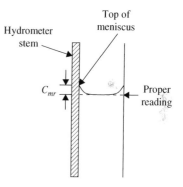

Figure 8.11 Determination of the meniscus correction, C_{mr}, of a hydrometer.

2. The dimension from the surface of the suspension to the center of buoyancy of the hydrometer is the effective reading depth of the hydrometer, H_r and changes with the hydrometer reading. The particle fall distance, H, used in Equation 8.4 depends on H_r but must be adjusted for the meniscus correction and the status of the hydrometer between readings. Use the following procedure to determine the relationship between H and the hydrometer reading. Two relationships are required: one for situations when the hydrometer remains in the suspension continuously and one for situations when the hydrometer is inserted only for the reading.

   a) Determine and mark the center of volume as the midpoint of the bulb.

   b) Using calipers, measure the distance between the center of buoyancy, c_b, and the maximum hydrometer reading, r_2. This is labeled $H_{r,2}$. Obtain a second

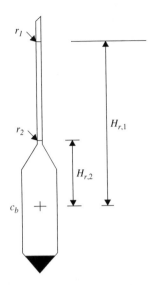

Figure 8.12 Calibration of reading, r, with depth, H_r, of a hydrometer.

measurement from the center of buoyancy c_b to the minimum hydrometer reading, r_1. This is labeled $H_{r,1}$. Refer to Figure 8.12.

c) The point-slope formula can now be used to compute the travel distance (H) of the particles for situations when the hydrometer remains in the suspension. Equation 8.21 provides this function:

$$H_m = H_{r,2} + \frac{(H_{r,1} - H_{r,2})}{(r_2 - r_1)} \times (r_2 - r_m + c_{mr}) \qquad (8.21)$$

Where:

H_m = Distance particles fall at reading m for situations when the hydrometer remains in the suspension continuously (cm)

H_r = dimension between the center of buoyancy and various readings on the hydrometer (cm)

r_m = specific gravity determined by the hydrometer at reading m (dimensionless)

c_{mr} = meniscus correction in units of specific gravity (dimensionless)

m = subscript indicating the reading number during the hydrometer test

d) Measure the volume of the hydrometer bulb (V_h), which includes everything up to the base of the stem. This can be done by either: a) filling a cylinder with clear water up to a calibration line, lowering the hydrometer into the cylinder and reading the displaced volume of water; or b) placing a partly filled beaker on the balance, taring the balance, lowering the hydrometer into the water to the base of the stem, and reading the mass of displaced water. Use the mass density of water at the appropriate temperature to calculate the volume of the hydrometer.

e) Measure the inside area of the cylinder (A_j) using either: a) the distance between two calibration lines, or b) partly filling the cylinder with water, marking the water level, add a known amount of water, and measuring the distance between the old and the new water levels. In either case the area is computed from by dividing the volume by the distance.

f) Equation 8.21 can be modified to account for the fact that insertion of the hydrometer into the suspension stretches the column of fluid. This leads to Equation 8.22, which is used to compute the travel distance of the particle for situations when the hydrometer is inserted into the suspension immediately before a reading and the removed until the next reading. The last term in the equation is often referred to as the immersion correction.

$$H_m = H_{r,2} + \frac{(H_{r,1} - H_{r,2})}{(r_2 - r_1)} \times (r_2 - r_m + c_{mr}) - \frac{V_h}{2A_j} \qquad (8.22)$$

Where:

H_m = Distance particles fall at reading m for situations when the hydrometer is inserted only for individual readings (cm)

V_h = Volume of the hydrometer bulb up to the base of the stem (cm^3)

A_j = cross-sectional area of the cylinder (cm^2)

g) Rather than use the above equations, it is equally acceptable to graph the two equations as shown in Figure 8.13.

3. Changes in fluid density: In the sedimentation test application, the calculation for the quantity of particles in suspension must be corrected for water density changes due to temperature, the presence of dispersing agent (an perhaps any other salt), and the meniscus reading location. This correction is essentially a zero shift in the

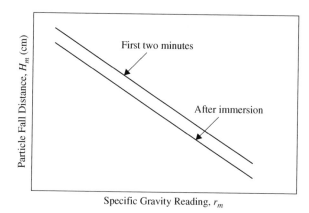

Figure 8.13 Plot of particle fall distance versus specific gravity reading, with and without immersion correction.

hydrometer scale. The scale on the hydrometer was set by the manufacturer at the calibration condition, which is normally distilled water at 20°C. Since dispersant agents have been added to the solution and the test temperature may be different than the calibration condition, a correction must be made to the reading obtained from the hydrometer scale. The combined density correction can be obtained by either: a) making a calibration measurement each time a slurry measurement is taken, or b) developing a general calibration relationship as a function of temperature. Either case requires a calibration cylinder that is composed of distilled water and dispersing agent in the same proportion as in the sedimentation test. Note that since both the sedimentation readings and the calibration readings are taken at the top of the meniscus, the meniscus correction is automatically accounted for in the combined fluid density correction.

a) When making individual sedimentation reading corrections, the calibration reading ($r_{w,m}$) is taken at the top of the meniscus of the calibration cylinder along with each sedimentation reading. This calibration cylinder must be located in the same temperature-controlled location as the sedimentation columns so that all cylinders are read at the same temperature.

b) The second option is to develop a temperature-based general calibration relationship for each hydrometer. A series of readings at the top of the meniscus, along with temperature, are made in the calibration cylinder. Be sure the solution is thoroughly mixed and temperature equilibrated at each measurement point. The hydrometer must be rinsed between readings as well. These points are graphed or fit with an equation. The temperature is then recorded with each sedimentation reading and used with the calibration relationship to obtain the composite correction.

Procedures are provided for performing three different particle size analyses. The hydrometer is used for the simple sedimentation analysis. It would be best performed on a fine-grained material having no coarse fraction. This will be helpful in evaluating the application of the procedure. The simple sieve should be performed on a relatively clean material (i.e. with little to no fines). This will make processing easier and reduce the amount of water and effort required to separate on the 75 μm sieve. The combined sieve and hydrometer is much more complicated. The material should contain an appreciable amount of plastic, fine-grained material. This will test the technical skill of the operator and illustrate the importance of controlling the various size fractions. For efficiency, the sample used for the combined analysis should be from the same source as used for the Atterberg Limits presented in the next chapter, since the grain size and Atterberg Limits are part of the Unified Soil Classification System presented in Chapter 10.

SPECIMEN PREPARATION

Specimen preparation is very specific to the material being used. Therefore, the specimen preparation techniques in this chapter are included within the procedures below. As a reminder, however, do not dry the soil prior to determination of the grain size analysis of fine-grained portions of soils. Use wet soil and estimate the dry mass of soil.

PROCEDURE

Simple Sedimentation

The following procedure contains two methods for the determination of the dry mass of the test specimen. For training activities, it is informative to perform both methods in order to learn the technique and to evaluate the experimental error in measuring the dry mass.

1. Obtain the equivalent of about 80 g of dry soil. Do not oven-dry the soil. Make a reasonable guess of the water content and use $M_{wet} = M_{dry} [1 + \omega_c]$.

2. Place the material on a glass plate.

3. Chop the soil into subcentimeter-sized pieces.

4. Blend this material and split in two portions.

5. Use one portion to measure the water content.

6. Measure the mass of clean plastic mixing container.

7. Place the second portion in the container and measure the mass with soil.

8. Place a small tare on the scale and zero.

9. Add 5.0 g of sodium hexametaphosphate to the tare and record mass to 0.01 g.

10. Add distilled water and the sodium hexametaphosphate to the bowl and mix the soil to a thick slurry (milk shake consistency).

11. Allow the solution to temper for at least 8 hours.

12. Transfer the mixture to a blender, increase the volume to about 500 mL with distilled water, and mix the slurry for 60 seconds. Be sure to rinse all the soil from the tempering container to the blender.

13. Transfer the dilute slurry to the 1000 ml cylinder and fill to the calibration mark with distilled water. Be sure to transfer all the soil from the blender to the cylinder.

14. Place in temperature bath or in relatively constant temperature area and allow time for temperature equilibration.

15. Mix the slurry thoroughly with plunging rod (or hand inversion) and begin sedimentation experiment by starting timer.

16. Obtain two sets of specific gravity readings (r_m) for the first two minutes of sedimentation with the hydrometer remaining in the suspension. Obtain readings at 15, 30, 60, 90, 120 seconds.

17. Measure the suspension temperature.

18. Always read the hydrometer at the top of meniscus (even when reading in a clear fluid).

19. Record the hydrometer reading to 0.2 of a division by estimating five increments between divisions. This resolution is necessary for the calculations.

20. Remix the suspension and obtain readings of specific gravity and temperature at 2, 4, 8, 16 minutes, and so on. Take readings for at least two days or until the fluid is clear, whichever comes first. If using a blank cylinder for the composite specific gravity correction, also collect these readings at the same times.

21. For each reading, place the hydrometer in the suspension about 10 seconds before the reading time, allow the hydrometer to stabilize, take reading, return to the wash water.

22. Insert and remove the hydrometer slowly to prevent mixing the upper portion of the suspension.

23. Between readings, be sure to cover the cylinder to prevent evaporation. It is important to keep the cylinder properly filled.

24. Once all the sedimentation readings have been taken, mix the suspension with the plunger and pour the slurry into an evaporating dish of known mass. Be sure to transfer all the slurry. Use a water jet to clean the inside.

25. Oven-dry for several days and obtain the final dry mass of soil and sodium hexametaphosphate.

The following procedure is for measuring the GSD of a coarse-grained moist material with maximum particle size smaller than 19 mm and without measurement of the GSD of the fine-grained fraction. Once again the procedure is written to measure the dry mass of the specimen by two methods for the purpose of checking calculations and evaluating errors.

Simple Sieve

1. Estimate the maximum particle size that is the smallest sieve that all the material will pass.

2. Use Table 8.2 to determine the required dry mass and use about three times the required value for this analysis.

3. Blend the material (break up lumps where obvious) provided and split to the appropriate specimen size. Two specimens will be required.

4. Use one specimen to determine the as-received water content.

5. Transfer the second portion to a large pan of known mass.

6. Measure the initial mass of the container and soil.

7. Cover the soil with water and add 5 g of sodium hexametaphosphate.

8. Mix thoroughly with a hard rubber paddle. Inspect for clumps.

9. Assemble a washing station using a 425 μm and 75 μm sieve and a collection bowl of known mass.

10. Transfer about 200 g to the 425 μm sieve and wash fine particles through the sieve.

11. Remove the 425 μm sieve and wash fine particles through the 75 μm sieve.

12. Transfer retained fraction on both sieves to an accumulation bowl.

13. Repeat steps 10 through 12 until all the material has been washed.

14. Add clean water to the accumulation bowl and mix by hand to be sure all the fines have been removed.

15. Decant water from the accumulation bowl, oven-dry the coarse-grained material, and measure the dry mass.

16. Oven-dry the fine-grained material and measure the dry mass plus sodium hexametaphosphate.

17. Select and clean a rack of sieves appropriate for the shaker and the soil distribution. The set must include the largest size that all material passes and the 75 μm sieve.

18. Inspect and record the size and mass of each empty sieve.

19. Assemble rack with a pan and pour soil into top sieve.

20. Shake for 10 minutes.

21. Disassemble rack and manually check that sieving is complete by tapping each sieve over a sheet of white paper.

22. If significant particles appear on paper, perform a second shaking cycle.

23. Determine the mass of each sieve and retained soil.

24. Check each sieve for overloading limit. A chart of the limit load is very helpful for quick comparison. If the limit is exceeded, repeat the sieving operation in two portions.

25. Empty and clean sieves.

Combined Sieve and Hydrometer

The following procedure is for measuring the GSD of a coarse-grained moist material with maximum particle size smaller than 19 mm and having a significant amount of fine-grained material. In this case, the procedure is modified to obtain sufficient information to check the dry mass.

1. Estimate the maximum particle size that is the smallest sieve that all the material will pass.

2. Use Table 8.2 to determine the required dry mass and use about three times the required value for this analysis.

3. Blend the material (break up lumps where obvious) provided and split to the appropriate specimen size. Two specimens will be required.

4. Use one specimen to determine the as-received water content.

5. Measure the moist mass of the material that will be used for the particle size analysis.

6. Separate the moist material on the 2 mm (#10) sieve.

7. Oven-dry, measure the mass, and sieve the portion retained on the 2 mm sieve. Use the sieving procedure described in the simple steps 17 to 24 with the 2mm sieve being the smallest size.

8. Split the moist material passing the 2 mm sieve into four portions.

9. Measure the moist mass of each portion.

10. Use the first portion for a hydrometer test as outlined above but starting from step 6.

11. Use the second portion for a water content measurement.

12. Use the third portion for a simple sieve analysis as outlined above but starting from step 7.

13. Oven-dry the fourth portion to obtain the dry mass.

Calculations

Hydrometer Analysis

1. Use Equation 8.4 to determine the particle size for each hydrometer reading.

2. Determine the dry mass (M_D) of the hydrometer specimen from the moist mass and the companion water content.

3. Use Equation 8.5 to determine the percent passing for each hydrometer reading.

4. Compute the alternate dry mass (M_{Da}) from the final mass measured in step 25 minus the added mass of sodium hexametaphosphate.

5. Compute the % mass error (E_m) as shown in Equation 8.23:

$$E_m = \frac{(M_D - M_{Da})}{M_D} \times 100 \tag{8.23}$$

Simple Sieve Analysis

1. Use the diameter marked on each sieve as the particle diameter.

2. Determine the dry mass (M_D) of the sieve specimen from the moist mass and the companion water content.

3. Use Equation 8.7 to determine the percent finer for each sieve size working from largest size to smallest.

4. Compute the alternate dry mass (M_{Da}) as the sum of the coarse-grained fraction measured in step 15 and the fine-grained fraction measured in step 16. Remember to subtract the 5 gm of sodium hexametaphosphate.

5. Compute the percent mass error (E_m) using Equation 8.23.

1. Determine the dry mass (M_D) using Equation 8.9.

2. Use the diameter marked on each sieve as the particle diameter.

3. Use Equation 8.10 to determine the percent finer for each sieve size in the first simple size analysis for the material larger than the 2 mm sieve.

4. Use Equation 8.12 to determine the percent finer for each sieve size in the second simple size analysis for the material between the 2 mm sieve and the 75 μm sieve.

5. Use Equation 8.4 to determine the particle size for each hydrometer reading.

6. Use Equation 8.13 to determine the percent finer for each hydrometer reading.

7. Compute the alternate dry mass (M_{Da}) as the sum of the coarse-grained fraction measured in step 7 plus the material used in the hydrometer in step 10 plus the oven-dried material in steps 11 and 13 plus the simple sieve material used in step 12.

8. Compute the percent mass error (E_m) using Equation 8.23.

For each sample tested, report the maximum particle size; the values of D_{60}, D_{30}, D_{10}, C_u, C_c; the percent passing the 75 μm (No. 200) sieve; and the error in dry mass. Include a plot of the grain size distribution curve for the material with the method of determination indicated (i.e., mechanical or sedimentation). Append the data sheets to the report.

Criteria for judging the acceptability of hydrometer test results have not been determined.

Sieve analysis does not produce independent measurements, in that multiple sieves are used, each making a measurement of the soil retained. When a grain of soil is retained on an upper sieve erroneously, it affects the mass retained on the underlying sieve. As a result, the full precision statement for grain size analysis is much more complicated than seen previously in this text.

ASTM E691 Standard Practice for Conducting an Interlaboratory Study to Determine the Precision of a Test Method does not provide a method to analyze such results. The precision statement included in D6913 was developed by determining the percent retained on each particular sieve by each laboratory, thereby analyzing the results of each sieve as a separate method. Therefore precision estimates will vary depending on the details of the grain size distribution and the sieve set used in the testing.

The ASTM Reference Soils and Testing Program conducted an ILS and developed precision estimates using D6913 and a poorly graded sand (SP) consisting of approximately 20 percent coarse sand, 48 percent medium sand, 30 percent fine sand, and 2 percent fines (Frederic Sand). The sieve set consisted of the following sieves: 4.75 mm, 2 mm, 850 μm, 425 μm, 250 μm, 150 μm, 106 μm, and 75 μm (Nos. 4, 10, 20, 40, 60, 100, 140, and 200). The precision statement provided below specifically applies to the material and sieve set used for the ILS. ASTM D6913 provides the equations to analyze data sets produced by testing other soils using different sieve sets.

Criteria for judging the acceptability of test results obtained by ASTM D6913, based on the ILS program for Frederick Sand (SP) using the specific sieve set and analysis described above, are given as follows

- *Within Laboratory Repeatability:* Expect your results for one trial on the SP soil described above to be within 2 percent retained on a given sieve as compared to a second trial on the same soil.
- *Between Laboratory Reproducibility:* Expect your results on the SP soil described above to be within 7 percent retained on a given sieve as compared to others performing the test on the same soil.

DETECTING PROBLEMS WITH RESULTS

There are a few obvious error indicators in the grain size analysis. The errors that are easiest to detect are having more than 100 percent passing in the hydrometer test, a large difference between the hydrometer and the sieve results at the 75 μm size fraction (i.e. the overlap zone between the two methods), and large quantities of material in the pan during sieve analyses involving separation into coarser and finer fractions. These are generally procedural errors in processing the materials, which may be systematic or may be due to handling unfamiliar materials. Requiring the extra effort to make sufficient measurements to complete the mass error calculation is the first step in finding solutions to such problems.

Several other methods of isolating the causes of errors are possible. Systematic errors due to equipment deficiencies can be identified by inspecting the sieves and the efficiency of the shaking action and checking that the initial hydrometer specimen dry mass is within a few percent of the final dry mass. Procedural and technique errors are best identified by performing the test on a coarse material with a known grain size distribution, such as calibrated glass beads or the SP reference soil (Frederick Sand) used in the ILS program, or a fine-grained material with known distribution of fines and no coarse-grained material. Erratic hydrometer curves would indicate poor technique inserting and reading the hydrometer. A good training exercise is to make measurements on distilled water at various temperatures. The insertion and reading should be accomplished in less than 7 seconds.

It is far more difficult to identify errors in size measurements. Calibration of sieves will eliminate one important source of error, but material processing deficiencies are material specific and most common with problematic materials. Unfortunately, exercising the laboratory procedures using standard materials is unlikely to uncover such deficiencies. Duplicate blind testing and comparison testing between various laboratories has the most potential to identify best practices.

In situations where a number of tests are performed on the same material, it becomes possible to use precision information to evaluate the results. If the standard deviation for one set of measurements exceeds 2 percent, then evaluate the techniques of the individual performing the test. The possible sources of problems are improper splitting and blending procedures, clogged sieves, poor control when placing the hydrometer in the suspension, incorrect recording of elapsed time for a hydrometer reading, or sloppy handling of the materials. If duplicate measurements exceed the within-laboratory repeatability provided above, sources of experimental error are likely related to insufficient shaking time, incomplete washing of fine soils, errors in the dry masses, or errors in the hydrometer readings. If the test results do not fall within the typical ranges or exceed the reproducibility limit, the likely cause of error is systemic, such as an error in the mass of a sieve, sieve opening sizes not meeting the criteria in ASTM E11, sieve shaker equipment out of calibration, loss of hydrometer specimen material during the course of the test, or hydrometer out of calibration or calibrated improperly.

REFERENCE PROCEDURES

ASTM D422 Particle-Size Analysis of Soils.
ASTM D6913 Particle-Size Distribution (Gradation) of Soils Using Sieve Analysis.

REFERENCES

Refer to this textbook's ancillary web site, www.wiley.com/college/germaine, for data sheets, spreadsheets, and example data sets.

Da Re, Gregory. 2000. "Physical Mechanisms Controlling the Pre-failure Stress-Strain Behavior of Frozen Sand," PhD thesis, Department of Civil and Environmental Engineering, Massachusetts Institute of Technology, Cambridge.

Larson, Douglas. 1992. "A Laboratory Investigation of Load Transfer in Reinforced Soil," PhD thesis, Department of Civil Engineering, Massachusetts Institute of Technology, Cambridge.

Young, Gretchen. 1986. "The Strength Deformation Properties of Smith Bay Arctic Silts," MS thesis, Department of Civil Engineering, Massachusetts Institute of Technology, Cambridge.

Chapter 9

Atterberg Limits

This chapter describes the procedures to perform the liquid limit, plastic limit, and shrinkage limit tests, which are collectively referred to as Atterberg Limits. Procedures are presented for determining the liquid limit by the multipoint method using the Casagrande cup, the plastic limit by hand rolling, and the shrinkage limit by the wax method. Other methods and variations over time are discussed briefly in the background section. The liquid limit and plastic limit are used as part of the Unified Soil Classification System (USCS) presented in Chapter 10.

SCOPE AND SUMMARY

Atterberg Limits are determined for fine-grained soil materials, which by ASTM D2487 Classification of Soils for Engineering Purposes (Unified Soil Classification System) is soil that has fifty percent or more mass passing the 0.075 mm (No. 200) sieve. The limit values are most appropriate for characterization of the fine-grained material. However, as a practical matter, the soil is separated on a 0.475 mm (No. 40) sieve and the testing is performed on the passing portion.

TYPICAL MATERIALS

Atterberg Limits are conceptual boundaries between various states of material behavior involving mixes of soil particles and water. They were developed by Swedish scientist

BACKGROUND

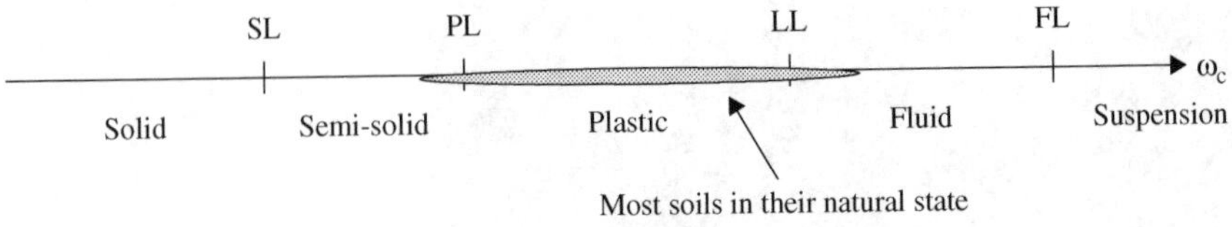

Figure 9.1 Atterberg Limits with respect to boundaries between material states.

Dr. A. Atterberg in 1911 to classify agricultural soils. The limits are represented by water content values corresponding to specific observations of behavior. Figure 9.1 illustrates a scale of increasing water content along with the various material behaviors possible for a particulate system.

The boundaries between the different states are somewhat vague and are defined in terms of simple index tests originally developed by Atterberg. The concept of defining transitional boundaries of behavior based on a measure of water content recognizes the importance of soil-water interaction for fine-grained materials. As the amount of water increases in the soil, the particle spacing increases and the interaction between adjacent particles will decrease, altering the mechanical behavior. The assemblage of particles changes from a densely packed solid to a very loosely packed liquid. Essentially, as the mixture gets wetter, the material becomes weaker. However, it is important to appreciate the fact that the rate of change in mechanical properties is strongly dependent on the size of the particles and the strength of the mineral surface charge. Hence, the values are important on a comparative basis but difficult to interpret as definitive borderlines between states.

Casagrande developed more formal standardized index tests for defining these boundaries. Descriptions of the tests are presented in a published paper (Casagrande, 1932) and remain relatively unchanged in the ASTM test methods referred to below.

The following definitions apply:

SL = Shrinkage Limit (ω_S). The shrinkage limit marks the boundary between a solid and a semi-solid. At this water content, the soil volume is at a minimum while maintaining a saturated state. In theory, the shrinkage limit is a well-defined condition but it does depend to some extent on the initial fabric of the material. The shrinkage limit can be determined by ASTM D4943 Shrinkage Factors of Soils by the Wax Method.

PL = Plastic Limit (ω_P). The plastic limit marks the boundary between semi-solid and plastic mechanical behavior; however, in reality the material slowly transitions between the two. At the water content corresponding to the plastic limit, the soil crumbles when rolled into a 3.2 mm (1/8 in.) diameter string. The plastic limit can be determined using ASTM D4318 Liquid Limit, Plastic Limit, and Plasticity Index of Soils.

LL = Liquid Limit (ω_L). The liquid limit marks the boundary between plastic and fluid-like behavior. Once again, this is a descriptive state rather than an absolute boundary. At the water content corresponding to the liquid limit, the soil becomes fluid under a standard dynamic shear stress. ASTM D4318 also covers the determination of the liquid limit.

FL = Fluidization Limit (ω_F). The fluidization limit is the boundary between the fluid state and a suspension. A suspension is a mixture of fluid and solid particles such that there are no interparticle contact forces. Particles in a suspension are essentially floating in the pore fluid. While this boundary can be defined, it has not yet been associated with a specific test. At this water content, there would be no effective stress within the material.

The symbols in parentheses above (e.g., ω_S) are sometimes used in practice to represent the limits. However, this text uses the capital letter designations to differentiate limits results from water content determinations. Limits are reported as integers without the percent designation (e.g., LL = 41), while water contents are typically presented to the nearest 0.1 percent and do include the percent symbol if reporting in percent (e.g. $\omega_C = 35.1\%$). Alternatively, the water content can be reported in decimal form (e.g., $\omega_C = 0.351$) as long as the units are made clear. Subsequent calculations using the limits must obey the rules of significant digits.

Atterberg limits are determined on specimens of remolded soils on the portion of particles finer than the 0.425 mm (No. 40) sieve (i.e., grains the size of fine sand and smaller). This is an arbitrary choice that makes processing the soil to eliminate the coarse particles relatively easy without including too many large, nonplastic particles in the measurement. This dilutes the measurement sensitivity to the fine-grained particles, but is justified compared to the additional effort that would be required to separate wet soil on the 0.075 (No. 200) sieve.

The plastic and liquid limit values are further used to define parameters referred to as the Plasticity Index (PI or I_p) and Liquidity Index (LI or I_L). The PI is the difference in water content between the liquid and plastic limits, as presented in Equation 9.1. The PI is important because it quantifies the range in water content over which the soil is in the plastic state. The PI is also a measure of how strongly the particles interact with the water.

$$PI = LL - PL \qquad (9.1)$$

Where:

PI = plasticity index (integer value without a percent symbol)

The LI locates a specific water content relative to the liquid and plastic limits while scaling to the Plasticity Index, and is given by Equation 9.2:

$$LI = \frac{\omega_C - PL}{PI} \qquad (9.2)$$

Where:

LI = liquidity index (decimal)
ω_C = water content (%)

The Liquidity Index is most often used with the natural water content (ω_N) to represent the field conditions relative to the Atterberg limits. Values around zero would mean the soil exists in nature around the plastic limit. When the Liquidity Index is above unity, the material is in the fluid range and likely to be sensitive.

Skempton (1953) developed a parameter called activity to quantify the relative importance of the clay fraction. Activity is defined as the ratio of plasticity index to the percentage of dry mass of material finer than 2 μm. Activity is expressed as shown in Equation 9.3:

$$\text{Activity} = \frac{PI}{\%\ \text{Mass} < 2\mu m} \qquad (9.3)$$

The Atterberg Limits provide significant insights about a soil. Most fine-grained soils exist in their natural state in a water content range that is slightly below the plastic limit to slightly above the liquid limit. The notable exceptions are arid and very deep deposits that can approach the shrinkage limit, and quick clays that exceed the liquid limit.

The plastic and liquid limits are a significant part of the USCS for the fine-grained portion of soils. A Casagrande plasticity chart, shown in Figure 9.2, is used to plot the values of the LL and PI in order to determine the group symbol for the fines. The plasticity chart is divided into several sections. The "U" line is considered the uppermost extent of limits for naturally occurring soils. The "A" line is separates silt (M) from clay (C). Liquid limits greater than 50 are high plasticity (H), while liquid limits less

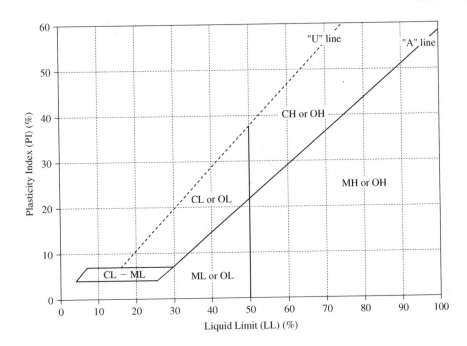

Figure 9.2 Casagrande Plasticity Chart.

than 50 are low plasticity (L). High plasticity and low plasticity are also referred to as fat and lean, respectively. Additionally, when used in conjunction with silt, the "H" can be referred to as "elastic." Most areas of the chart have either organic (OL or OH) or inorganic (CL, CH, ML, or MH) group symbols. A dual symbol of CL-ML is assigned to the area at the lower left area of the chart.

According to the current version of D2487, an organic soil is assigned the group symbol according to the value of the liquid limit, while the group name is assigned according to the position of the plotted results with respect to the "A" line. Specifically, an organic soil with a liquid limit less than 50 has a group symbol of OL and an organic soil with a liquid limit greater than 50 has a group symbol of OH, regardless of whether the results plot above or below the "A" line. Additionally, organic soils that plot above the "A" line are given the group name of organic clay and organic soils that plot below the "A" line are given the group name of organic silt, regardless of the value of the liquid limit.

To combat the confusion caused by allowing a group symbol to have two names, the authors prefer to differentiate organic soils in the various zones by using distinct names and symbols as follows: organic soils with a liquid limit less than 50 plotting in the OL zone of the chart below the "A" line are given the group name organic silt with the symbol OL, whereas when an organic soil with a liquid limit less than 50 and plotting above the "A" line is given the group name organic lean clay and a dual group symbol CL-OL. Likewise, for organic soils with a liquid limit greater than 50 percent (i.e., falling into the OH zone) with results plotting below the "A" line, the group name organic elastic silt and group symbol OH are given, while organic soils with a liquid limit less than 50 and plotting above the "A" line are given the group name of organic fat clay and group symbol of CH-OH.

The resulting group symbol of the fine portion of the sample is combined with the grain size results to classify soils. Further information on the classification can be found in the next chapter.

Atterberg Limits are commonly used in conjunction with correlations. Practice is generally most interested in the liquid limit and plastic limit. Correlations between the liquid limit or plasticity index and strength or coefficient of consolidation are provided in the chapters of this text addressing those subjects.

The profile shown in Figure 9.3 exemplifies a common application of plastic and liquid limits and natural water content obtained on samples over a depth range to evaluate

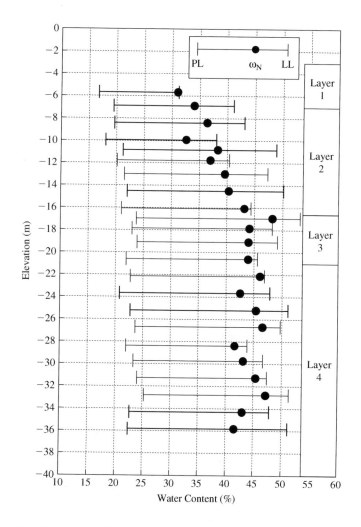

Figure 9.3 Common application of plastic limits, liquid limits, and natural water contents plotted versus depth to establish variability of deposit.

the uniformity of a deposit. Typically, the plastic limit is signified with the left-most bar, the liquid limit with the right-most bar, and the corresponding natural water content plotted as a symbol along the horizontal line.

The shrinkage limit is of particular importance when working with soils that experience volume change, particularly when they exist under constructed structures, such as pavements and concrete slabs. The shrinkage limit quantifies the minimum volume soil will occupy in an arid climate. Although individual agencies may have their own criteria on acceptable ranges of shrinkage limit for soils to be used in construction, a report by W. G. Holtz for the U.S. Bureau of Reclamation provides potential degrees of expansion relative to some index tests. This information is presented below as Table 9.1.

Table 9.1 Relationship of shrinkage limit and other soil properties to expected volume change.

Data from Index Tests[1]			Probable Expansion[2] (Percent Total Volume	
Colloid Content (% minus 0.001 mm)	Plasticity Index, PI (%)	Shrinkage Limit, SL (%)	Change, Dry to Saturated Condition)	Degree of Expansion
> 28	> 35	< 11	> 30	Very high
20 to 31	25 to 41	7 to 12	20 to 30	High
13 to 23	15 to 28	10 to 16	10 to 20	Medium
< 15	< 18	>15	< 10	Low

[1] All three index tests should be considered in estimating expansive properties.
[2] Based on a vertical loading of 7 kPa (1.0 lbf/in²).
Source: Holtz, 1959.

As mentioned previously, standard methods are available to measure the shrinkage limit, the plastic limit, and the liquid limit. Unfortunately, there is more than one method to measure each limit and some of the equipment has evolved over time. This has made it difficult to compare data from different sources and warrants a note of caution when using data from the literature. The problem is compounded by a lack of precision data for all of the various methods over time or benchmark data on reference materials.

The shrinkage limit test was originally performed using the mercury displacement method (ASTM D427). Due to the hazards inherent with mercury, the shrinkage limit is now measured using a wax-coated specimen and a water displacement technique. The wax method is now the ASTM standard and is used in this text.

The plastic limit test originated as a hand-rolling method. The labor-saving rolling machine was introduced in the 1990s. In addition, the technical literature is now making use of the fall cone as an improved method for determining the plastic limit. At present, only the rolling methods are recognized as standard methods for the plastic limit by ASTM D4318.

The liquid limit test has been the most dynamic of the three. The Casagrande device is the original standardized method but has been complicated by significant variations in the equipment used to run the test. The hardness of the base has been specified differently among individual standardization agencies, and those agencies have allowed different bases over time. The specified grooving tools have changed between standardization organizations and with time as well. In addition, a method has been developed using the fall cone to measure the liquid limit. While the fall cone is used as the standard liquid limit test method in many countries, the Casagrande device is the standard liquid limit method in ASTM D4318. For these reasons, the referenced standard should always be checked when using archival data.

Now that the application of Atterberg Limits have been covered, it is time to discuss the testing methods. In most cases, multiple methods will be discussed, while only the preferred method for each limit will be covered in the procedures section.

Liquid Limit by Casagrande Device (ASTM D4318)

The ASTM determination of the liquid limit is obtained using the Casagrande cup (1932) and flat geometry grooving tool. This equipment, along with water content tares, mixing spatulas, and a resilience tester, is shown in Figure 9.4.

This test is designed to simulate a miniature slope under impact loading. The energy required to fail the slope is measured for a range of water contents for the multipoint

Figure 9.4 Casagrande cup, flat grooving tool, mixing spatulas, water content tares, and resilience tester.

Classification[1]	Minimum Tempering Time, h
GW, GP, SW, SP	No requirement
GM, SM	3
All other soils	16

[1]According to USCS group symbols:
GW: Well-graded gravel, with or without sand
GP: Poorly-graded gravel, with or without sand
SW: Well-graded sand, with or without gravel
SP: Poorly-graded sand, with or without gravel
GM: Silty gravel, with or without sand
SM: Silty sand, with or without gravel
Source: After ASTM D1557-07, Table 2. Copyright ASTM INTERNATIONAL. Reprinted with permission.

Table 9.2 Minimum tempering times for soils with various USCS group symbols.

method (Method A). The energy is measured by the number of drops, or "blows", and given the symbol N. Failure is defined as closing the groove for a length of 13 mm.

The test is performed on material passing the 0.425 mm (No. 40) sieve that has been completely remolded to destroy the in situ structure. Disaggregate clumps of clay and silt, however, do not break down weakly cemented aggregates. The soil is mixed with distilled water and tempered for at least 16 hours. ASTM suggests mixing to about the 30 drop consistency, but the authors believe a 15 drop consistency is preferred because it provides more water for hydration and allows more control when adjusting the water content during the test. Tempering allows time for all the particles to have access to the water and completely hydrate. Tempering times may be shortened for some soils, as indicated in Table 9.2.

For each water content, the soil is placed in the cup in such a manner that the soil fills the same volume as water would if placed in the cup while in the impact (drop) position. Care must be taken such that no air bubbles are trapped in the soil pat. It is generally easiest to work the soil into the cup from the front lip and squeeze it to the back. Figure 9.5 depicts water placed to the specified level in the cup.

The grooving tool is used with the beveled edge at the front of the tool to create a groove down the center of the pat. The tool creates the geometry of the slope and controls

Figure 9.5 Casagrande cup with water placed to the specified level in the cup.

Atterberg Limits   **123**

Figure 9.6 Grooved soil pat.

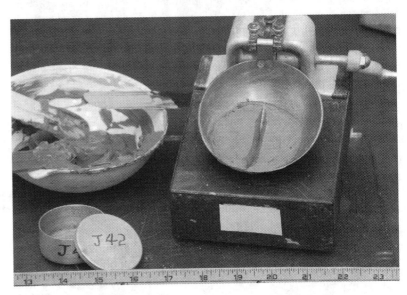

Figure 9.7 Groove closed by 13 mm (1/2 in.).

the volume of soil in the cup. The cut should be made from the back to the front of the cup. Care must be taken to keep the grooving tool at 90 degrees to the cup curvature at all times. Wipe the excess material from the front edge of the cup. A groove such as that shown in Figure 9.6 is made in the soil pat.

Turn the crank at a rate of 2 +/- 0.1 blows per second until the groove closes by 13 mm (1/2 in.), as shown in Figure 9.7.

For laboratory instruction purposes and to develop a consistent protocol, it is beneficial to repeat the blow count determination several times before taking material for a water content measurement. Scrape the material from the cup and remix it with the batch. Wipe the cup clean with a paper towel and repeat the process. Once two consecutive determinations yield the same (within one) blow count, remove a portion of soil for a water content determination. Remove a swipe of soil from across the groove with a spatula to obtain a representative specimen of at least 20 g for the water content determination.

Measurements of the number of blows to close the groove are required by ASTM D4318 for at least three different water contents; however, four or five determinations are preferable. Valid data should be limited to the range of 15 to 35 blows because the relationship between water content and number of blows is not linear at the extremes.

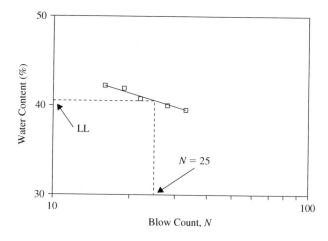

Figure 9.8 Flow curve for determining the liquid limit using the Casagrande cup.

The results are plotted on what is referred to as a flow curve. The flow curve has N on the x-axis as a log scale and the corresponding water content on the y-axis as a natural scale. The liquid limit is defined as the water content corresponding to closure at 25 drops, which is computed from a best fit line through the experimental data. Figure 9.8 demonstrates a typical flow curve and the determination of the liquid limit for Boston Blue Clay (BBC).

The determination of the liquid limit can also be obtained by use of the one-point method (ASTM D4318 Method B) and the same equipment described above. Soil is prepared at a water content close to 25 blow count material (i.e., 20 to 30). Visually identifying the consistency of paste that will fail at about 25 blows requires some experience with the specific material and the test method. At least two trials at that blow count are required. The blow count and the water content corresponding to that measurement are used to calculate the water content at 25 blows (i.e., the liquid limit) using Equation 9.4:

$$LL^n = \omega^n \left(\frac{N^n}{25} \right)^{0.121} \tag{9.4}$$

Where:

LL^n = one point liquid limit determined by measurement n
ω^n = water content for measurement n (%)
N^n = blow count at measurement n (dimensionless)

The method is not as reliable as the determination by the multipoint Casagrande method as the equation assumes a slope between the measurement point and the liquid limit. Not all soils have the same flow curve. Further, the method relies on one point of measurement instead of a best-fit line between at least three points.

The liquid limit is an index test and the quality of the result relies on careful control of the important parameters of the experiment and procedure. The energy per drop is delivered to a constant volume of soil with a specified geometry. The volume of soil is controlled by the shape of the cup, proper filling of the cup, and the material removed by the grooving tool. The grooving tool creates the geometry of the slope. The base width of the groove (distance for closure) is also controlled by the grooving tool. The amount of energy is controlled by the mass (cup plus soil) that falls, the height of the drop, the number of drops, and the hardness of the base. Figure 9.9 gives a schematic of the energy imparted to the control volume of the soil.

Usually, the back end of the grooving tool is dimensioned to properly set the drop height of the cup. The hinge of the cup is adjusted such that the strike point on the cup is 10 mm from the base when the cup is raised to its highest position.

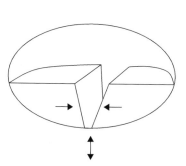

Figure 9.9 Schematic of the energy imparted to the control volume of soil in the Casagrande device.

Important considerations for the determination of the Liquid Limit:

- The material used for the base of the apparatus is a hard rubber and has been a source of considerable confusion. The base of the original Casagrande device was made of a product called Micarta No. 221A (Lambe, 1951, Appendix A, p. 152). This material was readily available because it was also used to make bowling balls and had a coefficient of rebound of about 85 percent. In the late 1980s, manufacturers could not obtain the Micarta material, and the base was replaced with a softer rubber having a coefficient of rebound of about 78 percent. The ASTM D4318 test method now specifies a base material having a coefficient of rebound of between 77 and 90 percent to allow both materials. The coefficient of rebound is defined as the ratio of the rebound height to the drop height (in percent) of a 7.94 mm (5/16 in.) diameter steel ball dropped on the base from a height of 254 mm (10 in.).
- In addition, the British Standard (BS1377) has always specified a less resilient (softer) base material. Using a softer base material will produce a higher interpreted value of the liquid limit. Based on publications by Norman (1958), Sampson and Netterberg (1985), and Sherwood and Ryley (1970), the liquid limit determined using the Casagrande cup with the harder base will be approximately 83 to 95 percent of the value determined using the softer base.
- Be sure the cup and grooving tool are in good condition, and the drop height of the cup is set to the proper distance.
- Clean and dry the cup after each measurement to be sure the interface remains consistent throughout the test. Allowing material to stick to the brass surface will strengthen the interface.
- Mix the soil well prior to tempering and between each determination. The strength of the mass of soil in the cup is determined by the weakest surface while the water content measures the average value. If the material has poorly hydrated clumps, the surface moisture will control the strength and introduce a bias in the results.
- Prevent entrapped air within the soil placed in the cup. These pockets of air create weak spots leading to early failure and nonuniform conditions.
- Over time, two different tools have been used to form the groove. The original tool shapes the slope, the groove width, and the height of the slope. A newer tool, often referred to as the curved tool, was easier to use but did not control the slope height. Only the original tool is used today since it was found to provide better control and more consistent results.
- Some soils tend to tear when making the groove. In such cases, make the cut with several passes of the tool, gradually increasing depth with each pass.
- Remove soil for water contents perpendicular to the groove. When testing lower plasticity soils, water can migrate down the slope during the test. Collecting material across the groove will give the most representative value.
- Work from wet to dry to improve the uniformity of the paste and maintain hydration of the particles.

Liquid Limit by Fall Cone Method (BS1377)

A fall cone method has also been developed to determine the liquid limit (Hansbo, 1957). The apparatus consists of a stainless steel cone with a polished surface. The cone has a 30° cone angle with a sharp point. The mass of the cone is controlled at 80.00 grams. A cone meeting these requirements is shown in Figure 9.10.

The procedure for this method is to place the soil in a specific cup, set the cone on the surface of the soil, release the cone for five seconds, and measure the penetration. The soil is then remixed and the measurement repeated. If the penetration measured in the two trials is different by less than 0.5 mm, the water content is determined. Trials are performed at a minimum of four different water contents and the results are plotted. The

Figure 9.10 Fall cone device for measuring the liquid limit.

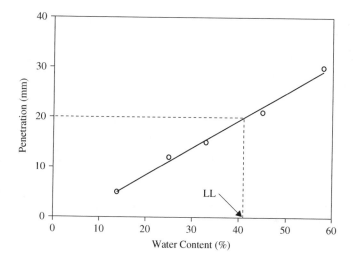

Figure 9.11 Determination of liquid limit using the fall cone.

liquid limit by the fall cone method is defined as the water content at a penetration of 20 mm, as demonstrated in Figure 9.11.

Many of the same important considerations given for the Casagrande method apply to the fall cone method as well.

The fall cone can be used to determine strength of soil (Hansbo, 1957). There have been a number of studies to determine the undrained strength of the soil corresponding to the liquid limit. The literature has suggested this value to be 1.7 kPa (Wroth and Wood, 1978).

Comparison between the Casagrande Method and the Fall Cone Method

In general, the fall cone method provides more consistent results since it is subject to fewer experimental and operator errors. However, the presence of large particles in the soil matrix can cause erroneous readings when using the cone method A comparison between the Casagrande method and the fall cone method, shown in Figure 9.12, is a simplification of the data shown in the referenced source (Wasti, 1987).

The methods typically agree well for soils with liquid limit values between 10 and 100.

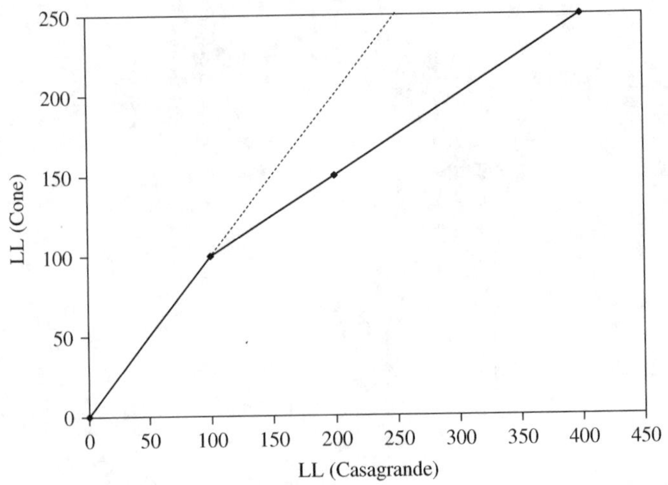

Figure 9.12 Comparison between liquid limit determined using Casagrande cup and fall cone. (Adapted from Wasti, 1987)

Figure 9.13 Soil threads rolled to 3.2 mm (1/8 in) without crumbling (left); soil threads rolled to the plastic limit (i.e., crumbling at 3.2 mm) (right).

Plastic Limit by Rolling (ASTM D4318)

The plastic limit is determined by rolling soil on a glass plate to a 3.2 mm (1/8 in.) diameter thread. Historically, rolling has been performed by hand. When the thread crumbles at 3.2 mm, the soil has reached the plastic limit. Refer to Figure 9.13 for an example of how soil threads appear when wetter than the plastic limit, and when at the plastic limit.

After rolling the soil thread to the plastic limit, the soil is placed in a tare for a water content determination and covered immediately to prevent moisture changes. The steps are repeated with another portion of soil, adding the soil to the tare until the mass of soil is at least 6 g. At least three trials are required and the average value is calculated, and reported to the nearest integer as the plastic limit.

A rolling device has been developed to control the diameter of the thread (Bobrowski and Griekspoor, 1992) and is included as optional equipment in the current ASTM standard.

Important considerations for the determination of the plastic limit include:

- The plate must be made of glass because it is nonabsorbing, providing better moisture control when rolling the soil threads.
- Maintain even pressure throughout the rolling process. Resist the tendency to increase the pressure as the soil gets drier, as this will cause premature breakage and therefore a higher determination of the water content.

Several papers (Sharma and Bora, 2003; Wasti, 1987; Feng, 2000; among others) have suggested the use of the fall cone to determine plastic limits to avoid the user-dependent errors associated with the rolling method by ASTM D4318. This would provide the ability to unify the LL and PL with a common method of testing. The fall cone technique to determine plastic limits would require a redefinition of the plastic limit and the proposal is to use 100 times the undrained strength at the liquid limit. As of the date of this writing, the method has not gained widespread use.

The method for determining the shrinkage limit is to measure the minimum volume of a dried soil pat, and then calculate the water content required to fully saturate that minimum volume. A portion of soil is prepared at about a 10 drop consistency paste and tempered. The soil is placed in a control volume and allowed to dry, possibly shrinking in volume. Care is taken to make sure the soil pat does not dry so quickly that the soil pat cracks. The dried soil pat is then immersed in a fluid and the volume of the displaced fluid is measured to determine the volume of the dried soil pat. Since the dried soil pat would absorb water and swell, for the wax method the soil pat is encased in wax of known density prior to submersion. Figure 9.14 presents the equipment necessary to perform a shrinkage limit test by the wax method.

The mercury displacement method mentioned previously is simpler since the dry soil pat will not absorb the mercury and therefore no steps need to be taken to seal the soil. However, the wax method is usually preferred to avoid the dangers and regulations associated with handling mercury.

Other important considerations for the determination of the shrinkage limit include:

- When a soil pat loses moisture rapidly, the soil may crack, resulting in an invalid test. If the soil pat cracks, the test must be redone and measures must be taken to slow the rate of moisture loss, such as drying the soil a humidity controlled room. However, slowing the rate of moisture loss in this way significantly extends the testing time, possibly by several weeks.
- During submersion of the wax encased soil pat in water, air bubbles may become attached to the wax surface or thread. The occurrences of these air bubbles can be reduced by making the wax coating smooth. When air bubbles do attach to the specimen or thread, brush them away or pop them, while taking care not to puncture the wax coating.

Figure 9.14 Equipment necessary to perform the shrinkage limit test by the wax method. Clockwise from upper left: wax pot, wax encased soil submerged in water within a beaker on a scale, ring stand to suspend the soil, spatula, tares, molds for containing the soil pat while drying, thread, wax pat formed in a control volume.

Table 9.3 Results of plastic and liquid limit determinations on natural soil compared to oven-dried soil.

	Bangkok clay (25% organic)			BBC
	LL	PL	PI	LL
Oven-dry	48	25	23	42
Natural	69	25	44	50

Source: Adapted from Ladd et al, 1971 for Bangkok Clay and personal database for BBC.

- Entrapped air bubbles in the wax, either during determination of the wax density or during coating of the dried soil pat, will cause significant errors in the test. If an air bubble is observed in the wax coating prior to solidifying, it may be popped and smoothed over. After solidifying, an air bubble must be cut out and replaced with wax. Take care to smooth out the surface so air bubbles will not be likely to attach to the wax surface during water submersion. During wax calibration, pour the wax slowly to minimize entrapment of air bubbles. Before solidifying, air bubbles can be popped. If air bubbles are observed after solidification and can not be cut out and replaced with fresh melted wax, prepare a new wax specimen.

Other important considerations for the determination of Atterberg Limits include:

- The soil must not be oven-dried prior to testing because some clay minerals (the more plastic minerals) will be altered irreversibly by the high temperature. The exception to this rule is when the ratio of the liquid limit after oven-drying to the liquid limit of undried soil is necessary to determine whether a soil is organic. A soil is classified as organic according to D2487 when this ratio is less than 0.75. The drying temperature can have a significant effect on the determination of limits, particularly for organic soils and soils containing halloysite. Table 9.3 presents selected results demonstrating the effect.
- When working soils through the No. 4 sieve in preparation for limit testing, do not break down weakly cemented aggregates within a finer matrix. Clumps of clay and silt, however, must be broken up to allow for sufficient hydration, creating a homogeneous specimen.
- Electrolytes in the water will influence the size of the double layer. Using distilled water to mix with the soil is preferred to tap water because it will slightly reduce the natural concentrations of electrolytes but will not add foreign ions found in tap water. Ideally, water from the site will be used to adjust the water content of soil for use in limits testing. Table 9.4 presents an example of the effect of dissolved salts on the determination of the plastic and liquid limit of Boston Blue Clay.

TYPICAL VALUES

The specified values of liquid and plastic limit according to USCS group symbol can be ascertained from the Casagrande chart in Figure 9.2. In addition, Table 9.5 lists some typical values for various clay deposits. As can be seen from Figure 9.13, the plastic and liquid limits can vary with depth range in a deposit, due primarily to differences in soil composition, as well as among geographic regions.

Among the Atterberg Limits, typical values for the shrinkage limit are most difficult to find in published literature. Besides rating systems, such as presented in Table 9.1, the data available is fairly limited. Table 9.6 presents some data available for the liquid, plastic, and shrinkage limits for soil minerals and soil deposits. The first source of information is for pure clay minerals only. The second set of values is for named soils. Note that the liquid and plastic limits for the named soils differ in some cases from the values provided for those soils above. The variations can be attributed to differences in deposit and geographic location of soils used for the study.

Equipment Requirements

General

1. Water content tares
2. Scale readable to 0.01 g with a capacity of at least 200 g for determination of water content
3. Desiccator
4. Water bottle
5. Distilled water
6. Mixing bowl
7. Cake frosting spatula

Liquid Limit By Casagrande Cup

1. Casagrande liquid limit device
2. Grooving tool

Plastic Limit By Hand Rolling

1. Glass plate
2. Example rod, approximately 3.2 mm in diameter

Shrinkage Limit By Wax Method

1. Calibrated wax
2. Wax melting pot
3. Shrinkage dish
4. Lubricant (such as vacuum grease)
5. Tap water (for calibration of the dish)
6. Capped cylindrical molding tube that will produce a specimen with a diameter of approximately 5 cm and a height of approximately 4 cm
7. Straight edge
8. Calipers with a measurement resolution of 0.01 mm
9. Equipment to measure volume by displacement method: string, beaker, and support hanger

Boston Blue Clay

Salt Concentration (g/L)	LL	PL	PI
0	30	20	10
35	40	20	20

Source: Adapted from Ladd, 1996.

Table 9.4. Example of the effects of dissolved salt on the determination of the liquid and plastic limit of Boston Blue Clay.

Table 9.5 Typical values of liquid and plastic limits for various clay deposits.

Soil Type	Liquid Limit (LL)	Plastic Limit (PL)
Boston Blue Clay (BBC)*	41	20
Bangkok Clay*	65	24
Atchafalaya Clay*	95	20
Maine Organic Clay*	65	31
Maine Clay**	30	20
San Francisco Bay Mud***	89	37
Mexico City Clay***	361	91
Harrison Bay Arctic Silt****	66	31

*After Ladd and Edgers, 1971, as appearing in Ladd et al., 1977.
**After Reynolds and Germaine, 2007.
***Mesri and Choi, 1984.
****Yin, 1985.

Table 9.6 Typical values of liquid, plastic, and shrinkage limits for soil minerals and soil deposits.

Soil Type	Liquid Limit (LL)	Plastic Limit (PL)	Shrinkage Limit (SL)
Montmorillonite*	100–900	50–100	8.5–15
Illite*	60–120	35–60	15–17
Kaolinite*	30–110	25–40	25–29
Mexico City Clay**	388	226	43
Boston Blue Clay**	41	25	19
Morganza Louisiana Clay**	104	75	14
Beverly Clayey Silt**	20	16	13

*Mitchell and Soga, 2005.
**After Lambe, 1951.

CALIBRATION

The Casagrande cup and associated equipment must be periodically calibrated and inspected for wear. In addition, for the shrinkage limit test, calibration of the dish and measurement of the wax density must be made.

Inspection of Wear (Casagrande Cup and Grooving Tool)

1. Base: The depression in the base at the point of contact between the base and cup must be less than 10 mm in diameter. If the base is worn larger than this, the base must be machined flat provided the dimensional tolerances are still met. Otherwise, the base must be replaced.

2. Cup: Using the grooving tool in the cup will cause a depression over time. The cup must be replaced when the depression is greater than 0.1 mm deep.

3. Cup Hanger: The mechanism must be replaced if the cup hanger pivot binds or if the lowest point of the cup rim can be moved from side-to-side by more than 3 mm.

4. Cam: The cam must be replaced when the cup drops before the cam follower loses contact with the cam.

5. Grooving Tool: The grooving tool must be retired from use when the width of the point does not meet the required tolerances of within 0.2 mm of 11 mm.

Adjustment of Height Drop (Casagrande Cup)

When the cup drop height is outside of the range of 0.2 mm of 10 mm, adjust the height of drop. The height can be adjusted in the laboratory using the following simple techniques. Note that the cup is assumed to be attached to the device during the adjustment.

1. Position a piece of tape such that the tape bisects the worn spot on the cup, with the tape parallel to the axis of the cup pivot, and with the tape on the pivot side of

the worn spot. If the cup does not have a worn spot, use carbon paper or tape with the sticky side up to determine the point of contact between the cup and the base.

2. Turn the crank until the cup is raised to its maximum height.

3. Slide the height gage under the cup from the front and observe whether the gage contacts the cup or the tape.

4. If the tape and the cup are both simultaneously contacted, the height of drop is ready to be checked. If not, adjust the cup slider until simultaneous contact is made.

5. Check the adjustment by turning the crank at two revolutions per second while holding the gage in position against the tape and cup.

6. If a faint clicking sound is heard without the cup rising from the gage, the adjustment is correct.

7. If no noise is heard or if the cup rises from the gage, readjust the height of drop.

8. If the cup rocks on the gage during this checking operation, the cam follower pivot is excessively worn and the worn parts must be replaced.

9. Remember to remove the tape after adjusting the drop height.

The shrinkage limit dish must be calibrated to determine the volume. The following series of steps and contained terminology have been written to be generally consistent with ASTM D4943.

Calibration of Shrinkage Limit Dish

1. Gather the shrinkage limit dish, a glass plate, lubricant, and a supply of water that will be used to fill the dish. All materials must be equilibrated to room temperature prior to calibration.

2. Place a light film of lubricant on the face of the plate and along the inside of the dish. Make sure there is a sufficient amount of lubricant to create a seal between the plate and the dish, but not so much as to cause clumping of the lubricant.

3. Obtain the mass of the lubricated glass plate and lubricated dish (M_c) to 0.01 g.

4. Add equilibrated water to the dish until overfilled.

5. Place the lubricated face of the place over the top of the dish, without trapping any air below the plate. This creates the control volume.

6. Remove any water outside the control volume.

7. Obtain the mass of the glass plate, dish, lubricant, and contained water (M_{wc}) to 0.01 g.

8. Calculate and record the mass of water (M_w) in the control volume to 0.01 g using Equation 9.5:

$$M_w = M_{wc} - M_c \qquad (9.5)$$

Where:

M_w = mass of water (g)
M_{wc} = mass of plate, dish, lubricant, and contained water (g)
M_c = mass of glass plate, dish, and lubricant (g)

9. Record the mass density of water (ρ_w) for the applicable temperature. Refer to Appendix B for the mass density of water versus temperature.

10. Calculate and record the volume of the dish (V_m) to 0.01 cm^3 using Equation 9.6.

$$V_m = \frac{M_w}{\rho_w} \qquad (9.6)$$

Where:
V_m = volume of the dish (cm^3)
ρ_w = mass density of water (g/cm^3)

11. Repeat the dish calibration process. If the resulting volumes differ by more than 0.03 cm³, repeat the process until the two measurements are within this range.

12. Average the values obtained using the two successful trials and record the result to 0.01 cm³.

Measurement of the Wax Density

The wax density must be measured in order to determine the volume of the wax seal.

1. Gather the cylindrical molding tube, cap, lubricant, and melted wax.

2. Place a thin layer of lubricant on the inside of the molding tube. The lubricant will act as a mold release.

3. Cap the cylindrical molding tube.

4. Pour melted wax into the molding tube, making sure there are no trapped air bubbles. Allow the wax to cool completely.

5. Remove the cap and extrude the wax specimen, taking care not to deform the specimen.

6. Trim both ends of the wax specimen.

7. Measure the height of the specimen at four locations to 0.001 cm. Average the four measurements and record the value as h_{wp}.

8. Measure the diameter of the specimen at four locations to 0.001 cm. Average the four measurements and record the value as d_{wp}.

9. Calculate the volume of the wax plug (V_{wp}) to 0.01 cm³ using Equation 9.7:

$$V_{wp} = \frac{\pi \cdot d_{wp}^2 h_{wp}}{4} \qquad (9.7)$$

Where:
V_{wp} = volume of the wax plug (cm³)
d_{wp} = diameter of the wax plug (cm)
h_{wp} = height of the wax plug (cm)

10. Obtain and record the mass of the wax plug (m_{wp}) to 0.01 g.

11. Calculate the mass density of the wax (ρ_x) to 0.01 g/cm³ using Equation 9.8:

$$\rho_x = \frac{m_{wp}}{V_{wp}} \qquad (9.8)$$

Where:
ρ_x = mass density of the wax (g/cm³)
m_{wp} = mass of the wax plug (g)

SPECIMEN PREPARATION

For classroom efficiency, the sample used for Atterberg Limits should be from the same sample as used for the grain size analyses, since the plastic limit, liquid limit, and grain size analyses are part of the Unified Soil Classification System presented in Chapter 10.

Do not oven-dry the soil prior to the determination of a standard liquid limit test. Performing a companion liquid limit test on soil that has previously been oven-dried is required for assessing whether a soil is organic within the USCS.

Prepare the soil by working the material through the 0.475 mm (No. 40) sieve and obtaining the natural water content. Mix soil with enough water to obtain about 600 g to 700 g of paste at a consistency of approximately 15 drops. Generally, 15-drop consistency is similar to soft-serve ice cream. Adjust the water content by spreading the material on a glass plate to dry or adding water and mixing in the bowl. Be sure the paste has a uniform consistency before proceeding.

Separate at least 150 g of the soil for the shrinkage limit determinations. Add distilled water until the soil has about a 10-drop consistency. Be sure the paste has a uniform consistency before proceeding.

Cover the bowls and place in a humid room for about 24 hours to temper.

Separate at least 40 g of the tempered soil at fifteen drop consistency just prior to starting the liquid limit determination and set aside to dry for the plastic limit determinations. The soil can be separated into three portions and placed directly on a paper towel to assist in drying.

PROCEDURE

The determination of the liquid and plastic limits will be performed in general accordance with ASTM D4318, while the reference method for the shrinkage limit is ASTM D4943.

Liquid Limit (Casagrande Cup)

1. Place tempered soil in a clean, calibrated Casagrande cup to maximum depth of 1/2 inch. The soil should form a flat, horizontal surface with the bottom lip of the cup. This volume can be checked by filling the cup with water while it is in the strike position. Be sure to remove entrapped air and to prepare a smooth surface.

2. Groove soil with ASTM type tool and keep the tool perpendicular to the cup at the point of contact.

3. Turn crank at 2 blows per second until groove closes for a length of 0.5 inches and record the number of blows.

4. Remove soil from cup and return it to the dish.

5. Mix soil in dish and repeat steps 1, 2, 3, and 4 until two consistent blow counts ($\pm$ 1) are measured.

6. Obtain the water content of soil by removing about 10 g of paste perpendicular to and across the closed groove.

7. Obtain four separate water content determinations between 15 and 35 blows by drying the soil slightly and repeating steps 1 through 6.

Plastic Limit

1. Roll one-third of the soil set aside for the plastic limit test into a 3.2 mm (1/8 in.) strand on the glass plate.

2. Gather the material into a ball.

3. Repeat steps 1 and 2 until the strand shows signs of crumbling when it reaches 3.2 mm in diameter. This is the plastic limit.

4. Place in a water content tare and cover. Repeat steps 1 through 3 until at least 6 g of soil are collected for a water content determination.

5. Measure the water content for the plastic limit determination.

6. Repeat steps 1 through 5 for each of the remaining two-thirds.

Shrinkage Limit

Only one shrinkage limit determination is required according to D4943. However, for laboratory instructional purposes, about three determinations on the same soil will provide information to detect problems with results.

1. Record the identifying information for the shrinkage limit dish.

2. Place a thin layer of lubricant on the inside of the shrinkage limit dish.

3. Obtain the mass of the shrinkage limit dish and lubricant (m) to 0.01 g.

4. Place approximately one-third of the volume of the 10-drop-consistency tempered soil in the dish. Tap the dish against a padded, firm surface to force the soil to flow and level out. Take care not to trap any air bubbles in the soil pat.

5. Repeat step 4 for two additional portions of soil.

6. Using a straight edge, strike off the extra soil from the top. Remove any particles adhering to the outside of the dish.

7. Immediately obtain the mass of the dish and wet soil (m_w) to 0.01 g.

8. Air-dry the soil until the soil color turns light. If the soil pat cracks during drying, restart the test and slow the rate of moisture loss.

9. Once the color of the soil has turned from dark to light, use a forced draft oven at 110 +/− 5°C to dry the soil to constant mass.

10. Obtain the mass of the dish and oven-dried soil (m_d) to 0.01 g.

11. Remove the soil pat from the dish and tie a piece of thread around the soil.

12. Hold the pat of soil by an end of the thread and immerse the pat in melted wax. Make certain that the whole soil pat is covered in wax and that no air bubbles are trapped in the wax.

13. Remove the pat from the wax and allow the soil and wax to cool completely.

14. Determine the mass of the wax encased soil pat in air (m_{sxa}) to 0.01 g.

15. Place a water bath on a balance and zero the reading on the balance.

16. Suspend the wax-encased soil pat from a hanger placed beside the balance, submerging the wax-encased soil pat in a water bath, which is on the balance. Make certain that no air bubbles are attached to any portion of the submerged pat or thread. Record the mass (m_{wsx}) to 0.01 g.

Calculations

1. Liquid Limit: Plot the water contents against log of blows, draw the flow curve, and select the liquid limit as the intersection of this curve and the 25 blow line. Report the nearest whole number as the liquid limit (LL).

2. Plastic Limit: Calculate the average of the three plastic limit trials. Report the nearest whole number as the plastic limit (PL).

3. Shrinkage Limit: The following series of calculations and contained symbols have been written to be generally consistent with ASTM D4943.

   a. Calculate the mass of the dry soil pat using Equation 9.9:

$$m_s = m_d - m \tag{9.9}$$

   Where:
   m_s = mass of oven-dry soil pat (g)
   m_d = mass of dish and oven-dry soil pat (g)
   m = mass of dish and lubricant (g)

   b. Calculate the initial water content of the soil in the dish using Equation 9.10:

$$w_C = \frac{m_w - m_d}{m_s} \times 100 \tag{9.10}$$

   Where:
   w_C = initial water content of the soil in the dish (%)
   m_w = mass of dish and wet soil (g)

c. Calculate the volume of the wax and wax encased soil pat using Equation 9.11:

$$V_{dx} = \frac{m_{wsx}}{\rho_w}$$ (9.11)

Where:
V_{dx} = volume of the wax and wax encased soil pat (cm³)
m_{wsx} = mass of water displaced by the wax encased soil pat (g)

d. Calculate the mass of the wax using Equation 9.12:

$$m_x = m_{sxa} - m_s$$ (9.12)

Where:
m_x = mass of wax (g)
m_{sxa} = mass of the wax encased soil pat in air (g)

e. Calculate the volume of wax using Equation 9.13:

$$V_x = \frac{m_x}{\rho_x}$$ (9.13)

Where:
V_x = volume of the wax (cm³)

f. Calculate the volume of the oven-dry soil pat using Equation 9.14:

$$V_d = V_{dx} - V_x$$ (9.14)

Where:
V_d = volume of the oven-dry soil pat (cm³)

g. Calculate the shrinkage limit using Equation 9.15:

$$SL = \omega_C - \frac{(V_m - V_d)\rho_w}{m_s} \times 100$$ (9.15)

h. Report the average of the three measurements to the nearest whole number as the shrinkage limit (SL).

Report

Report the liquid limit, plastic limit, shrinkage limit, plasticity index, liquidity index, and natural water content along with identifying sample, testing, and project information.

PRECISION

Criteria for judging the acceptability of test results obtained by test method D4318 (Liquid and Plastic Limits) are given as follows as based on the interlaboratory study (ILS) conducted by the ASTM Reference Soils and Testing Program. Note that the wet preparation method and the multipoint liquid limit method were used to produce these results.

- *Within Laboratory Repeatability:* Expect the standard deviation of your results on the same soil to be on the order of 0.7 for the liquid limit and 0.5 for the plastic limit.
- *Between Laboratory Reproducibility:* Expect the standard deviation of your results as compared to others on the same soil to be on the order of 1.3 for the liquid limit and 2.0 for the plastic limit.

Criteria for judging the acceptability of test results obtained by test method D4943 (Shrinkage Limit) are given as follows as based on the AASHTO Materials Reference Laboratory (AMRL) Proficiency Sample Program conducted on a CL material.

- *Within Laboratory Repeatability:* Expect the standard deviation of your results on the same soil to be on the order of 0.8.
- *Between Laboratory Reproducibility:* Expect the standard deviation of your results as compared to others on the same soil to be on the order of 1.4.

This information is consistent with data contained in a Bureau of Reclamation report (Byers, 1986), which indicates it is reasonable to expect that two tests properly performed on the same soil by the same person in the same laboratory with the same equipment within a short period of time will be within 2 points of each other.

DETECTING PROBLEMS WITH RESULTS

For the liquid and plastic limits, the Casagrande plasticity chart is useful for a first check on the reasonableness of results. If the limits plot above the "U" line, there is likely a problem with the results. Check the calculations of the plastic and liquid limit values.

If the standard deviation for one set of measurements exceeds the criteria provided above, then evaluate the techniques of the individual performing the test. If duplicate measurements exceed the within laboratory repeatability estimates, sources of experimental error are likely related to errors in counting the blow counts, insufficient rolling time for the plastic limit, incorrect water content determinations, or entrapped air for the shrinkage limit. If the test results do not fall within the typical ranges or exceed the reproducibility limit, the likely cause of error is systemic, such as insufficient water content specimen size, incorrect closing distance of the groove in the soil, improper volume of soil in the Casagrande cup, erroneous density of the wax, or equipment out of calibration.

Several methods of isolating the causes of errors are possible. Systematic errors due to equipment deficiencies can be identified by methods such as verifying that the proper fall height of the cup is measured at the point of contact between the base and the cup, confirming the coefficient of rebound of the base, or recalibrating the volume of the shrinkage limit dish. Procedural and technique errors are best identified by performing the test on a soil with known values of limits.

REFERENCE PROCEDURES

ASTM D4318 Liquid Limit, Plastic Limit, and Plasticity Index of Soils.

ASTM D4943 Shrinkage Factors of Soils by the Wax Method.

REFERENCES

Refer to this textbook's ancillary web site, www.wiley.com/college/germaine, for data sheets, spreadsheets, and example data sets.

Atterberg, A. 1911. Uber die physikalische Bodenuntersuchung und uber Die Plastizität der Tone, *Internationale Mitteilungen für Bodenkunde*, vol. 1.

Bobrowski, L. J., Jr., and D. M. Griekspoor. 1992. "Determination of the Plastic Limit of a Soil by Means of a Rolling Device," *Geotechnical Testing Journal*, GTJODJ, *15*(3), 284–287.

BS1377. *Methods of Test for Soils for Civil Engineering Purposes*, British Standards Institution, London.

Byers, Jack G. 1986. *Alternative Procedure for Determining the Shrinkage Limit of Soil*, Report No. REC-ERC-86-2. Bureau of Reclamation, Denver, CO.

Casagrande, A. 1932. "Research on the Atterberg Limits of Soils," *Public Roads*, *13*(8).

Feng, T. W. 2000. "Fall-cone Penetration and Water Content Relationship of Clays," *Géotechnique*, *50*(2), 181–187.

Hansbo, S. 1957. "A New Approach to the Determination of the Shear Strength of Clay by the Fall-cone Test," *Royal Swedish Geotechnical Institute Proceedings*, Stockholm, *14*, 5–47.

Holtz, W. G. 1959. "Expansive Clays—Properties and Problems," Bureau of Reclamation Report No. EM-568, Denver, CO.

Ladd, Charles C. 1996. "Soil Structure and Environmental Effects: Effects of Mineralogy and Environmental Factors," *1.361Advanced Soil Mechanics,* Class Notes.

Ladd, C. C., and L. Edgers. 1972. "Consolidated-Undrained Direct-Simple Shear Tests on Saturated Clays," *Research Report R72-82*, No. 284, Department of Civil Engineering, Massachusetts Institute of Technology, Cambridge, MA.

Ladd, C. C., R, Foott, K. Ishihara, F. Schlosser, and H. G. Poulos. 1977. "Stress-Deformation and Strength Characteristics," State-of-the-Art Report, *Proceedings of the IX International Conference on Soil Mechanics and Foundation Engineering*, Tokyo, 421–494.

Ladd, C. C., Z-C Moh, and D. G. Gifford. 1971. "Undrained Shear Strength of Soft Bangkok Clay," *Proceedings of the 4th Asian Regional Conference on Soil Mechanics and Foundation Engineering*, Bangkok, vol. 1, 135–140.

Lambe, T. W. 1951. *Soil Testing for Engineers*, John Wiley and Sons, New York.

Mesri, G. and Y. K. Choi. 1984. "Time Effects on the Stress-Strain Behavior of Natural Soft Clays," Discussion, *Géotechnique*, *34*(3), 439–442.

Mitchell, J. K. and K. Soga. 2005. *Fundamentals of Soil Behavior*, John Wiley and Sons, New Jersey.

Norman, L. E. 1958. "A Comparison of Values of Liquid Limit Determined with Apparatus Having Bases of Different Hardness," *Géotechnique*, 8(2), 79–84.

Reynolds, Richard T. and John T. Germaine. 2007. "Benefits and Pitfalls of Multistage Embankment Construction," *Proceedings of the 7th International Symposium on Field Measurements in Geomechanics (ASCE GSP 175)*, September 24–27, 2007, Boston, MA.

Sampson, L. R. and F. Netterberg. 1985. "The Cone Penetration Index: A Simple New Soil Index Test to Replace the Plasticity Index," *Proceedings of the Eleventh International Conference on Soil Mechanics and Foundation Engineering*, August 12–16, San Francisco, CA, 1041–1048.

Sharma, Binu and Padma K. Bora. 2003. "Plastic Limit, Liquid Limit and Undrained Shear Strength of Soil: Reappraisal," *Journal of Geotechnical and Geoenvironmental Engineering*, 129(8), 774–777.

Sherwood, P. T., and Ryley, M. D. 1970. "Investigation of a Cone-Penetrometer Method for the Determination of the Liquid Limit," *Géotechnique*, 20(2), 135–136.

Skempton, A. W. 1953. "The Colloidal Activity of Clay." *Proceedings of the Third International Conference on Soil Mechanics and Foundation Engineering*, Zurich, vol. 1, pp. 57–61.

Terzaghi, K. 1925. "Simplified Soil Tests for Subgrades and Their Physical Significance," *Public Roads*, October.

Wasti, Yildiz. 1987. "Liquid and Plastic Limits as Determined from the Fall Cone and the Casagrande Methods," *Geotechnical Testing Journal*, GTJODJ, *10*(1), 26–30.

Wroth, C. P., and D. M. Wood. 1978. "The Correlation of Index Properties with Some Basic Engineering Properties of Soils," *Canadian Geotechnical Journal*, *15*(2), 137–145.

Yin, Edward Yen-Pang. 1985. "Consolidation and Direct Simple Shear Behavior of Harrison Bay Arctic Silts," MS Thesis, Department of Civil Engineering, Massachusetts Institute of Technology, Cambridge.

Chapter **10**

Soil Classification and Description

SCOPE AND SUMMARY

This chapter details the procedures to describe soils using the Visual-Manual method and to formally classify the soil according to the Unified Soil Classification System (USCS). Combining the description and the classification will provide the information necessary for a complete identification. The Visual-Manual procedures are comprised of mainly qualitative observations (such as can be used in the field with a minimum of equipment), and provide a method to estimate the USCS grouping. On the other hand, the USCS relies on quantitative laboratory testing obtained through Atterberg Limits and Grain Size Distribution testing. These values are then used to definitively compute the classification. The chapter provides techniques for both fine-grained and coarse-grained soils.

TYPICAL MATERIALS

Soil classification is applicable to all geotechnical materials. The USCS system is most applicable to naturally occurring materials; however, the system can be applied to man-made assemblages of particles with some modifications.

Soils exhibit many different types of behaviors. In fact, soils are often divided into groups based on general behavior characteristics. For example, consider the differences you experience while walking on a sandy beach versus trying to walk across a clam-digging mud flat. These two materials are not totally different but they do behave differently under your foot pressure and hence it would be appropriate to put them in different groups.

There is a distinction between classification and description, although the terms are sometimes casually interchanged in practice. The formal division of soils into types or groups according to prescribed characteristics is termed classification. Many different materials will fit into a given type or group. On the other hand, description pertains to individual observations about the material without quantitative results. One would expect a description to be more subjective and more specific (i.e., with a description, two individual soils could be differentiated).

Classification and description of soils are of utmost importance in geotechnical engineering. Quality soil identification is necessary on every project. The identification process should begin as soon as contact is made with the material, keeping in mind the varied audience that will be using the description. A good description distinguishes the soil from all others and starts the process of determining how a particular soil will behave as compared to other soils within an engineer's experience base. From a complete description, the soil can be informally classified into a group (with a group symbol and group name if using the USCS).

Performing a proper Visual-Manual description takes experience and practice. When dealing with a new material it is always best to apply the entire procedure systematically. However, local experience should be used to shortcut the process and focus on observing variations from a baseline description rather than wasting time repeating needless steps. While the Visual-Manual procedure yields a USCS classification, it is important to self-calibrate with the formal USCS classification.

Soil descriptions and classifications are crucial to effective engineering, but are often skipped or neglected today. Identification of a sample is invaluable to the engineering process in many situations: the technician handling the soil can detect when sample labeling errors have occurred; the field engineer is able to relay reliable information to the office engineer based on soil description; time can be saved tailoring a testing program according to sample classification; context is available to the engineer based solely on subsurface information, lab results, and analyses; and site information is synthesized as it comes in, giving the engineer a feel for future projects.

Numerous soil classification systems are in use. They vary from being founded on a specific application, such as the Corps of Engineers Frost Susceptibility Classification System, to applying to a broad range of soil types and applications. Some of the more common classification systems are United States Bureau of Public Roads, Airfield Classification, U.S. Bureau of Soils Triangle Textural Classification, MIT Classification, and Burmister. A paper by Arthur Casagrande (1948) discusses many of these classification systems, as well as providing a tremendous amount of valuable detail on soil characteristics. Details concerning the Burmister classification system can be found in a 1951 paper written by Donald Burmister. There are many similarities between the Airfield Classification system, described in depth by Casagrande (1948), and the USCS system.

In this textbook, the USCS is used as a basis because it applies to the range of soils encountered for most types of engineering projects. Another advantage of the USCS is that the Visual-Manual procedure uses compatible criteria, but can be used in the field or without laboratory testing. The combination of a formal and informal procedure provides a very powerful tool to gain specific, identifying information as well as allowing the field engineer or office engineer to use test results to develop the ability to visually identify soils. The USCS was developed in 1952 by the Bureau of Reclamation and the Corps of Engineers (Lambe and Whitman, 1969). The USCS bases the classification on both a grain-size component and a consistency of fines component. The classification consists of a group symbol and a group name. The group name provides more specificity than the group symbol.

The USCS is formalized in ASTM D2487 Standard Practice for Classification of Soils for Engineering Purposes (Unified Soil Classification System). The practice is simple, relevant to all soil types, incorporated widely in practice, and relatively fast to use. In addition to a soil description, a USCS classification requires quantitative grain size and Atterberg Limits data, except limits are not required if the material contains less than 5 percent fines. The system is based on two simple principles: size distribution of the grains is important for coarse-grained material, and the interaction of the grains with water is most important for fine-grained material. Materials are then separated into fractions based on mass percentages. One obvious shortcoming of the USCS is the fact that it ignores the importance of particle geometry for coarse-grained materials.

The USCS uses objective groupings for classification. There are three tiers relative to grain size (coarse, fine, peat), which is further broken into six classes: G, S, M, C, O, Pt (gravel, sand, silt, clay, organic, peat) and four adjectives: W, P, H, L (well-graded, poorly-graded, high plasticity, low plasticity). Two letter combinations of these symbols are then used to identify each group. There are also dual symbols assigned as described below. The percentage of soil particle sizes based on dry mass is used to determine the tier. Then, if in the coarse tier, the percentage of material in the various size ranges are used to determine the class. Majority always rules and a tie (equal parts) is given to the smaller fraction. As an example, if a material is 51 percent greater than the No. 200 sieve, and 30 percent of the bulk material is gravel size particles, the soil is in the coarse-grained tier and the gravel class.

Borderline groups are indicated by two group symbols separated by a slash (e.g., CL/CH) and can be used when a soil classification is close to two group symbols. This is commonly used in the Visual-Manual procedure due to a lack of quantitative information, but can also be used with the USCS classification. When using borderline group symbols with the USCS, the first group symbol in a borderline case is assigned as the designation resulting from the classification and the second is the symbol for the soil that is close. Borderline symbols are distinct from dual symbols. Dual symbols are indicated by two group symbols separated by a dash (e.g., SP-SM) and are used when either a coarse sample has 5 to 12 percent fines, or when the plotted results of the plastic and liquid limit testing of the fines are in the cross-hatched area of the plasticity chart, labeled CL-ML. Combined symbols are something entirely different and consist of two class symbols. Combined symbols are used for coarse-grained materials having more than 12 percent fines to indicate that the plasticity of the fines are more important than the size distribution of the coarse-grained material (e.g., GM).

Group Names also provide cross-tier or cross-class recognition depending on the quantity of minor material present in a mix. The material is ignored if less than 5 percent. The term "with" is used for soil mixes having more than 15 percent coarse material (e.g., silt with sand or poorly-graded gravel with sand) or having 5 to 12 percent fine material (e.g., well-graded gravel with clay) to warn of presence. When greater than 12 percent fines are present in a coarse material, or 30 percent or greater coarse material is present in a fine material, the adjective status is used (e.g., silty gravel or sandy fat clay), acknowledging the fact that this fraction is important to behavior.

Fine-grained materials are distinguished based on plasticity or interaction with water. High-plasticity clays (CH) are sometimes referred to as "fat" clays, while low-plasticity clays (CL) are referred to as "lean" clays. Notice that the term "clay" is defined in this context as plotting in the plasticity chart on or above the A-line. Silts that do not exhibit plasticity (which is typical for silt-sized particles) have the group symbol ML and are typically referred to as just "silt." Silts that do exhibit plasticity are assigned the group symbol MH and are referred to as "elastic silt." In this context, "silt" is defined as material plotting below the A-line.

The distinction between silt and clay is somewhat confused by the fact that the profession has several definitions for each term. In Chapter 8, "Grain Size Analysis," the terms "silt" and "clay" were based on a specific particle size. When referring to the composition of an individual grain, "clay" refers to a particular group of mineral structures. Illlite is an example of a clay mineral. One could have a silt-sized particle of the clay mineral illite. In the USCS classification system and the Visual-Manual procedure, the distinction between silt fines and clay fines is based on the plasticity of the minus No. 40 fraction. This introduces a "performance-based" definition for the terms "clay" and "silt." It is therefore important to provide context when referring to material as clay or silt, and organic lean clay and organic silt for the OL group names.

Organic, fine-grained soils in the USCS are indicated when the liquid limit of the oven-dried soil is less than 75 percent of the liquid limit of undried soil. As described in Chapter 7, "Organic Content," other more deterministic methods are available for determining the organic content of a material, such as digestion or loss on ignition; however, the USCS does not use these other methods. Organic soils are divided into high (OH) and low (OL) plasticity based on the liquid limit. Unfortunately, the USCS uses the same group name for OH (organic clay) and OL (organic silt) soils plotting above and below the A-line. This text uses different names and group symbols to distinguish the soils falling into the various zones, as described in the previous chapter. In summary, the authors prefer to use "organic fat clay" and "organic elastic silt" for the OH group names, and "organic silt" or "organic lean clay" for the OL group names. The terms "clay" and "silt" are used in the name when the limits plot above or below the A-line, respectively, in the same manner as for inorganic soils. Additionally, the authors prefer to use dual symbols for organic soils when the results plot above the A-line. When the liquid limit is less than fifty the group symbol is "CL-OL," and when it is greater than fifty the group symbol is "CH-OH."

Figure 10.1 provides an overview of the USCS. Details concerning the individual elements presented in the chart follow later in the text.

The first step in both the USCS and the Visual-Manual procedures for both fine-grained and coarse-grained soils is the general description of the soil. Many companies will have their own preferred standardized form and order of a description, as well as requesting that certain aspects are skipped in the interest of time, or perhaps will vary the list depending on project type or location. The person describing the soil should note the results of the observations, and then use these results to form the concise text describing the sample in the order and form requested by the company.

A number of tests and observations can be performed in the field, or in the lab with little sample preparation or equipment. Most of these tests yield immediate results, which make them invaluable tools for soil identification. ASTM D2488 Standard Practice for Description and Identification of Soils (Visual-Manual Procedure) provides these manual procedures to identify soils. When working with intact samples the first observations should focus on the fabric and structure of the material. For coarse-grained material the focus is on the grains, connectivity of grains, and distribution of sizes. These descriptors are easily obtained by direct observation. For fine-grained material the focus is on evaluating the level of plasticity. The procedures used in the standard practice, along with one additional method are collectively referred to as "quick tests."

That said, both the ASTM practice for the Visual-Manual procedure and the ASTM practice for USCS include the following categories as part of the soil description:

- Range of Particle Sizes: The distribution of particles in a soil is probably the most important characteristic relating to material performance. Therefore, it is prominent in the USCS and the first item to evaluate when doing a Visual-Manual description. The range of particle sizes is listed as a percentage of dry mass by grain size as provided in Table 10.1. A soil is coarse-grained if more than 50 percent of the material is retained on the No. 200 sieve (0.075 mm opening); otherwise the soil is fine-grained. This should provide a reasonable estimate of the particle size distribution and allow one to assign a USCS group symbol for coarse-grained materials. With some experience, the size fractions can be determined by eye to within 5 percent.

Figure 10.1 USCS Classification Chart.

Field Identification Procedures (Excluding particles larger than 3 in. and basing fractions on estimated weights)				Group Symbols [a]	Typical Names	Information Required for Describing Soils	Laboratory Classification Criteria
Coarse-grained soils — More than half of material is *larger* than No. 200 sieve size [b] (The No. 200 sieve size is about the smallest particle visible to naked eye)	Gravels — More than half of coarse fraction is larger than No. 4 sieve size (For visual classification, the 1/4 in. size may be used as equivalent to the No. 4 sieve size)	Clean gravels (little or no fines)	Wide range in grain size and substantial amounts of all intermediate particle sizes	GW	Well graded gravels, gravel-sand mixtures, little or no fines	Give typical name; indicate approximate percentages of sand and gravel; maximum size; angularity, surface condition, and hardness of the coarse grains; local or geologic name and other pertinent descriptive information; and symbols in parentheses. For undisturbed soils add information on stratification, degree of compactness, cementation, moisture conditions and drainage characteristics. Example: *Silty sand*, gravelly; about 20% hard, angular gravel particles 1/2-in. maximum size; rounded and subangular sand grains coarse to fine, about 15% non-plastic fines with low dry strength; well compacted and moist in place; alluvial sand; (*SM*)	$C_U = \dfrac{D_{60}}{D_{10}}$ Greater than 4; $C_C = \dfrac{(D_{30})^2}{D_{10} \times D_{60}}$ Between 1 and 3
			Predominantly one size or a range of sizes with some intermediate sizes missing	GP	Poorly graded gravels, gravel-sand mixtures, little or no fines		Not meeting all gradation requirements for GW
		Gravels with fines (appreciable amount of fines)	Silty fines (for identification procedures see ML below)	GM	Silty gravels, poorly graded gravel-sand-silt mixtures		Atterberg limits below "A" line or PI less than 4 / Above "A" line with PI between 4 and 7 are *borderline* cases requiring use of dual symbols
			Plastic fines (for identification procedures, see CL below)	GC	Clayey gravels, poorly graded gravel-sand-clay mixtures		Atterberg limits above "A" line with PI greater than 7
	Sands — More than half of coarse fraction is smaller than No. 4 sieve size	Clean sands (little or no fines)	Wide range in grain size and substantial amounts of all intermediate particle sizes	SW	Well graded sands, gravelly sands, little or no fines		$C_U = \dfrac{D_{60}}{D_{10}}$ Greater than 6; $C_C = \dfrac{(D_{30})^2}{D_{10} \times D_{60}}$ Between 1 and 3
			Predominantly one size or a range of sizes with some intermediate sizes missing	SP	Poorly graded sands, gravelly sands, little or no fines		Not meeting all gradation requirements for SW
		Sands with fines (appreciable amount of fines)	Nonplastic fines (for identification procedures, see ML below)	SM	Silty sands, poorly graded sand-silt mixtures		Atterberg limits below "A" line or PI less than 5 / Above "A" line with PI between 4 and 7 are *borderline* cases requiring use of dual symbols
			Plastic fines (for identification procedures, see CL below)	SC	Clayey sands, poorly graded sand-clay mixtures		Atterberg limits below "A" line with PI greater than 7

Determine percentages of gravel and sand from grain size curve. Depending on percentage of fines (fraction smaller than No. 200 sieve size) coarse grained soils are classified as follows: Less than 5% — GW, GP, SW, SP; More than 12% — GM, GC, SM, SC; 5% to 12% — *Borderline* cases requiring use of dual symbols.

Fine-grained soils — More than half of material is *smaller* than No. 200 sieve size		Identification Procedures on Fraction Smaller than No. 40 Sieve Size					
		Dry Strength (crushing characteristics)	Dilatancy (reaction to shaking)	Toughness (consistency near plastic limit)			
	Silts and clays liquid limit less than 50	None to slight	Quick to slow	None	ML	Inorganic silts and very fine sands, rock flour, silty or clayey fine sands with slight plasticity	Give typical name; indicate degree and character of plasticity, amount and maximum size of coarse grains; colour in wet condition, odour if any, local or geologic name, and other pertinent descriptive information, and symbol in parentheses. For undisturbed soils add information on structure, stratification, consistency in undisturbed and remoulded states, moisture and drainage conditions. Example: *Clayey silt*, brown; slightly plastic; small percentage of fine sand; numerous vertical root holes; firm and dry in place; loess; (*ML*)
		Medium to high	None to very slow	Medium	CL	Inorganic clays of low to medium plasticity, gravelly clays, sandy clays, silty clays, lean clays	
		Slight to medium	Slow	Slight	OL	Organic silts and organic silt clays of low plasticity	
	Silts and clays liquid limit greater than 50	Slight to medium	Slow to none	Slight to medium	MH	Inorganic silts, micaceous or diatomaceous fine sandy or silty soils, elastic silts	
		High to very high	None	High	CH	Inorganic clays of high plasticity, fat clays	
		Medium to high	None to very slow	Slight to medium	OH	Organic clays of medium to high plasticity	
Highly Organic Soils		Readily identified by colour, odour, spongy feel and frequently by fibrous texture			Pt	Peat and other highly organic soils	

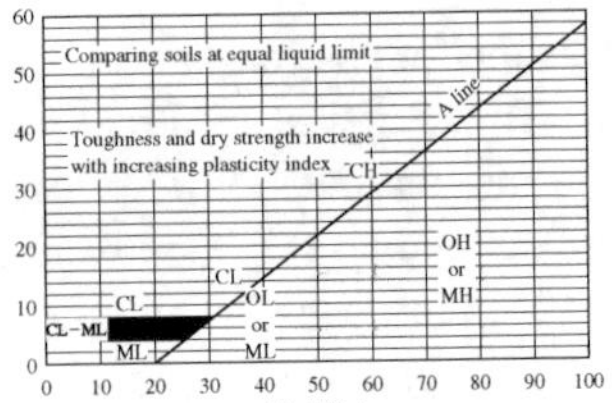

Use grain size curve in identifying the fractions as given under field identification.

Plasticity chart for laboratory classification of fine grained soils. Comparing soils at equal liquid limit — Toughness and dry strength increase with increasing plasticity index. (A line; Plasticity index vs Liquid limit; CL-ML, ML, CL, OL, CH, OH or MH)

From Wagner, 1957.

[a] *Boundary classifications.* Soils possessing characteristics of two group are designated by combinations of group symbols. For example GW – GC, well graded gravel-sand mixture with clay binder.

[b] All sieve sizes on this chart are U.S. standard.

Field Identification Procedure for Fine Grained Soils or Fractions

These procedures are to be performed on the minus No. 40 sieve size particles, approximately 1/64 in. For field classification purposes, screening is not intended, simply remove by hand the coarse particles that interfere with the tests.

Dilatancy (Reaction to shaking):
After removing particles larger than No. 40 sieve size, prepare a pat of moist soil with a volume of about one-half cubic inch. Add enough water if necessary to make the soil soft but not sticky.
Place the pat in the open palm of one hand and shake horizontally, striking vigorously against the other hand several times. A positive reaction consists of the appearance of water on the surface of the pat which changes to a livery consistency and becomes glossy. When the sample is squeezed between the fingers, the water and gloss disappear from the surface, the pat stiffens and finally it cracks or crumbles. The rapidity of appearance of water during shaking and of its disappearance during squeezing assist in identifying the character of the fines in a soil.
Very fine clean sands give the quickest and most distinct reaction whereas a plastic clay has no reaction. Inorganic silts, such as a typical rock flour, show a moderately quick reaction.

Dry Strength (Crushing characteristics):
After removing particles larger than No. 40 sieve size, mould a pat of soil to the consistency of putty, adding water if necessary. Allow the pat to dry completely by oven, sun or air drying, and then test its strength by breaking and crumbling between the fingers. This strength is a measure of the character and quantity of the colloidal fraction contained in the soil. The dry strength increases with increasing plasticity.
High dry strength is characteristic for clays of the CH group. A typical inorganic silt possesses only very slight dry strength. Silty fine sands and silts have about the same slight dry strength, but can be distinguished by the feel when powdering the dried specimen. Fine sand feels gritty whereas a typical silt has the smooth feel of flour.

Toughness (Consistency near plastic limit):
After removing particles larger than the No. 40 sieve size, a specimen of soil about one-half inch cube in size, is moulded to the consistency of putty. If too dry, water must be added and if sticky, the specimen should be spread out in a thin layer and allowed to lose some moisture by evaporation. Then the specimen is rolled out by hand on a smooth surface or between the palms into a thread about one-eight inch in diameter. The thread is then folded and re-rolled repeatedly. During this manipulation the moisture content is gradually reduced and the specimen stiffens, finally loses its plasticity, and crumbles when the plastic limit is reached.
After the thread crumbles, the pieces should be lumped together and a slight kneading action continued until the lump crumbles.
The tougher the thread near the plastic limit and the stiffer the lump when it finally crumbles, the more potent is the colloidal clay fraction in the soil. Weakness of the thread at the plastic limit and quick loss of coherence of the lump below of plastic limit indicate either inorganic clay of low plasticity, or materials such as kaolin-type clays and organic clays which occur below the A-line.
Highly organic clays have a very weak and spongy feel at the plastic limit.

Figure 10.1 USCS Classification Chart.
(Lambe and Whitman, 1969. Copyright John Wiley and Sons. Reprinted with permission.)

Table 10.1 USCS grain size boundaries.

Grain Type	Sieve Size		Grain Size (mm)	
	Lower Bound	Upper Bound	Lower Bound	Upper Bound
Boulders	12 in.	—	300	—
Cobbles	3 in.	12 in.	75	300
Gravel	—	—	—	—
Coarse	0.75 in.	3 in.	19	75
Fine	No. 4	0.75 in.	4.75	19
Sand	—	—	—	—
Coarse	No. 10	No. 4	2	4.75
Medium	No. 40	No. 10	0.425	2
Fine	No. 200	No. 40	0.075	0.425
Fines	—	No. 200	—	0.075

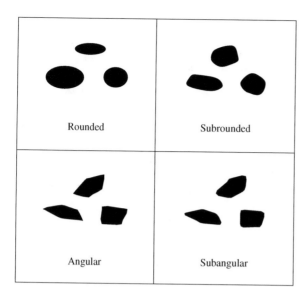

Figure 10.2 Soils with different angularities.

Descriptor	Criteria
Angular	Particles have sharp edges and relatively plane sides with unpolished surfaces.
Subangular	Particles are similar to angular description but have rounded edges.
Subrounded	Particles have nearly plane sides but have well-rounded corners and edges.
Rounded	Particles have smoothly curved sides and no edges.

Table 10.2 Criteria for describing angularity.

Source: After ASTM D2488-06, Table 1. Copyright ASTM INTERNATIONAL. Reprinted with permission.

- Maximum Particle Size: In addition to the percentages of soil sizes, the maximum particle size must be listed. Keep in mind that the sampling method will likely limit the maximum particle size. For example, a split spoon sampler will not collect a particle larger than about 45 mm. On the other hand, the maximum particle size dictates the amount of material required to do a proper particle size analysis or water content determination.
- Angularity: Angularity describes the shape of individual particles, and is rated on a scale from rounded to angular. Describing these characteristics is appropriate for the coarse fraction of a soil material. Angularity can be ascertained by eye for most sand and gravel-sized particles; however, in some cases a magnifying glass will be helpful. The particles in a sample will normally include a range of angularities, often varying with particle size. The description should encompass the observed variability without creating an excessive burden. Figure 10.2 shows examples of the degrees of angularity, while Table 10.2 gives word descriptors.
- Shape: Shape pertains to the relative dimensions of the individual particles. Particles are described as regular, flat, elongated, or flat and elongated. Table 10.3 lists the criteria for the descriptors. Relative dimensions are normally estimated rather than carefully measured with an instrument. As with angularity, shape will vary within a sample and over the particle size range. Figure 10.3 shows a picture of various shaped particles meeting these criteria.

- Color: Color must be described in conjunction with the moisture state of the soil, because the color will change dramatically as the soil changes moisture state. In most situations, the color will change from dark to light as the soils dries. For

Descriptor	Criteria
The particle shape shall be described as follows where length, width, and thickness refer to the greatest, intermediate, and least dimensions of a particle, respectively.	
Regular	All dimensions are within a factor of 3.
Flat	Particles with width/thickness > 3.
Elongated	Particles with length/width > 3.
Flat and Elongated	Particles meet criteria for both flat and elongated.

Source: After ASTM D2488-06, Table 2. Copyright ASTM INTERNATIONAL. Reprinted with permission.

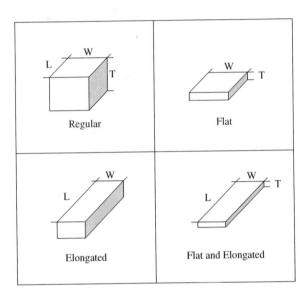

Figure 10.3 Various shaped particles.

Table 10.4 Criteria for describing moisture condition.

Description	Criteria
Dry	Absence of moisture, dusty, dry to the touch.
Moist	Damp but no visible water.
Wet	Visible free water; usually soil is below water table.

Source: After ASTM D2488-06, Table 3. Copyright ASTM INTERNATIONAL. Reprinted with permission.

standardized coloring, the Munsell® soil charts are often used to provide a common color standard, thus allowing comparison of descriptions between organizations.

- Odor: Soils with an organic component will typically smell "earthy," or in extreme cases will smell like sulfur (rotten eggs) or petroleum products. Care must be taken to distinguish whether a petroleum-like odor actually comes from petroleum or from organics. Soils that are almost entirely organic matter are considered to be Peat (PT) and the USCS and Visual-Manual procedures do not particularly describe them well. Rather, the appearance of the peat should be noted including color; odor; presence of roots, weeds, or other matter; reaction to chemicals; and level of decomposition. Then, if required, laboratory methods can be employed to fully classify these materials instead. Once such method is ASTM D4427 Classification of Peat Samples by Laboratory Testing.

- Moisture Condition: The moisture condition of the soil at the time of description is described in accordance with Table 10.4. The descriptions are most useful if the sample is described soon after sampling. The description should include the sample condition (e.g., as received, intact, bulk, and so on).
- HCl Reaction: The reaction of the soil in the presence of a dilute (1N) hydrochloric acid (HCl) solution is a useful method to detect the presence of carbonates. ASTM Test Method D2488 states precautions to be used when handling acids and when mixing with water. The test is performed by placing a few drops of the solution on the soil and observing the reaction. The reaction is described according to Table 10.5.
- Consistency: Consistency applies to intact, fine-grained specimens and is related to the stiffness of the soil. The scale ranges from very soft to very hard. Criteria for the consistency ratings are provided in Table 10.6.
- Cementation: Cementation is bonding between individual particles by a material other than the particle itself. The presence of cementation is evaluated by assessing the resistance of the soil to breaking down under pressure. Cementation is primarily applicable to coarse-grained soil and is an especially important aspect of residual soils. Evaluate cementation by placing a lump of soil in the palm of the hand and working with finger pressure. The rating system ranges from weak to strong, and is presented as Table 10.7.

Description	Criteria
None	No visible reaction.
Weak	Some reaction, with bubbles forming slowly.
Strong	Violent reaction, with bubbles forming immediately.

Table 10.5 Criteria for describing the reaction with HCl.

Source: After ASTM D2488-06, Table 4. Copyright ASTM INTERNATIONAL. Reprinted with permission.

Description	Criteria
Very soft	Thumb will penetrate soil more than 25 mm (1 in.).
Soft	Thumb will penetrate soil about 25 mm (1 in.).
Firm	Thumb will indent soil about 6 mm (0.25 in.).
Hard	Thumb will not indent soil but readily indented with thumbnail.
Very hard	Thumbnail will not indent soil.

Table 10.6 Criteria for describing consistency.

Source: After ASTM D2488-06, Table 5. Copyright ASTM INTERNATIONAL. Reprinted with permission.

Description	Criteria
Weak	Crumbles or breaks with handling or little finger pressure.
Moderate	Crumbles or breaks with considerable finger pressure.
Strong	Will not crumble or break with finger pressure.

Table 10.7 Criteria for describing cementation.

Source: After ASTM D2488-06, Table 6. Copyright ASTM INTERNATIONAL. Reprinted with permission.

- Structure: Structure is often referred to as macrofabric and is a description of the mixture of materials in the deposit. Structure can only be observed using intact samples or when viewing a cross section in an excavation, such as the wall of a test pit. When working with intact samples, the material should be cut on a vertical plane with a wire saw or split with a knife. The material must be examined in a moist (preferably natural) state initially, and then observed while drying for changes and variations. A soil may appear to be homogeneous clay when moist or dry, but zones of different material will become significantly lighter during drying, indicating siltier layers. Radiography can show layering and inclusions and is discussed in Chapter 11, "Background Information for Part II." Table 10.8 lists the structural descriptions, along with the criteria. Figure 10.4 shows a picture of an intact soil sample cut using the technique described above and allowed to dry to show zones of different grain size.

- Hardness: Determine the hardness of individual coarse particles by hitting with a hammer. Particles that do not break are considered hard. For particles that do break, a simple description of what happens to particles is sufficient (e.g., gravel-size particles cleave under blow of a hammer).

Table 10.8 Criteria for describing structure.

Description	Criteria
Homogeneous	Same color and appearance throughout.
Stratified	Alternating layers of varying material or color with layers at least 6 mm thick; note thickness.
Laminated	Alternating layers of varying material or color with the layers less than 6 mm thick; note thickness.
Fissured	Breaks along definite planes of fracture with little resistance to fracturing.
Slickensided	Fracture planes appear polished or glossy, sometimes striated.
Blocky	Cohesive soil that can be broken down into small angular lumps that resist further breakdown.
Lensed	Inclusion of small pockets of different soils, such as small lenses of sand scattered through a mass of clay; note thickness.

Source: After ASTM D2488-06, Table 7. Copyright ASTM INTERNATIONAL. Reprinted with permission.

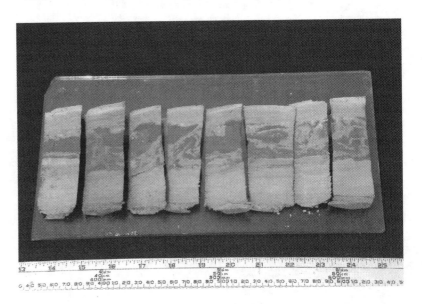

Figure 10.4 Sample cut and allowed to dry to show structure.

- Sensitivity: While not a part of the ASTM Visual-Manual classification procedure, sensitivity is an important observation when working with intact, fine-grained materials. Quantitatively, sensitivity is the ratio of the intact strength to the strength in the remolded state and at the same water content. Refer to Chapter 11, "Background Information for Part II" for further information on measurements of sensitivity. A reasonable estimate of sensitivity for classification purposes can be determined by taking a slice (about 1 cm thick) of material, applying pressure between the thumb and index finger, and sliding the thumb and finger in the opposite directions to shear the material. Sensitivity is then reported on a scale of "insensitive" to "quick" (sometimes referred to as "quick clays"). Table 10.9 lists the sensitivity descriptions, along with the criteria.

The following observations are used to evaluate the plasticity of the fine fraction. For the Visual-Manual procedure, plastic and liquid limit tests are not run, but rather estimates of plasticity are made by performing a few "quick tests." They are all easy to perform, require little equipment, and provide insights about the nature of the fine particles. The tests are performed when 10 percent or more of the material is judged to pass the No. 200 sieve. In the interest of time, the tests are actually performed on the material passing the No. 40 sieve. While this does dilute the measurements, it makes the separating task much more manageable as compared to using the No. 200 sieve.

- Dry Strength: Dry strength describes the crushing characteristics of a 12 mm (0.5 in.) ball of material once it has been allowed to dry from about the plastic limit consistency. The dry strength increases with plasticity. It is more sensitive to the type of particle (mineral) that to the size of the particle. Figure 10.5 shows the process of conducting the dry strength quick test. Table 10.10 lists the criteria for describing dry strength.

Description	Criteria
Insensitive	Deforms continuously under constant pressure.
Sensitive	After yielding, material deforms with less pressure.
Very Sensitive	After yielding, material is much softer and deforms under slight pressure.
Quick	After yielding material is unable to hold shape.

Table 10.9 Criteria for describing sensitivity of intact, fine-grained soil.

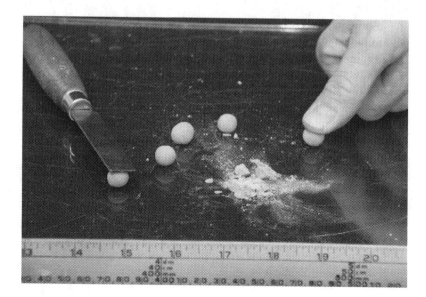

Figure 10.5 The dry strength quick test.

- Dilatancy: Dilatancy refers to the reaction of a pat of soil in the hand to shaking. It is important to thoroughly mix the soil with just enough water to get the pat to about the liquid limit consistency. The test will give a false reading if too much water is added because the soil will fluidize. Silt has a positive (quick) reaction (i.e., water appears on surface upon shaking, and then disappears when the fingers are stretched to shear the pat). Clay has no reaction. Figure 10.6 shows the process of conducting the dilatancy quick test. Table 10.11 presents the dilatancy criteria.

Table 10.10 Criteria for describing dry strength.

Description	Criteria
None	The dry specimen crumbles into powder with mere pressure of handling.
Low	The dry specimen crumbles into powder with some finger pressure.
Medium	The dry specimen breaks into pieces or crumbles with considerable finger pressure.
High	The dry specimen cannot be broken with finger pressure. Specimen will break into pieces between thumb and a hard surface.
Very High	The dry specimen cannot be broken between the thumb and a hard surface.

Source: After ASTM D2488-06, Table 8. Copyright ASTM INTERNATIONAL. Reprinted with permission.

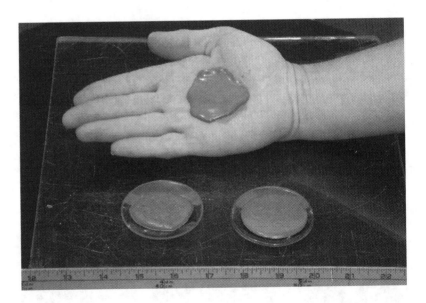

Figure 10.6 The dilatancy quick test.

Table 10.11 Criteria for describing dilatancy.

Description	Criteria
None	No visible change in the surface appearance.
Slow	Water appears slowly on the surface during shaking and does not disappear or disappears slowly upon squeezing.
Rapid	Water appears quickly on the surface during shaking and disappears quickly upon squeezing.

Source: After ASTM D2488-06, Table 9. Copyright ASTM INTERNATIONAL. Reprinted with permission.

- Toughness: Toughness is a measure of the strength of a thread of soil at the plastic limit. Toughness is estimated while rolling a 3 mm thread on a smooth surface. Toughness increases with the plasticity index. A bit of experience is required to develop a sense of the relative amounts of pressure required to roll the thread. Table 10.12 shows the toughness criteria.
- Plasticity: Plasticity is based on observation of the nature of the plastic limit thread. Plasticity is evaluated over the range of nonplastic to highly plastic based on the characteristics of the soil thread while performing the plastic limit test. Table 10.13 provides an overview of the criteria used to make the assignment. The plastic limit is the water content at which the material is at the transition point between semi-solid and solid.

The following two tests (dispersion and the "tooth test") are not required by ASTM D2487 or D2488. The dispersion test is included as an optional tool called the "Jar Method" in Appendix X.4 of D2488.

- Dispersion: Dispersion (or sedimentation) can be performed to assist in estimating the percentages of grain size for soils consisting of mainly fine sand or silt. High-plasticity clay particles do not drop out of suspension in a short enough time span to make this technique practical in the field for these materials. It is also very difficult to properly hydrate plastic clays using manual methods. These clumps of clay can be very misleading. A test tube is a convenient way to perform this "mini" sedimentation test in the field. Place soil in the test tube so that approximately 2 cm of the test tube is filled with soil. Add water to the test tube so that the water level is about 2 cm from the top of the test tube. Place a finger or thumb over the open end of the test tube and agitate until the soil and water are well mixed. Stop agitation and allow the soil particles to settle out.

Table 10.12 Criteria for describing toughness.

Description	Criteria
Low	Only slight pressure is required to roll the thread near the plastic limit. The thread and the lump are weak and soft.
Medium	Medium pressure is required to roll the thread to near the plastic limit. The thread and lump have medium stiffness.
High	Considerable pressure is required to roll the thread to near the plastic limit. The thread and the lump have very high stiffness.

Source: After ASTM D2488-06, Table 10. Copyright ASTM INTERNATIONAL. Reprinted with permission.

Table 10.13 Criteria for describing plasticity.

Description	Criteria
Nonplastic	A 1/8 in. (3 mm) thread cannot be rolled at any water content.
Low	The thread can barely be rolled and the lump cannot be formed when drier than the plastic limit.
Medium	The thread is easy to roll and not much time is required to reach the plastic limit. The thread cannot be rerolled after reaching the plastic limit. The lump crumbles when drier than the plastic limit.
High	It takes considerable time rolling and kneading to reach the plastic limit. The thread can be rerolled several times after reaching the plastic limit. The lump can be formed without crumbling when drier than the plastic limit.

Source: After ASTM D2488-06, Table 11. Copyright ASTM INTERNATIONAL. Reprinted with permission.

Figure 10.7 shows the results of sedimentation quick tests performed on several soils.

Table 10.14 provides timeframes for which various particle sizes fall out of suspension and are deposited in sequence at the bottom of the test tube. Note that fine particles will "fluff" (i.e., water will hold the fine particles away from each other), whereas sands and gravels will settle until point-to-point contact. Therefore, the approximate volumes of the coarser fractions can be determined by the relative layer thickness in the tube, whereas the volumes of clay and silt will be considerably less than appears from estimating layer thicknesses along the tube. An estimate of the dry mass must be made based on the relative volumes of material settling out in the time ranges prescribed below. Watch for clumping, which will make fine particles settle out more quickly than they should.

- Tooth test: The tooth test can be used to discern whether fine-grained soils are greater than clay size (0.002 mm). A soil particle greater than 0.002 mm will feel gritty on the tooth. This method is not recommended unless the person performing the test is certain there are no contaminants in the soil. As a reminder, a person with normal vision can detect the fine-grained boundary, which is at 0.075 mm.

Additional comments may be warranted, depending on the preferences of the company or client, such as presence of foreign matter, construction debris, petroleum products, anything notable while obtaining the sample, geological interpretation of the material, or other information specific to the project or application should be noted.

After collecting information from the various Visual-Manual procedures, it is time to determine the USCS group symbol and group name. ASTM D2487 and D2488 provide

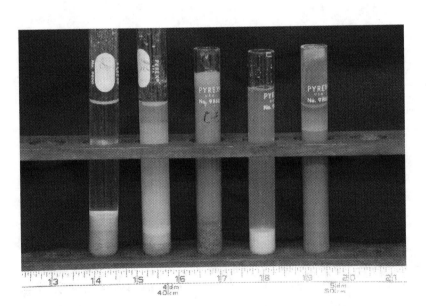

Figure 10.7 Sedimentation quick tests performed on several soils.

Table 10.14 Soil type associated with settling time in a field dispersion test

Settling Time	Size (mm)	Soil Type Settled Out of Suspension
< 1 seconds	0.4	Medium sand and larger
20 seconds	0.075	Fine sand and larger
5 minutes	0.02	Coarse silt and larger
60 minutes	0.010	Most of the silt and larger
> 4 hours	0.002	Only clay in suspension

flow charts to help guide the classification process. The charts from D2487 (USCS) are reprinted below for classifying fine-grained soils, organic fine-grained soils, and coarse-grained soils as Figure 10.8, Figure 10.9, and Figure 10.10, respectively. These flow charts are based on numerical results from the grain size and Atterberg Limits testing. They are very similar to those in D2488 (Visual-Manual) and can be used relatively easily with the Visual-Manual procedure. Of course, the flow charts in D2488 can be used directly instead, but it would be a bit redundant to reprint both sets here.

When classifying a fine-grained material, the results from the quick tests are very helpful in assigning a group symbol. Table 10.15 can assist in combining the observations and arriving at one group symbol. Remember that borderline symbols, such as CL/ML or CL/CH, can be used with the Visual-Manual procedure when a soil exhibits properties of both group symbols.

Based on the group symbol, one enters into the group symbol column of Figure 10.8 and proceeds to the right based on the estimated coarse-grained fractions to obtain the appropriate group name.

When classifying an organic fine-grained material, the task is less rigorous than for inorganic materials. One simply assigns the combined OL/OH group symbol and proceeds to determine the group name accounting for the coarse-grained fraction. This can be done using any of the rows in Figure 10.9 and replacing the noun silt or clay with "soil." It is possible (but not required) to use the quick tests to separate high and low plasticity organic soils.

When classifying a coarse-grained soil, use the estimates of the size fractions and enter the left most column of Figure 10.10. Proceed to the right to obtain both the group

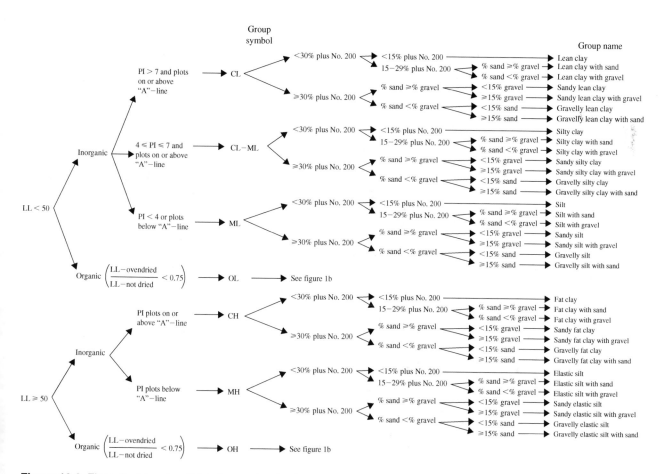

Figure 10.8 Flow chart for classifying fine-grained soils using the USCS.

(ASTM D2487-06, Figure 1. Copyright ASTM INTERNATIONAL. Reprinted with permission.)

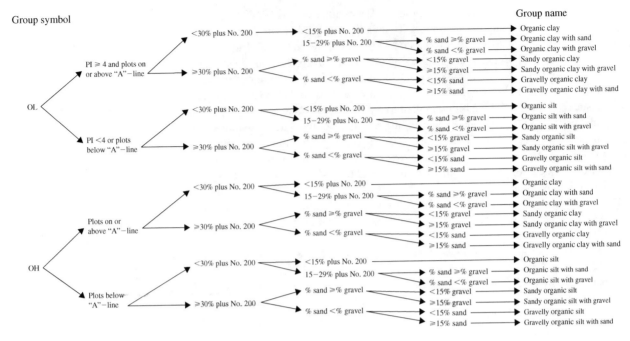

Figure 10.9 Flow chart for classifying organic fine-grained soils using the USCS.

(ASTM D2487-06, Figure 2. Copyright ASTM INTERNATIONAL. Reprinted with permission.)

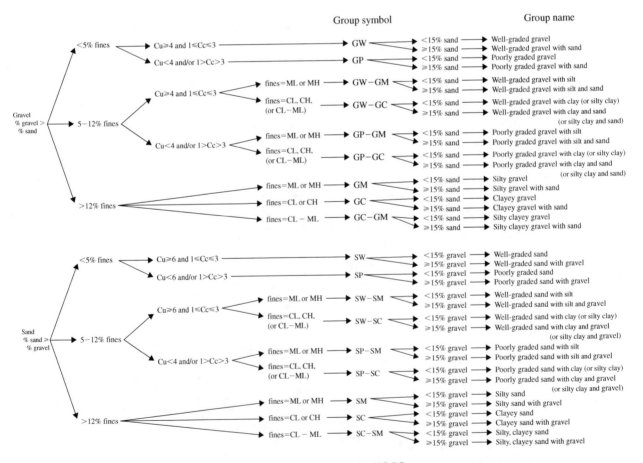

Figure 10.10 Flow chart for classifying coarse-grained soils using the USCS.

(ASTM D2487-06, Figure 3. Copyright ASTM INTERNATIONAL. Reprinted with permission.)

symbol and the group name. When the plasticity of fine-grained material is required, use information from the quick tests and Table 10.15 to obtain the appropriate fines classification.

The final step in the identification process is to compose a concise unique description of the sample using the observations and classification.

As stated previously, the form of a description or classification may vary largely from company to company or from project to project. Table 10.16 provides a checklist (adapted from ASTM D2488) to be used as a tool while describing soils. The list is generally in the order of making the determination and is consistent with the sequence provided in the procedures section. An entry should be provided for every row using "N/A" for any item which does not apply. This is important because it distinguishes between a forgotten item as compared to recognition that the item is not applicable.

Soil Symbol	Dry Strength	Dilatancy	Toughness	Plasticity
ML	None to low	Slow to rapid	Low	Nonplastic to low
CL	Medium to high	None to slow	Medium	Low to medium
MH	Low to medium	None to slow	Low to medium	Medium to high
CH	High to very high	None	High	High

Table 10.15 Identification of inorganic fine-grained soils from manual tests.

Source: After ASTM D2488-06, Table 12. Copyright ASTM INTERNATIONAL. Reprinted with permission.

Table 10.16 Checklist for description of soils.

General Information

1. Your name or initials and the date
2. Source of material: in situ, bulk, tube, etc.
3. Local name
4. Geologic interpretation

For intact samples

5. Structure: stratified, laminated, fissured, slickensided, lensed, homogeneous
6. Sensitivity: insensitive, sensitive, very sensitive, quick
7. Consistency: very soft, soft, firm, hard, very hard
8. Cementation: weak, moderate, strong

For all samples

9. Color (in moist condition)
10. Odor (mention only if organic or unusual)
11. Moisture: dry, moist, wet
12. Unusual material in sample
13. Maximum particle size or dimension
14. Percent of cobbles or boulders, or both (by volume)
15. Reaction with HCl: none, weak, strong
16. Percent of gravel, sand, or fines, or all three (by dry weight)
17. Particle-size range

   Gravel: fine, coarse

   Sand : fine, medium, coarse

18. Particle angularity: angular, subangular, subrounded, rounded
19. Particle shape (if appropriate): regular, flat, elongated, flat and elongated

(continued)

Table 10.16 (*continued*)

20. Hardness of coarse sand and larger particles

For Fines (material passing #40 sieve)

21. Dilatancy: none, slow, rapid

22. Sedimentation: fine sand, silt, clay

23. Toughness: low, medium, high

24. Plasticity of fines: nonplastic, low, medium, high

25. Dry strength: none, low, medium, high, very high

26. Organic fines: yes or no

Summary

27. Group symbol

28. Group name

29. Additional comments: presence of roots or root holes, presence of mica, gypsum, etc., surface coatings on coarse-grained particles, saving or sloughing of auger hole or trench sides, difficulty in augering or excavating, and so on.

Source: Modified from ASTM D2488-06, Table 13. Copyright ASTM INTERNATIONAL. Reprinted with permission.

An example resulting description is:

10% fine sand, 90% fines: no dilatancy, medium toughness, medium plasticity, medium dry strength, sensitive.

Firm, moist, blue-gray, lean clay (CL). No odor. Maximum particle size 1 mm. Occasional fine sand lenses. GLACIOMARINE DEPOSIT.

CALIBRATION

Calibration is not required for classification according to the USCS or description according to the Visual-Manual procedure. However, the two methods should be used in conjunction with one another to "calibrate" the person performing the Visual-Manual procedures. It is helpful to have a collection of particles in "standard" sizes and shapes for training activities and an example of high- and low-plasticity clay.

Equipment Requirements

1. Magnifying glass
2. Dilute HCl solution (one part 10N HCL with three parts water)
3. Dilute Hydrogen Peroxide solution (3 percent is available in drug stores; optional)
4. Hammer
5. Equipment for Grain Size Distribution, if not already performed
6. Equipment for Atterberg Limits, if not already performed
7. Test tube
8. Water
9. Ruler
10. Spatula
11. Color chart (Munsell® preferred; optional)

Specimens for classification can be obtained directly from the sampling apparatus (such as split spoon or push tube), jar sample, or a bag sample of bulk materials. In fact, the specimens are best classified with no preparation and the in situ conditions preserved as much as practical. Sample quantity is often limited by practical field constraints. However, it is good practice to adhere to the general guidelines provided for a grain size analysis. A reasonable minimum dry mass would be about 200 times the mass of the largest individual particle when working with coarse-grained materials and about 150 g of fine-grained material.

The classification analysis and description procedure will be performed in general accordance with ASTM Standard Practices D2487 and D2488. Procedures are slightly different for intact and bulk samples. The order of observation is important and the following procedures (and Table 10.16) have been arranged in the preferred order of operation. In most engineering applications, the samples will arrive at the laboratory in the moist condition. One should always collect structure specific information on intact samples before proceeding with the bulk characteristics. Complete all the individual observations and evaluations, determine the USCS group symbol, and then proceed to write a concise, final description.

More than one soil should be used for classification and description in this laboratory. Ideally, having an intact fine-grained tube sample and a large bulk coarse-grained sample serves to illustrate the extremes. It is also important to experience the difference between plastic and nonplastic fine-grained material. However, if that is not possible, it is suggested that a soil with a wide range of grain sizes (such as a clayey sand) is used, so that the student gains experience with multiple aspects of these procedures.

General sample description:

1. Record the sample identification information, boring or test pit number, sample number, type of sample (tube, bag, bucket), local name and geologic interpretation (if available), the date, and initials of the person describing the soil.

2. If the sample is from a tube, extrude a section of soil and obtain a representative sample. Refer to Chapter 11, "Background Information for Part II," for more information on processing tube samples. If the sample is from a bag or bucket, process the material and choose a representative sample. Refer to Chapter 1, "Background Information for Part I" for more information on processing bulk samples and selecting representative samples.

For intact, fine-grained samples (50% or more passing No. 200 sieve):

3. Cut a vertical slice off the edge of the sample (about 1 cm thick) using a wire saw. Describe the layering with reference to Table 10.8. Lay the slice on a glass plate and set aside to air dry. Check the slice periodically for changes (new fine layers) in layering as it dries.

4. Use a knife to cut vertically into the sample and then twist the knife to split the sample in two. The fractured surface will help identify three dimensional features of structure outlined in Table 10.8.

5. Cut a horizontal slice (about 1 cm thick) of the sample. Bend this slice and observe how it breaks (fractures or bends). Next compress the edge of the slice between your fingers and observe the behavior. Use this information to describe sensitivity according to Table 10.9.

6. Measure the consistency according to Table 10.6 on one of the cut surfaces.

For intact, coarse-grained samples (>50% retained on No. 200 sieve):

7. Observe extent of layering and general surface features. (Note: this is only possible when working on an exposed vertical surface in the field.)

8. Examine material for signs of cementation, clusters of particles, and so on. Use Table 10.7 to describe the cementation. Check that cementation persists in the presence of water by inundating (flooding with water) some of the material.

For every sample:

9. If large differences in layers are observed, separate as appropriate before proceeding with the bulk characterization.

10. Describe the color of the as-received material. It is often useful to add a color description of the fully dried material. Refer to standardized color charts (such as the Munsell® soil color charts), if requested.

11. Describe any odor. The odor can be enhanced by sealing a small amount of the material in a container for several minutes.

12. Describe moisture condition with respect to Table 10.4.

13. Identify and describe any unusual materials.

14. Record the maximum particle size.

15. Separate the large particles (usually greater than 19 mm) and estimate the percentage of the total mass.

16. Proceed to work with the finer fraction.

17. If there is concern for carbonate products, check the reaction to a dilute HCl solution with respect to Table 10.5.

18. Work the soil between the hands to separate the particles by size and estimate the percentage (by dry mass) of soil particles within the ranges given in Table 10.1. If in the field and observing a test pit excavation, include the percentage of cobbles and boulders visible in the hole by volume. Use the dispersion procedure if further assistance is needed with visually estimating percentages of the finer fraction particle sizes.

19. If a significant amount of the material is fine-grained, then separate the sample on the No. 40 sieve. Dry the coarser fraction to make it easier to work with.

20. Describe the particle angularity of the coarse particles with respect to Table 10.2 and Figure 10.2. Note that the angularity may vary with particle size.

21. Describe the particle shape of the coarse particles with respect to Table 10.3. Note that the particle shape may vary with particle size.

22. Describe the hardness of coarse particles as described in the background section of this chapter.

23. Add additional comments such as listed as number 28 in Table 10.16.

24. If less than 5 percent of the dry mass passes the No. 200 sieve, proceed to step 34.

25. Obtain a sample of material finer than the No. 40 sieve.

For fine-grained samples (50% or more passing No. 200 sieve on material finer than No. 40 sieve):

26. Chop up the material into relatively small pieces (about 1/5 cm cubes) and thoroughly mix. Remove a reasonably-sized sample and obtain the water content according to Chapter 2.

27. Separate about 50 g of material and mix with sufficient water to create a liquid limit consistency.

28. Separate a second 50 g subsample and place this on a paper towel to dry toward the plastic limit.

29. Using a spatula, create a smooth pat of the liquid limit consistency material in the palm of your hand. Cup your hand and tap the back with your other hand. Observe the surface of the pat for changes in texture and record the results of the dilatancy test with respect to Table 10.11.

30. Continue to add water to the pat, creating a thin slurry. Transfer the slurry into a test tube, fill with water, and mix thoroughly. Perform the sedimentation test and estimate the distribution of size fractions.

31. Take some of the material off the paper towel and roll it into a consistent lump. If the material prefers to stick to surfaces (adhesive) rather than stay in a lump (cohesive), it is too wet. Roll the cohesive lump into a thread on a flat surface. While rolling, observe the toughness with respect to Table 10.12 and the plasticity with respect to Table 10.13. Record the results.

32. Use some of the thread to make several 12 mm diameter balls. Allow these balls to completely dry (in the sun or in a lab oven) and then perform the dry strength test and record the results with respect to Table 10.10.

33. Decide whether the fines are organic or inorganic. Organic soils can be identified visually by a dark brown to black color, organic odor, presence of peat fibers, and a spongy feel when at the plastic limit. Organic material will react (slowly) when covered with a 3 percent solution (pharmacy grade) of hydrogen peroxide.

Visual-Manual Classification:

34. Based on the qualitative information collected in the preceding steps, it is now possible to obtain an estimated USCS group symbol and group name. The decision trees provided in Figures 10.8, 10.9, and 10.10 are taken from D2487 (USCS Classification), which utilizes the quantitative results from a grain size distribution curve and Atterberg Limits. Similar decision trees (based on estimated values) can be found in D2488 for use with the Visual-Manual procedure.

35. If greater than 50 percent by dry mass is larger than the No. 200 sieve size, the material is a coarse-grained soil. Use the flow chart in Figure 10.10 and the estimated percentages of the various particle size fractions to obtain the appropriate group symbol and name. In place of numerical values of uniformity, use judgment to decide if all particle ranges are adequately represented. In order to be well-graded, the sample must have a wide range (two orders of magnitude) in particle sizes and representation of all intermediate sizes. Anything else would be poorly graded. For borderline situations it is acceptable to use dual symbols when performing the visual manual classification.

36. If the material is coarse-grained and has 10 percent or more fine-grained material, it is necessary to evaluate the plasticity of the fines (Table 10.15) and either use a dual symbol (10 percent fines) or a combined symbol ($\geq$ 15 percent fines). The dual symbol recognizes that sufficient fines are in the sample to impact behavior (e.g., GW-GM). The combined symbol further elevates the importance of the fines, suggesting the coarse grain distribution is no longer as important (e.g., GM). Note that these percentages are different than used in Figure 10.10 for USCS classification because the estimation is assumed to be in increments of 5 percent.

37. If 50 percent or more by dry mass passes the No. 200 sieve size, the soil is fine-grained. If the fines are judged to have considerable organic material, ASTM D2488 uses a combined OL/OH Organic Soils designation rather than separating out plasticity. However, if desired, use Table 10.15 to choose between OL (low plasticity) and OH (high plasticity) and then the flow chart in Figure 10.9 to obtain the group name. If the fines are inorganic, use Table 10.15 to assign a group symbol and the flow chart in Figure 10.8 to obtain the group name.

USCS Classification:

1. Quantitative measurements are required in order to perform the formal USCS classification.

2. Perform a sieve analysis on the soil using the procedures provided in Chapter 8, "Grain Size Analysis."

3. Perform the Liquid and Plastic Limits test on the material passing the No. 40 sieve using the procedures provided in Chapter 9, "Atterberg Limits."

4. If there is concern that the material is organic, perform a second Liquid Limit test on material passing the No. 40 sieve and oven-dried according to the procedures of Chapter 2, "Phase Relations."

5. Use the results of the sieve analysis and Atterberg Limits tests and the appropriate flow chart (Figure 10.8, Figure 10.9, or Figure 10.10) to determine the group symbol and name for the soil.

Report

Report the description, the results of the Visual-Manual Tests, and the resulting group name and group symbol by D2488 for each of the soils provided. Add the results of the grain size distribution, Atterberg Limits testing, and USCS Classification by D2487 for the same soils. Comment on any differences between the two methods.

PRECISION

Since classification is a practice instead of a test method, precision cannot be determined. However, the trained eye can generally judge the percentage of grain size to within about 5 percent.

DETECTING PROBLEMS WITH RESULTS

If the sample descriptions from the two methods do not match for the same soil, a personal recalibration for performing the Visual-Manual procedures is likely the cause. Otherwise, verification of the laboratory grain size distribution and plastic and liquid limit tests should be performed.

REFERENCE PROCEDURES

ASTM D2487 Classification of Soils for Engineering Purposes (Unified Soil Classification System).

ASTM D2488 Description and Identification of Soils (Visual-Manual Procedure).

REFERENCES

Refer to this textbook's ancillary web site, www.wiley.com/college/germaine, for data sheets, spreadsheets, and example data sets.

Burmister, Donald M. 1951. "Identification and Classification of Soils," *Symposium on Identification and Classification of Soils,* ASTM STP 113, 3–24.

Casagrande, Arthur. 1948. "Classification and Identification of Soils," *Transactions of the American Society of Civil Engineers,* 113, 901–991.

Lambe, T. W., and R. V. Whitman. 1969. *Soil Mechanics,* John Wiley and Sons, New York.

Part II

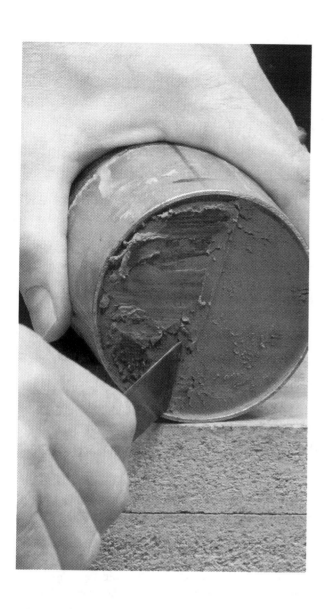

Chapter 11

Background Information for Part II

Part I of the text concentrated on the characterization of soil. The index properties are independent of how the particles are initially packed together, so the tests are performed on completely remolded material. Part II progresses into the measurement of compaction and engineering properties of soils. The tests covered in Part II are:

- Compaction Test Using Standard Effort
- Hydraulic Conductivity: Cohesionless Materials
- Direct Shear (DS)
- Strength Index of Cohesive Materials
- Unconsolidated-Undrained Triaxial Compression (UU)
- Incremental Consolidation by Oedometer (OED)

With the exception of the compaction test, these measurements depend on the initial arrangement of the particles and the density of the specimen. Intact soils have a unique fabric, structure, and density of the particles. These parameters take on special importance when engineering in-situ performance. Therefore, careful consideration must be given to the topic of intact sampling, sample handling, specimen selection, specimen trimming, and specimen reconstitution.

SCOPE AND SUMMARY

This chapter does not address the various sampling tools in detail or the rationale used for field sampling.

Specimen preparation involves the various procedures that are employed to create material of the proper test geometry from the sample. This can be done by trimming an intact sample, or by fabricating a specimen from bulk material. In either case, the process requires well-defined goals and careful attention to detail. This chapter provides an overview of the various methods available for specimen preparation, as well as detailed instructions for processing intact samples and reconstituting bulk samples for mechanical testing.

Moving forward in technology to the measurement of engineering properties of soils requires a considerable increase in the level of sophistication in the laboratory. Experiments will generally require simultaneous readings of several parameters, relatively high reading frequencies, and long testing durations. As a general rule, testing times scale with the inverse of the hydraulic conductivity. For fine-grained soils, tests will often take days or weeks, and can even require months to complete. These technical requirements, along with incentives to have computer-assisted testing capabilities and the availability of economical electronic devices, have caused dramatic changes in the laboratory environment over the past two decades. Laboratories make routine use of transducers for most measurements and record data with automated systems. The final part of this chapter provides an introduction to the operation, calibration, and application of transducers common in the geotechnical laboratory, and a description of the components of data acquisition.

INTACT SAMPLING

In the first chapter, disturbed sampling methods were discussed. This section provides information on intact sampling of fine-grained materials. Intact sampling is intended to collect a sample while preserving as many of the in-situ characteristics as possible. These samples are never truly undisturbed because the sampling process must cause some change to the in-situ condition, whether stress state, temperature, or the like. The amount of disturbance depends on many factors, some of which can be quantified (magnitude of the stress change), while others are poorly known procedural anomalies (overheating the sample during transport).

Intact sampling methods all share one common feature: the sample is collected at some depth from the ground surface in a hole with limited information about the area around the sample. This brings into question the extent to which the sample is representative of the area. Countless intact sampling technologies have been developed to obtain samples from the variety of materials encountered in the field. These tools vary in the sample quality and quantity they are able to produce, and the difficulty of tool operation.

Samplers can be separated into three distinct groups: block samplers, penetration samplers, and coring samplers. Block sampling is performed very close to the ground surface or at the base of an excavation, and has a very limited application. The samples are literally carved by hand and contained in a tight-fitting box for transport. Block samples are generally the highest-quality samples. Penetration samplers force a rigid tube into the deposit and then extract the sample and tube simultaneously. These devices collect a sample at the base of a predrilled hole and are operated on the end of a drill pipe. They cause small to excessive disturbance, depending on the geometry and operation. Coring samplers also operate on the end of a drill pipe at the base of a predrilled hole. These devices cut the material from the perimeter of the sample as the material fills the sampling tube. These methods are slower than penetration samplers, but have the potential to collect excellent-quality samples.

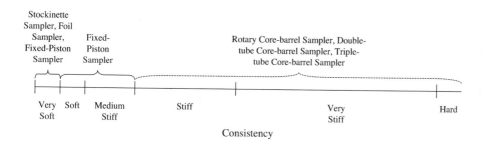

Figure 11.1 Scale of consistency of homogeneous soils with sampling method recommended by the Corps of Engineers. (Adapted from Corps of Engineers, 2001)

Sampling in a borehole is accomplished using hand techniques at depths of up to a few meters, conventional truck-mounted drill rigs at depths of up to a few hundred meters, and by massive drilling platforms at depths of up to several kilometers. A very important and specialized aspect of subsurface drilling is control of the borehole stability.

Selection of a particular sampling technology depends on cost, availability, and material being sampled. Most samplers are interchangeable and compatible with common drill rig technology. The exact details of the sampling operation are very specific to the sampler itself. Many of these are discussed in the Corps of Engineers (COE) manual titled *Geotechnical Investigations* (2001). The sampling tool must be matched to the type and consistency of material to be sampled. The following discussion is limited to fine-grained materials. When the soils are homogeneous, the most appropriate sampling tool depends on the consistency (strength) of the material. Figure 11.1 is a schematic scale depicting recommendations provided by the COE.

The foil-type sampler is pushed through the deposit to the sampling location, at which point the core barrel is extended to collect a sample. This technology is only applicable to very soft clays. The fixed-piston type sampler is penetrated into the soil from the bottom of a borehole. The piston helps draw the material into the tube and holds it in place during extraction. The piston is operated by either hydraulic or mechanical means. The mechanical sampler has a second internal drill rod connected to the piston head. This provides more control and helps troubleshoot problems. The hydraulic sampler is easier to operate and much faster. Both work equally well under good conditions. When using a thin-walled sample tube, the push samplers have an upper limit of medium clays (strength of about 300 kPa). Thick walled tubes increase this limit to stiff clays (about 600 kPa). Coring sample technology uses a rotating cutter around the sample tube. Drilling fluid circulates through the cutter to wash out the removed material. The tube protects the sample from the outside rotating cutter and the fluid circulation. These samplers can be used from soft soil to very hard rock. The cutters increase in aggressiveness from flat blades, to carbide tips, to diamond-impregnated epoxies.

Heterogeneous deposits are the most difficult to sample, and the challenge increases as the contrast in layering increases. The COE recommends that heterogeneous soils be sampled using samplers different than indicated above in some cases for homogeneous soils. A modified fixed-piston sampler, also known as the foil or stockinette sampler, is suggested for varved clays, while the Pitcher sampler, which is a double-tube core-barrel sampler, is recommended for sampling deposits that alternate between soft soils and rock.

Push sampling technology is most often used in practice and is therefore discussed in further detail. ASTM D1587 Standard Practice for Thin-Walled Tube Sampling of Soils for Geotechnical Purposes presents the procedures and considerations necessary for intact sampling. Some of the most important issues for intact sampling follow.

The selection of type of materials for the tubes depends on the consistency of the deposit and the intended duration of sample storage. Tubes are made of mild steel, galvanized steel, stainless steel, epoxy-coated steel, brass, or plastic liners. Brass and stainless steel tubes can last for several years without corrosion. Brass is much weaker, which limits the ability to sample stiff soils. Plain steel tubes are much cheaper but will rust within a few weeks of storage. The rust quickly penetrates the sample and renders the material untestable. Galvanized steel tubes can be stored for a period of time in between these estimates. Epoxy coating or plastic liners are used for environmental sampling and essentially eliminate chemical interactions. The storage estimates are based on the tubes having moisture-tight seals, as described later in this section. Access to oxygen will accelerate corrosion of tubes, even if they are manufactured from stainless steel.

Hvorslev (1949) performed considerable work on factors contributing to sampling disturbance. The geometry of push samplers is important both to the quantity of material recovered and to the quality of preservation of mechanical properties. The four important features relative to the geometry of a sample tube are the inside clearance ratio, area ratio, tip geometry, and length.

Push samplers are designed to cut a sample that is slightly smaller in diameter than the inside diameter of the sample tube. This is done to reduce side wall friction as the sample enters the tube. It is also common practice to lubricate the inside of the sample tube with a spray lacquer. Allow the lacquer to dry for at least a few hours before using. The Inside Clearance Ratio (ICR) characterizes the size of this gap and is defined in Equation 11.1:

$$ICR = \frac{D_i - D_e}{D_e} \times 100 \tag{11.1}$$

Where:

ICR = Inside Clearance Ratio (%)
D_i = inside diameter of the tube (mm)
D_e = inside diameter of the cutting edge (mm)

ASTM D1587 requires that the ICR must be less than or equal to 1 percent for an intact sampling tube, unless specified otherwise. While an adequate ICR may improve sample quality, it also works against recovery. This is because the sample must bond to the inside of the tube as the sampler is extracted from the deposit. For soft soils, ASTM D1587 indicates that a wait time of 5 to 30 minutes after pushing a sample is usually sufficient to allow the soil to swell inside the tube and bond. Reducing the ICR is often a good solution to poor recovery in nonswelling materials.

As the sampler is pushed into the deposit, soil must be displaced to make room for the tube. The area ratio (AR) is used to characterize this aspect of the sampler geometry and is defined by Equation 11.2:

$$AR = \frac{D_w^2 - D_e^2}{D_e^2} \times 100 \tag{11.2}$$

Where:

AR = Area Ratio (%)
D_w = outside diameter of the tube

The area ratio is generally regarded as one of the most important factors in intact sampling. Using sampling tubes with smaller area ratios provides better potential sample quality. However, at very small area ratios, the tube becomes fragile. Hvorslev recommended that a thin-walled tube should have an area ratio of less than 10 to 15 percent, although various organizations will permit higher maximum ratios. ASTM D1587 does not provide specific limits on the area ratio for a thin-walled sampler. The ICR contributes to the area ratio of the tube. For comparison, a coring-type sampler removes the material to make room for the sampler and would have an AR equal to zero. This type of device is able to provide the highest possible sample quality.

The geometry of the penetrating end of the sample tube is also an important factor. The cutting edge is designed to guide the material displaced by the tube away from the sample. This allows the sampled material to move directly into the tube. In addition, the area ratio is reduced as the cutting edge is sharpened. The criteria for acceptable tip sharpness is still a matter of research, but sharper is clearly better. Typical cutting-edge angles are in the range of 9 to 15 degrees. The sampling process wears down the end of the tube and the dimensional criteria will not be met for a second round of sampling. For this reason, intact sampling tubes are *not* reusable.

Hvorslev indicated that the lengths of intact sampling tubes should be limited to 10 to 20 times the diameter for cohesive soils. ASTM D1587 suggests lengths of 0.91 m (36 in.) for either 50.8 mm (2 in.) or 75.2 mm (3 in.) diameter tubes and 1.45 m (54 in.) for 127 mm (5 in.) diameter tubes, which are consistent with Hvorslev's recommendations. Both Hvorslev and D1587 recognize that the length is typically limited by the practicalities of drilling and the ability to properly handle certain lengths of sample. Longer tubes are often employed in offshore applications to provide more sample in situations that require extremely long times to extract the sample (called tripping the hole). Although this practice is common, the sample quality question has not been properly addressed.

Access to the sampling depth is most often accomplished by drilling a borehole, which is usually 1.5 to 2 times the diameter of the sampler. ASTM D1587 limits the inside diameter of the borehole to 3.5 times the outside diameter of the thin-walled sample tube. The borehole is drilled from the surface using a drill bit to chew up the material while circulating a fluid (mud) through the center of the bit and up the sides of the hole.

Drilling mud is water mixed into a slurry with either the clay from the foundation, bentonite powder, or bentonite powder plus a weighting agent. Bentonite mud can achieve a mass density of 1.3 g/cm^3 while weighting agents can increase this to 2.0 g/cm^3. The use of drilling mud has the advantage that it compensates the horizontal total stress in the bottom of the hole as soon as the hole is advanced. The mud lubricates the bit, provides liquid to soften the material, and carries the cuttings back to the ground surface. It will also penetrate and seal the side walls of boreholes in sandy deposits. In addition, mud helps hold the sample in the tube. Drillers' experience working with mud can help optimize the process considerably.

Creating the borehole changes the state of stress in the immediate area, including at the bottom of the hole where the sample will be taken. The drilling mud applies a hydrostatic pressure to the walls and bottom of the borehole. Two catastrophic limiting conditions exist due to the change in stress. If the mud pressure is too low, the walls of the hole will squeeze in and the bottom will heave up. This is a condition of borehole collapse, possibly trapping expensive drilling equipment in the hole, and will prevent further borehole advancement. If the mud pressure is too high, then the soil around the borehole will fracture (called hydraulic fracture) and a crack will develop perpendicular to the minor principle stress. This will cause loss of the drilling fluid, preventing further advancement of the hole, and possible failure of the material to be sampled.

A properly designed drilling mud should balance the horizontal total stress (or the minor principal stress) at the sampling location. This will reduce the average stress and remove the in-situ shear stress. Since the pressure gradient with depth is different in the hole and in the ground, the mud density must be varied as the hole is advanced. With deep holes when using only mud, it becomes impossible to avoid hydraulic fracture at the upper section of the borehole while at the same time balancing the stress at the sampling location. This requires the upper section of the borehole to be reinforced with steel casing, which is cemented in place.

Once the borehole is advanced to the desired sampling depth, the drill bit must be removed from the hole. During the tripping process, it is essential that the borehole remains completely full of mud. Next, the sampler is lowered into the hole and advanced according to appropriate practices for the sampling equipment. After advancement of the sampling apparatus into the soil, a period of time should be allowed for setup. This allows the clay to expand into the tube, helping to prevent the sample from dropping out of the tube during extraction. Two forces must be overcome in order to extract the sampler from the formation: the tensile strength of the soil at the bottom of the tube and the suction developed as the tube moves upward. The soil is normally failed at the end of the tube by rotating the tube several revolutions to shear the base, although some special samplers use a cutting wire or cutting blades. The sampler is then slowly extracted until the suction is released and then raised to the surface. Specialty devices also exist to vent the head of the sampler.

Once the sample tube is removed from the sampler, it should be handled with a level of care appropriate for soil consistency. Sensitive or weak samples should be maintained in the upright position. Mark the tubes with the exploration and sample number, depth, and which end of the tube is the shallower depth (i.e., "top"). The depth reference is the bottom of the borehole, which is equal to the top surface of the sample. Make sure that these tube labels are at least 6 inches from the ends of the tube so they are not covered by the seal. Clean the bottom and top soil surfaces down to intact material. Tools are available to reach inside the tube since the top surface will be 15 cm inside the top of the tube. Measure the length of recovered material and the mass of the tube and soil. Record the recovery, which is the ratio of the recovered sample length to the length of penetration. Ideally this is 100 percent, and low numbers indicate a problem that must be identified and solved. Finally, measure the index strength of the material at the bottom of the tube using a hand-operated shear vane, such as made by Torvane®, or a pocket penetrometer. Remove the soil from the testing location for a companion water content. This information provides immediate feedback to guide or modify the sampling program.

The sample must be sealed to prevent moisture loss and slow oxidation. Tubes can be sealed using various methods. O-ring packers are two disks of hard plastic, separated by a large o-ring, and attached to each other using a specifically designed screw assembly. The packer just fits into the tube, and as the screw is tightened to bring the two disks of plastic closer together, the o-ring is squeezed out to create the seal. Packers are reusable, but must be inspected for scratches on the o-rings and replaced when necessary. Packers must be maintained by regreasing the o-rings every time the packer is removed and replaced.

The tubes can be sealed using a wax mixture. The wax can be pure paraffin; however, this wax is very brittle and will shrink away from the edges of the tube or crack easily, destroying the seal. Paraffin is inexpensive and can purchased at many hardware or grocery stores. Bees' wax and microcrystalline are very expensive but more flexible. A combination of 50 percent paraffin and 50 percent bees' wax is a good mix. However, the authors' preferred mix is 50 percent paraffin with 50 percent petroleum jelly. The reasons for the preference are that the hardened mixture can be cut with a wire saw, both are readily available in a pinch if needed, and the components of the mixture are fairly inexpensive when purchased in bulk.

To place the wax seal, first remove soil to allow for a 1.25 cm to 2.5 cm void. Create a level soil surface on each end of the tube and use a damp cloth to clean the inside surface of the tube. The seal will not be successful if any soil remains on the tube surface. Support the tube in a vertical position. Pour the wax into the designated space and allow it to cool, then place a plastic tube cap over the end of the tube. Finally, wrap a layer of flexible tape (such as electrical tape) around the tube cap and tube to secure the cap and to apply another level of seal. Tubes sealed in this manner can be stored for several years. Refer to Figure 11.2 for a schematic of a tube sealed with wax, a tube cap, and tape.

Recommended handling and transportation procedures for soil samples are covered in ASTM D4220 Preserving and Transporting Soil Samples. Tubes for testing intact soil properties are designated as Group C for intact soil tubes and Group D for fragile or highly sensitive intact soil samples within ASTM D4220. Store the samples on site in a cool, shady location. Never leave the samples in direct sunlight. In general, sample tubes should not be left in temperatures below 10°C or above 30°C. During transport

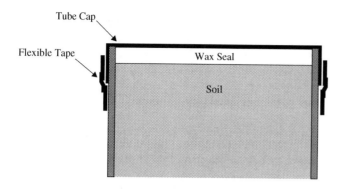

Figure 11.2 Sample tube sealed with wax, a tube cap, and flexible tape.

back to the laboratory, tubes should be in the upright position, cushioned from jarring, and harnessed in place to prevent toppling over. Companies usually devise a crate specifically designed for this purpose. Be aware that without cushioning, the sharp cutting edge of the sampling tube can cut through the bottom cap during transport. Samples of medium consistency or stronger are routinely shipped by commercial carriers in packing tubes. ASTM D4220 provides guidance for the design of transport containers for soil samples.

When samples are received at a large laboratory, they are probably put through a laboratory-specified logging process. The necessary information from the tube is cross-checked with a chain of custody form, and then this information, along with the field observations, is entered into a database. The person logging the tube should note any tube damage or other irregularities, especially whether the seals are intact and in working order. Once logged, the tube is either moved on directly for processing test specimens or put into storage. Tubes should be stored in a cool, relatively constant temperature location. A dark area is not required, but sunlight must be avoided. Properly sealed tubes will last longer in a dry room than in a humid room because humidity promotes rusting from the outside.

The Arthur Casagrande Lecture given by Charles C. Ladd and Don J. DeGroot, titled "Recommended Practice for Soft Ground Site Characterization," describes preferred options for intact sampling for cohesive soils, as well as field testing techniques and suggested laboratory testing for various analyses. The Corps of Engineers *Geotechnical Investigations Manual* (2001) provides guidance on sampling cohesionless materials.

PROCESSING INTACT SAMPLES

Processing a sample requires a sequence of decisions and procedures to remove the soil from the tube in preparation for creating a test specimen. Two levels of decisions are required for this to happen. First, the engineer must decide on project test needs, giving consideration to the forthcoming analyses and the material availability. Preliminary sample information concerning recovery, field classification, index strength, and density are essential to this early decision making. Radiography (discussed next) is extremely valuable at this stage of the project. These decisions generally lead to a series of test requirements for individual sample tubes. The second level of decision concerns where to locate the specimen within the individual sample tube. Radiography is even more important at this stage of testing because it provides an image of the sample without actually opening the tube.

Intact samples can be divided into three categories. The first category contains materials that "hold their shape" when in contact with air and without lateral support. These materials are held together by either true cohesion (interparticle bonds) or negative pore pressures (matric suction). This includes soft to stiff fine-grained materials, most silts, many fine sands, and all cemented deposits. The second category consists of very soft, fine-grained materials that are so weak that they deform significantly under several centimeters of self-weight. Soils in this category will be sampled using one of the many intact sampling methods and must be handled very carefully to preserve the geometry as best as practical. Finally, very stiff materials requiring core-type sampling are cut to specimen size in the laboratory using specialized equipment and lubricating fluids. The special procedures necessary for handling and testing very soft or very stiff materials are beyond the scope of this text.

Radiography

Radiography is a nondestructive method that measures the intensity of an electron beam after it passes through the sample. Normally, the procedure produces a two dimensional image of the three dimensional object. Advanced equipment has been used to create three dimensional images, but this method has not been used in routine geotechnical practice. The radiography technique used for geotechnical work is essentially the same as used in the medical profession, except the beam is much more powerful for use on soils.

Radiography should be a standard practice for soil collected using intact sampling methods. It provides a method to preview all the intact material available on a project before opening a single tube. This eliminates a large uncertainty for the project. It also allows precise locating of test specimens to focus on specific material types or avoid bad spots in the sample. ASTM has a practice on the subject titled D4452 X-Ray Radiography of Soil Samples. This document provides guidance on how to design the setup to hold the tube samples. Industrial X-ray inspection companies (for welds and so on) have the proper radiography equipment to make the image. The expense of radiography is approximately 25 percent of a consolidation test, which is relatively small compared to the cost of acquiring the samples and running sophisticated tests on disturbed or nonrepresentative soil specimens.

Radiography is readily capable of showing the following:

1. Voids, cracks, and separations due to air pockets in the samples
2. Variations in soil types, especially granular versus cohesive materials
3. Macrofabric features resulting from bedding planes, varves, fissures, shear planes, and so on
4. Presence of "inclusions" such as sand lenses, stones, shells, calcareous nodules, peaty materials, drilling mud, and the like
5. Variations in the degree of sample disturbance, ranging from barely detectable curvature of the bedding planes adjacent to the sample edges to gross disturbance as evidenced by a completely contorted appearance and large voids and cracks (most often occurring at the ends of the tube). Radiography will not detect distortion disturbance in uniform materials.

Many of these features are difficult to identify from visual inspection of the extruded samples, at least without trimming or breaking the sample apart. Hence, radiography provides a nondestructive means for selecting the most representative and least disturbed portions of each tube for engineering tests. It is not unusual to find that the sample quality is variable within a tube and to have zones of high or low quality. Layering is sometimes difficult to detect unless the radiation direction is parallel to the layers, so when there is a concern about the presence of layers, the sample can be radiographed at several angles. Combining the radiograph with field logging information helps in planning the overall testing program based on the amounts of suitable material. Such information can be essential for projects having a limited number of expensive tube samples, as occurs with offshore exploration.

Figure 11.3 shows a schematic of the equipment used for radiographing 7.6 cm (3 in.) diameter sample tubes at MIT. The soil samples are typically 60 cm in length and are radiographed in at least 22.9 cm (9 in.) increments. The X-ray source must be far enough away from the tube to reduce the deviation of the beam angle as it passes through

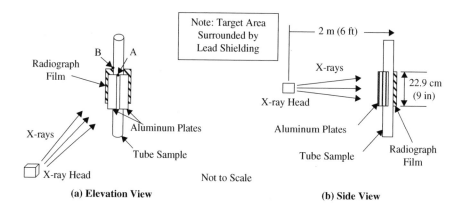

Figure 11.3. Schematic diagram of a method used to X-ray undisturbed samples.

the sample. The system illustrated uses an excitation voltage of 160 kV and current of 3.8 ma exciting a beryllium target to generate the radiation. Exposure times range from 3 to 5 minutes depending of the material density for a 7.6 cm steel sample tube.

Since the tubes are cylindrical, X rays that strike the center of the tube (point A) must travel through 0.5 cm (0.2 in.) of steel and 7.1 cm (2.8 in.) of soil, while those hitting point B penetrate much less soil. Therefore, aluminum plates of varying thickness are positioned in front of the specimen such that all X rays will penetrate an approximately equal mass. Abrupt changes in the thickness at the edges of the aluminum plates cause vertical lines in the photographs. Lead shielding placed around the tube to reduce scattered radiation results in a black background. Lead numbers and letters are attached at 1-inch intervals along the length of the tube to provide distance reference marks. The tubes must be X-rayed in segments to minimize the divergence of the X-ray beam.

The X-ray image is recorded on plastic plate films. Much like those used in a doctor's office, negatives are difficult to work with in the laboratory. Therefore, the image is printed onto paper, yielding a photographic positive. On the positive print, dense objects appear dark and voids appear white. The final product is about 5 percent larger than the actual size of the soil sample.

The image produced on the radiograph is an integration of all the material along the line from the X-ray source to the film. Changes in darkness depend on the relative absorption capacity of the materials being penetrated (i.e., steel, soil, air, shells, and the like). As a result, some features do not cause a sufficient change in absorption capacity and hence cannot be seen on the X-ray photograph. Other features are only visible when the X-rays penetrate at the correct orientation. For example, an inclined crack filled with air within the sample will not be seen unless the X-ray path is parallel to the crack orientation. In general, changes in absorption capacity (which is generally equated to density) as small as 5 percent can be observed.

Figure 11.4 provides three example radiographic positives to illustrate the usefulness of this technique. The first image (a) shows a section of a sample of Boston Blue Clay illustrating the fineness of the layering in this sedimentary deposit. The alternating bands of dark and light are caused by alternating silt and clay-rich layers. The second image (b) illustrates the extreme shear distortion that can be created when pushing a thick-walled sampler into a layered deposit. The material is relatively uniform. The light curved lines are cracks. These cracks are symmetrical to the tube, which is a sure sign of a sampling-created feature. This material is severely disturbed due to the sampling process. The third image (c) is of another layered material showing a few light lines distributed randomly throughout the sample. These light lines are internal disturbance cracks, a relatively common occurrence in tube sampling. In this case, the material between markers 0 and 3 is uncracked and is the best location for an engineering test. This level of test material selectivity would not be possible without the X-ray.

Selecting Representative Specimens

Specimens are prepared for the engineering tests (strength, deformation, and hydraulic conductivity) from short sections of the tube sample. When radiographs are available, it is possible to trim the specimens from exact targeted locations in the sample. Specimens should be located to avoid the inclusions, cracks, and disturbed areas. The consolidation tests are usually given the highest priority and located in the most uniform areas of the tube, with preference given to the more plastic materials. The other tests (strength, hydraulic conductivity, and so on) are given lower priority, especially if the material will be consolidated in the test apparatus.

It is necessary to establish a frame of reference on the tube. This is done by marking the outside of the tube with permanent ink indicating the orientation of the X-ray and the distance from the bottom of the tube. Reference numbers should be included every 10 to 15 cm to allow removal of sections of the tube while preserving the precise location within the tube. This allows the material to remain in the tube for further storage and the tube to be tested in increments over extended periods of time.

Cutting Tubes

Hydraulic jacking of the entire tube is routinely practiced in commercial laboratories. This is an unfortunate practice that has a negative impact on the quality of test results. For most soils, the bond between the sample and tube becomes very significant with time. A large axial force is required to overcome this bond and this force causes significant sample disturbance. This disturbance is obvious if one observes the distortion and cracking of the sample as it exits the tube. Aside from the added disturbance, consider the sample illustrated in Figure 11.4c. Jacking this sample out of the tube will close these cracks, rendering the damage unknown prior to testing, but will affect the test results.

A much more reasonable practice is to cut the tube in "test length" sections and extrude the material in small segments. This dramatically reduces the force required to remove the sample and hence reduces the added disturbance. The practice also makes it possible to store the rest of the sample for later testing. In the absence of X-rays, this technique comes with the risk that the selected section will contain disturbed or even untestable material. In this case, it is better to know the state of the material than to unknowingly make measurements on nonrepresentative material. The following paragraphs present a recommended practice.

The tubes are cut approximately 1.5 to 2 cm above and below the targeted specimen location. Tubes containing cohesive materials are cut with a horizontal band saw.

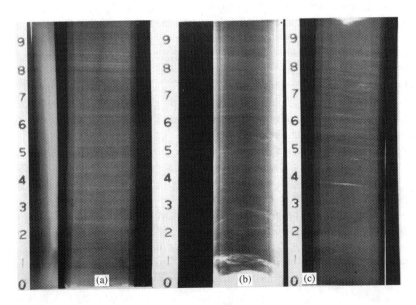

Figure 11.4 Three examples of sample radiographs: a) layering in sedimentary clay deposit; b) significant shear distortion; c) isolated testable material in a generally disturbed sample.

Blade life can be extended by lubricating with oil, but this increases the mess for little benefit. Vibration from the band saw does not cause disturbance of plastic clays, but is a problem for clays and silts with a low plasticity index. For vibration-sensitive material, the tube is cut with a standard hand-rotated tube cutter. Distortion of the tube is prevented by reinforcing the section with a two-piece split collar clamped around the tube above and below the cut location. This technique was developed by Richard S. Ladd for cutting sand samples, and is very effective. Refer to Figure 11.5 through Figure 11.7 for a tube being cut using a band saw and Figure 11.8 for a tube being cut with a tube cutter.

Make sure to handle the tube sections carefully as burrs may be present due to the cutting process. After the tube has been cut, use a metal file to smooth the outside of the tube, as shown in Figure 11.9.

The density of soil can be determined on a tube section. Obtain the mass of the section of tube. The band saw will create burrs on the cut surface of the tube and the tube cutter will roll the metal creating a slightly smaller inside diameter. A reaming tool should be used to remove either burrs on the inside edge of the tube or the rolled edge,

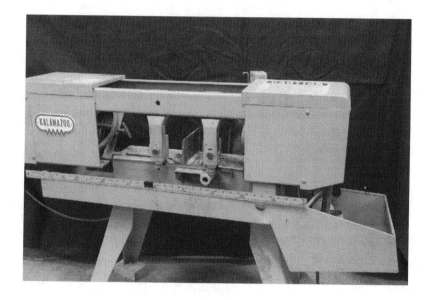

Figure 11.5 Band saw used to cut tube sections.

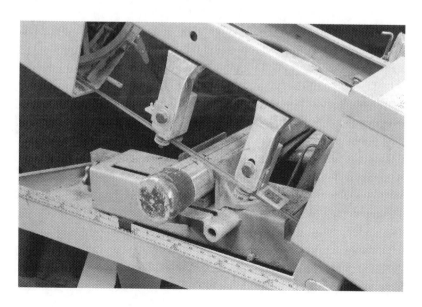

Figure 11.6 Tube set up in band saw, ready to begin cutting.

Figure 11.7 Tube being cut in band saw.

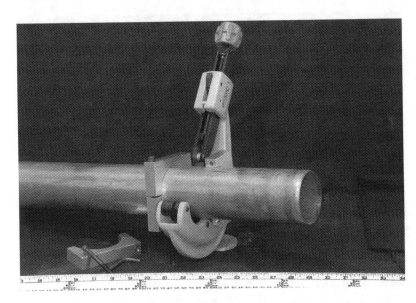

Figure 11.8 Tube being cut with tube cutter.

Figure 11.9 Metal file smoothing the outside edge of the cut tube section.

Figure 11.10 Deburring tool reaming the inside edge of the cut tube section.

Figure 11.11 A straight edge is used to remove soil and debris from the surface, creating a level and clean soil surface.

creating a smooth inside surface, as shown in Figure 11.10. Measure the length at three locations along the tube, approximately 120 degrees apart.

Once the soil is extruded from the metal section, obtain the mass of the empty tube. Calculate the mass of the soil. Measure the inside diameter at three locations approximately 120 degrees apart at both ends of the tube. Calculate the volume of the tube and then the density of the soil section.

Strength index and water content measurements along the length of the tube provide useful information when comparing engineering test results. These measurements can be performed on the end of the remaining unused tube segments. As shown in Figure 11.11, use a straight edge to remove the soil with shavings from the tube cutting and deburring operations. Make the soil surface clean and level.

Next, perform handheld shear vane measurements on the end of tube. The process of making handheld vane shear measurements is demonstrated in Figure 11.12 through Figure 11.14. Refer to Chapter 15, "Strength Index of Cohesive Materials," for further information on the handheld shear vane test.

Figure 11.12 Handheld shear vane inserted into the level, clean soil surface.

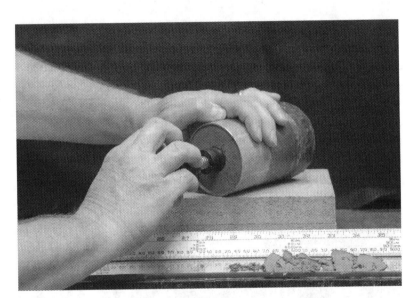

Figure 11.13 Handheld shear vane being twisted to make the measurement.

Figure 11.14 Handheld shear vane after measurement has been taken.

Use a straight edge to remove 0.5 to 1 cm of material from the end of the tube and use this material for a water content measurement associated with the index strength. Mark the locations of the cuts, the handheld shear vanes, and the water contents on sample log. Level the soil surface, clean the inside of the tube, and seal the end of the tube with wax.

Once the wax has hardened, the section can be placed against the adjacent section of tube that has not been tested, provided the other end has been prepared similarly with a wax seal. Place the two sections of tube end to end, tape with electrical tape as a moisture seal, and then reinforce with duct tape to secure the two sections together. The tube section can now be returned to storage for later use.

Experience has shown that the bond between the soil and the tube can be sufficiently strong that simply pushing on the end of a short tube section will cause significant disturbance. The following technique has been developed to break the adhesive bond between the tube and soil prior to extrusion. This is done by penetrating the soil with a hollow 1 mm diameter rigid tube at a location along the inside perimeter wall of the tube. A 0.5 mm steel wire is passed through the hollow tube that is then extracted from the soil. The wire is tensioned and used to core the soil along the inside perimeter of the tube, and the soil section can be extruded from the tube with slight hand pressure. The section of soil can then trimmed to proper testing size using equipment designed for each specific test.

Start with a tube section that has been deburred at both ends around the outside and inside surfaces. The soil surface should be clean and flat. Select a hollow tube that is longer than the tube section and open on one end. Plug the hollow tube with a solid rod to prevent soil from entering during insertion. Force the hollow tube through the sample, just inside the tube shell. Take care to maintain alignment of the hollow tube so it stays along the outside perimeter of the sample. The goal is to avoid penetrating any of the soil that will be used for the test specimen. Refer to Figure 11.15 for a picture of the sample tube section after insertion of the hollow tube.

For soft clays, slight hand force will easily insert the hollow tube. For stiffer clays, silts, and sandy mixtures, it can be quite difficult and may require the aid of a tool. At times, the authors have inserted the hollow tube only a short way by hand (making sure alignment is maintained), then supported the sample tube with both hands with the vertical axis perpendicular to the bench while pushing down on the tube section and hollow tube, using the lab bench to force the tube through the soil.

Once the hollow tube has been inserted, remove the solid rod center and set the tube section aside, taking care that the tube section cannot roll off the bench. Use a vice

Removing Samples from Tube Section

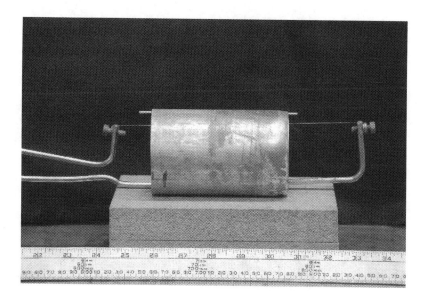

Figure 11.15 Tube section after hollow tube has been inserted. Note that the tube extends past both ends of the tube section.

firmly mounted to the lab bench to secure one end of a thin steel wire (such as piano wire). Pick up the tube section again and thread the other end of the thin wire through the hollow tube. Pull the hollow tube out, taking care not to enlarge the hole. Set the hollow tube aside. Refer to Figure 11.16 for a picture of this process.

Secure the free end of the wire in the jaws of a vice grip. Squeezing the wire between two washers will prevent cutting the wire. Hold the handles of the vice grips in one hand and the tube in the other. (See Figure 11.17.) Tension the wire against the bench vice and rotate the tube while sliding it back and forth along the wire. Be sure the wire stays firmly against the inside of the tube. (See Figure 11.18.)

The wire will break the bond between the soil and the tube. Rotate the wire around the inside of the tube at least twice. Release the vice grips and slide the tube section away from the bench vice, removing the wire.

Place the tube section on top of a machined cylindrical base or similar object with a diameter slightly smaller than the interior diameter of the sample tube. Push the sides of the tube down, maintaining the direction of push that the soil was subjected to during sampling. (See Figure 11.19 and Figure 11.20 for photos of this process.)

Figure 11.16 Tube section after piano wire has been threaded through the tube and the tube has been removed and set aside.

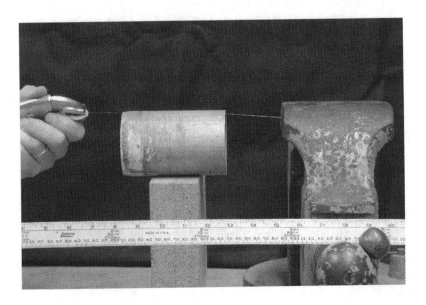

Figure 11.17 Piano wire secured to vice on one end and held with two washers and vice grips on the other end.

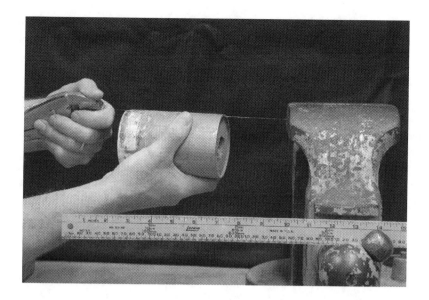

Figure 11.18 Tube section being rotated while piano wire is forced against the tube.

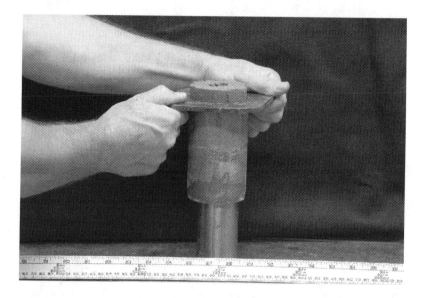

Figure 11.19 After breaking the bond between the soil and the tube, the tube section can be placed on a cylindrical base and the soil can be pushed out of the tube. Maintain the same direction of movement of soil through the tube as the soil experienced during sampling.

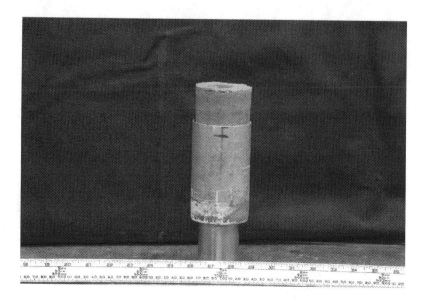

Figure 11.20 Soil in the process of being extruded from the tube. This picture shows the soil only partially extruded for demonstration purposes. Soil should be fully extruded, then transferred to a plate to prepare for trimming.

Carefully transfer the soil section to supporting plates. Never pick up a soft soil specimen by hand, but rather use flat plates or other methods to manipulate the block. The appropriate methods will depend largely on the stiffness of the soil. A freshly cut surface in soft clay will adhere to any flat surface. When working with such materials, cover each surface with wax paper or plastic wrap to prevent adhesion.

Trimming Intact Samples into a Confinement Ring

Several engineering tests require the sample to be trimmed into a rigid specimen container. The consolidation test is a perfect example. The specimen must fit tightly into the container to satisfy the necessary boundary conditions. This method of preparing specimens is only applicable to soft to medium-stiff soils. Since it is not possible to trim harder materials into rigid containers, those specimens must be prepared by coring.

In general, moisture loss is a concern when trimming soft, intact specimens. Trimming should be conducted in a humid environment or performed quickly enough to prevent significant evaporation. In dry climates, a humid room or environmental enclosure will be necessary. Do not touch the specimen with your fingers. Indentations caused by finger pressure are locations of disturbance and will lead to water redistribution within the soil.

Trimming cohesive soils and very low hydraulic conductivity cohesionless materials into a testing ring can be performed using one of two techniques: the cutting ring technique or the shaping tool technique.

The cutting ring technique uses a cutting ring that fits together with the specimen ring or has the cutting edge as part of the ring. Begin the process by squaring the ends of the section perpendicular to the axis of the sample. Ensure that the section of soil has at least an extra 3.5 mm thickness (but not more than 6 mm) on all surfaces to allow for trimming. If the sample is significantly larger than the diameter or width of the trimming ring, use a wire saw and miter box to trim the sides to within 6 mm of the cutting ring.

Place the soil section on a rigid plate covered with a piece of wax paper. Always make sure that wax paper separates the soil and rigid surfaces during the entire specimen preparation procedure. Then place the cutting ring on the soil, as shown in Figure 11.21.

Normally, the soil and the ring are put in a trimming frame to maintain alignment and provide stability during the trimming process. Make sure that the trimming assembly is aligned with the perpendicular axis of the soil section (i.e., that it does not tilt during advancement). Figure 11.22 provides a picture of a typical alignment device.

With this trimming technique, the cutting ring should shave off a thickness of less than 1 millimeter of material directly adjacent to the specimen. The majority of the cutting is performed with a spatula or knife well in advance of the cutting ring. The spatula is used to cut a taper leading up to the cutting ring. Next, the assembly is advanced in increments so small that the soil cut away does not cover the beveled portion of the cutting tool. Figure 11.23 shows an increment of advancement of the cutting ring into

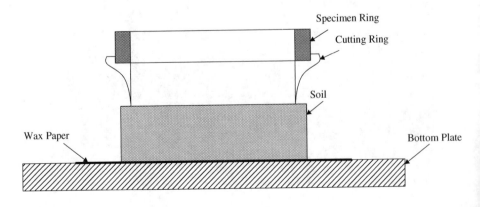

Figure 11.21 The cutting ring technique of trimming a sample.

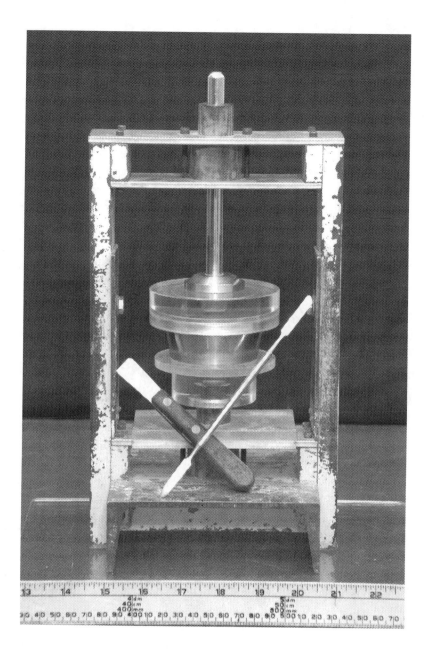

Figure 11.22 Alignment device to aid in trimming a specimen into a ring.

the soil. Take care not to tear the soil by advancing the cutting ring in increments too large. When trimming the soil away, trim in front of the blade to avoid cutting into the specimen.

Continue the iterative cutting procedure until the cutting ring has advanced into the soil a sufficient depth to create a specimen 3 to 5 mm thicker than that required for the test. Remove the cutting ring and soil from the alignment device. If using a separate cutting ring, advance the soil through the specimen ring until several millimeters protrude beyond the top of the ring. Place a rigid plate (without wax paper) on the soil and cut off the excess material with a wire saw. While cutting the surface, apply a slight force to the plate to keep the material in compression and saw back and forth with the wire using the ring surface as a guide. Clean the wire saw and make a second pass through the cut while lifting up on the plate. The soil should stick to the plate and be cleanly removed from the specimen. Use this removed material for a water content of the trimmings.

If the test will require the specimen to be depressed slightly into the ring (referred to as a "recess"), this cut surface will not be final. Otherwise, use a large sharp knife to cut

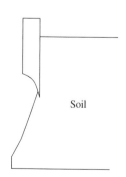

Figure 11.23 Advancing the cutting ring into the specimen.

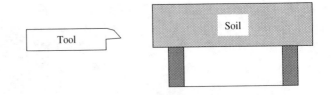

Figure 11.24 Initial setup for trimming soil using a shaping tool.

the surface flat with the ring. The wire saw will always create a slightly curved surface and should always be finished with a straight edge. Cover this surface with wax paper or a filter fabric and a rigid plate. Flip the specimen over and repeat the process on the opposite surface. This will require removal of the separate cutting ring. This surface will be prepared to straight-edge flatness. If a recess is required, then cover the surface with the filter fabric and use a recess spacer to advance the soil out the other end of the ring. Repeat the cutting process and create the final flat surface. Measure the specimen dimensions and mass.

This method provides total support for the specimen during the trimming process. However, since the soil is trimmed directly into a ring, voids and inclusions cannot be seen on the sides of the specimen. Encountering an inclusion that extends partly behind the ring is almost impossible to remove and often requires restarting by selecting an entirely new specimen. This is another good reason to have a radiograph

The second trimming technique involves a shaping tool to trim the soil. The shaping tool is designed to fit against the specimen ring and cut the sample to the inside diameter of the ring. The advantage of this method is that the surface is exposed before it enters the ring. This provides access to cut out inclusions or fill voids before the material enters the ring. Use the same dimensional tolerances and handling procedures described for the cutting ring technique. Place the soil on top of the specimen ring, as shown in Figure 11.24. Some trimming devices have a retractable pedestal to support the soil during the trimming operation.

Use the tool to scrape away the soil just above the ring as shown in Figure 11.25. After completing one rotation around the specimen, push down on the soil with slight force, advancing the soil by a small increment into the ring.

Repeat the process until the soil is completely in the ring. After completing the side trimming process, perform the same sequence of steps as described previously to prepare the top and bottom surfaces of the specimen. Measure the specimen dimensions and mass.

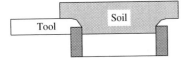

Figure 11.25 Scraping excess soil away prior to advancing the soil into the specimen ring in increments.

Trimming Intact Samples with a Miter Box

The second category of specimen preparation involves trimming the sample to some prescribed geometry using a miter box. This is used for a wide variety of test applications involving cylindrical and cubic geometries. For example, the triaxial test uses a cylindrical specimen trimmed using a miter box. The method is limited to soft to medium stiff materials. Very soft soils will slump during trimming and must be trimmed using a method that provides continuous lateral support. The trimming equipment for very soft soils is specialized, and the discussion is beyond the scope of this text. It will not be possible to cut stiff soils with a wire saw or hand knife. Such materials are machined with special cutters and lubricating fluids. Trimming techniques for very soft or stiff soils are beyond the scope of this text but equipment and procedures are available for each group.

The trimming process should be performed in a humid environment or very quickly to avoid moisture loss. As with the previous techniques, these samples should be manipulated using wax paper and rigid plates to avoid causing more disturbance.

The steps required to create cubic geometries are presented in Chapter 2. This section applies to cylindrical specimens. The process begins with a block of soil that is larger in all dimensions than the final specimen. Place the sample in an orthogonal miter

Figure 11.26 Soil being rough-trimmed in a miter box.

Figure 11.27 Specimen being final-trimmed in a cylindrical miter box.

box, such as shown in Figure 11.26, and create a flat surface perpendicular to the axis of the sample tube. This will establish the direction of the major axis of the specimen.

It will be necessary to reduce the diameter if it is greater than 8 mm larger than the final specimen diameter. Place the newly cut surface on a wax-paper-lined rigid plate with dimensions similar to the sample. Align the sample in the orthogonal miter box. Use the wire saw and the alignment edges to reduce the diameter to about 8 mm larger than that of the final specimen. Rotate the specimen about 15 degrees between each cut. Set the trimmed material aside for index testing.

Transfer the rough-cut specimen to a cylindrical miter box. Trim the surface with the wire saw. Most miter boxes have two settings: one to reduce the diameter to slightly larger than final size, and one for the final size. When trimming the final surface, move the wire saw in both the vertical and horizontal direction in one smooth motion and wipe off the wire after each cut. This process assures the soil is cut with clean wire throughout the cut. Rotate the cylinder by about 5 degrees for each cut. If the cylinder is rotated clockwise, the wire should move from left to right so the wire is cutting into the soil. The final surface should be smooth and clean of cuttings, as shown in Figure 11.27.

Stiffer materials will cause the wire to deflect away from the soil. This cuts a slightly oversize specimen. In this situation, a long, straight knife can be used to repeat the final trimming step to remove the excess material.

Wrap the cylinder in a piece of wax paper that has been cut to a few millimeters less than the final specimen length. Put the specimen in a split tube and gently tighten the clamp to hold the specimen in place. Use the wire saw to trim each end flush with the edges of the split tube. Finish the surfaces with the knife. This will produce a cylinder with flat and parallel ends, such as shown in Figure 11.28. Remove the soil from the mold, measure the dimensions and mass, and proceed with the testing. The split tube should be inspected on a regular basis for wear and alignment.

Remolding Intact Samples

Intact specimens can be remolded for various purposes, such as determining the remolded strength using an unconfined compression test. As presented in Chapter 1, the term "remolded" signifies modifying the soil by shear distortion (such as kneading) to a limiting destructured condition without significantly changing the water content and density.

The first step in the remolding process is to take the intact specimen, usually after failure in the previous test, and immediately place the specimen in a plastic bag, remove most of the air in the bag, and seal it. This must be done quickly to prevent changes in water content. Additional material from the adjacent portion in the tube judged to be sufficiently similar to the intact specimen may be added to ensure there is sufficient material for the remolded specimen. Work the soil between the fingers to remove all structure. Do not entrain air during the kneading process. Several minutes of kneading will be required to destructure the material thoroughly.

Open the plastic bag and remove the material. Select a mold with a plastic liner and work the soil into the lined mold using a rod. Take care not to trap any air in the specimen. Again, work quickly to avoid changes in water content. Once the mold is filled, trim the ends of the specimen. Remove the soil from the mold, measure the dimensions and mass, and proceed with the testing.

RECONSTITUTING SAMPLES

At times, soil is formed in the laboratory to particular conditions, usually specified by water content or density, or both. This process is called reconstitution. Reconstituted material is generally made from a bulk source using a controlled process. Reconstitution is most often necessary when working with free-draining coarse-grained soils

Figure 11.28 Using a split mold to hold the specimen while performing a final trim on the ends.

to circumvent problems associated with sample disturbance. This material would be collected in bulk from the field. Processing bulk samples is covered in Chapter 1, "Introduction and Background Information for Part I." Such materials are dominated by grain-to-grain contact and the fabric is easily changed during the sampling process. Fine-grained materials are also reconstituted for comparative studies relating to soil behavior. A number of techniques are used to reconstitute materials. Techniques vary depending on soil characteristics and the need to simulate particular field conditions. Reconstitution methods can be used to create either a large sample (used for processing into specimens) or an individual test specimen.

There are many factors to consider when choosing a method to reconstitute specimens. The specimen is being manufactured from bulk material by a mechanical process. This process will create the fabric and internal stress state. Each method creates a particular arrangement of particle-to-particle contacts. Ideally, the process mimics the field conditions and the most important features. Consideration should be given to the potential to create non-uniformities, variations in density, layering interfaces, segregation of fines, or otherwise altering the grain size distribution within the specimen. Finally, the method of preparation must consider damage caused by impacting the particles. The following sections provide an overview of the various methods.

Dynamic compaction can be used to create a specimen using a specified amount of energy per unit volume. The method is applicable to both fine-grained and coarse-grained soils. The material is compacted at a specified water content and the resulting density will depend on the response of the soil to the energy input. The specimen is compacted in layers and will always have a density gradient within each layer. The process often results in interface boundary irregularities. Dynamic compaction also creates high internal stress levels. Chapter 12, "Compaction Test Using Standard Effort," covers dynamic compaction in detail for the purpose of measuring the density versus water content response of soil. This relationship is obtained using standardized energy levels selected to simulate field compaction equipment. Standard energy levels can be used to create an individual test specimen and to link the results to the moisture/density curve. It is equally acceptable to use alternative energy levels to create the test specimen. For nonstandard dynamic compaction energy levels, it is most important to use consistent procedures in order to achieve reproducible specimens. Figure 11.29 presents a general illustration of dynamic compaction.

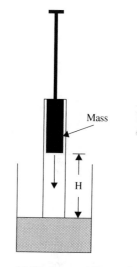

Figure 11.29 General illustration of dynamic compaction.

Kneading compaction is also an energy-based method. Kneading compaction provides more shearing action than dynamic compaction to densify the material. It is most applicable to fine-grained soil. Wilson (1970) presents the kneading compaction method in detail for the purpose of measuring density versus water content relationship at a specified energy level using equipment called the Harvard Miniature device. Unfortunately, this method has never been developed into an ASTM standard test method but is used in practice to simulate field compaction with a Sheep's Foot Roller.

The material is compacted at a range of specified water contents with the resulting density depending on the soil response. Kneading compaction uses a spring-loaded rammer to apply a load to the soil surface. The piston is 12.7 mm (0.5 in) in diameter and the spring is available with the choice of two different force levels (20 lbf and 40 lbf). The specimen is compacted in 3 (or 5) layers with 25 blows per layer. The method can also be used to produce specimens of different size or energy level. As with dynamic compaction, the specimen contains internal density variations and interface discontinuities. Refer to Figure 11.30 for a schematic of this method. The specimen size is small (33.34 mm in diameter and 71.5 mm in height), thereby requiring relatively little soil volume, and produces a specimen ideal for a triaxial test.

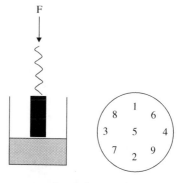

Figure 11.30 Schematic of the Harvard Miniature Compaction Test.

One of the simplest methods to reconstitute a laboratory specimen to a certain water content and density is to use static compaction. Static compaction can be used with most soils, but is limited to water contents that are low enough to maintain continuous air voids. The material is placed in a rigid mold at the specified water content. One or two mandrels are used to control the specimen volume. The apparatus is put

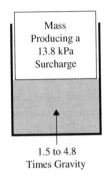

Figure 11.31 Reconstituting specimens using static compaction: single (left); double (right).

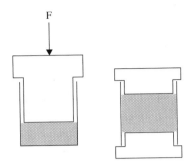

Mass Producing a 13.8 kPa Surcharge

1.5 to 4.8 Times Gravity

Figure 11.32 Schematic of the variables to be controlled when using the vibrating table to produce reconstituted coarse-grained specimens.

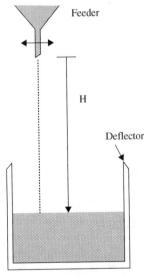

Feeder

H

Deflector

Figure 11.33 Reconstituting specimens using slow pluviation.

in a press and loaded with sufficient force to seat the mandrels. There is no control on the energy input and hence the method is not used to measure the density versus water content relationships. The method often creates non-uniform specimens with considerable density gradients due to side wall friction. The double mandrel method reduces this density variation, as does compacting the specimen in layers. Refer to Figure 11.31 for a schematic of this method.

Dry, coarse-grained soils can be prepared to a specified density using the techniques discussed in Chapter 4, "Maximum Density, Minimum Density." If the desired density is not at the maximum or minimum density, the experimental details can be altered to achieve slightly different results. This will require a bit of trial and error, and is not able to produce a density largely different from the index densities.

Vibratory compaction can be used to produce specimens of coarse-grained materials at a controlled water content and density. Based on the vibratory table method to measure the maximum density (ASTM D4253), the amplitude and/or frequency of the vibration as well as the vertical load can be changed to obtain the desired density. Refer to Figure 11.32 for a schematic of the vibrating table set up. This approach is best used when the desired density is relatively close to the maximum index density. It is difficult to obtain uniform specimens with vibratory compaction.

Slow pluviation can be used to create very uniform specimens of dry, poorly graded coarse-grained soils when the desired density is close to the maximum index density. The material is placed in a container (hopper) at a specified distance, H, above the specimen surface. The feed door is opened and the material is allowed to rain down in a slow, constant stream. The hopper is continuously traversed across the specimen, depositing a thin layer of material with each pass. The process is continued until the specimen mold is overfilled by about a centimeter. The top surface is formed with a straight edge and the specimen is ready for testing. It is essential that the top surface of the mold be beveled to deflect particles away from the specimen rather than bouncing off the mold rim and falling into the specimen. The method is prone to segregation of particle sizes when the size range becomes large. Refer to Figure 11.33 for a schematic of one possible slow pluviation set up.

Rather than constructing a fancy hopper system (which is required for large samples), a specimen-sized container with properly sized holes in the bottom can be used to deposit the material. The pepper shaker is then agitated by hand to rain the particles into the specimen. The holes should be slightly larger than the maximum particle size so the material will not flow out to the container but all particles will pass through the holes when the container is shaken. For the reasons explained previously in Chapter 4, the slow deposition provides little interference to the particles moving around on the surface, resulting in a dense particle packing. More information on the procedure is included in Chapter 4, along with a picture of the equipment.

The drop method can be used to create very uniform specimens of dry, coarse-grained soils at a density near the minimum index density. The drop method is performed by placing the soil in the hopper and releasing the trap door, allowing the soil to drop en masse into a mold of known volume placed below. The top surface is trimmed off and the specimen is ready for testing. It is important to have a deflector on the top of the mold to direct particles away from the specimen rather than bouncing off the surface and into the mold. The material can have a rather wide size range but cannot contain a significant fraction (<2 percent) of fines since fine-grained materials will most likely be expelled from the specimen. Fast pluviation causes high interference between particles along with an air cushion that causes the particles to settle in a very loose-packed condition. This method often yields a lower density than the minimum index density. Figure 11.34 shows the drop method for fast pluviation.

Multiple Sieve Pluviation (MSP) can be used to create very uniform specimens of dry, coarse-grained materials with relative densities between 10 and 85 percent. Soil is placed in a container with a sliding plate positioned to close the holes in the bottom. The slider plate is slid sideways and the grains are rained at a uniform rate over multiple sieves. Like the other pluviation methods, the top of the mold must be beveled. The rate of deposition is controlled by how much the slider plate has been moved. Deposition must occur over the entire specimen at the same time. The method varies the rate and height of deposition to control the hindrance and impact force, and thus varying the resulting specimen density. Multiple sieves are used to distribute the particle paths, resulting in a more uniform raining density. The sieves are selected such that the openings are about one sieve size larger than the maximum particle size in the soil. Segregation of fines will occur with this method, and therefore must only be used for relatively uniform, coarse soils. Figure 11.35 shows a schematic setup of MSP.

The three pluviation methods are normally performed under atmospheric conditions. Drag forces are the major cause of segregation in the various methods. The equipment can be modified to allow pluviation under a vacuum, thus eliminating the drag and greatly extending the applicability to larger size ranges.

The undercompaction procedure is a hybrid method that combines kneading compaction with density control. See Ladd (1977) for further discussion on this technique. The procedure is applicable to any material and any density. Water contents can be almost any value, but must have continuous air voids in the compacted condition. When working with coarse-grained materials, a low water content (2 percent) can be used to bulk up the material. This is very useful when fabricating loose specimens.

The specimen is compacted by tamping the surface with a rod to a specified height. (See Figure 11.36.) The sample must be constructed using a minimum of five layers. The procedure achieves a specimen of uniform density by adjusting for the fact that compaction of the upper layers induce additional compaction of the layers below. Unfortunately, there are a number of experimental variables and little information to guide the decision making. The number of layers, the molding water content, the target density, and the undercompaction of the bottom layer all effect the uniformity of the specimen. A reasonable starting value for the undercompaction of the bottom layer is about 4 percent for most soils.

The percent undercompaction in the layers varies linearly from the undercompaction set for the bottom layer to zero for the top layer. When performing shear tests, the geometry of the deformed specimen can be used as feedback to adjust the value of undercompaction for the bottom layer.

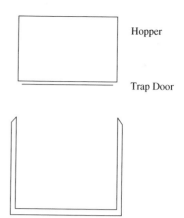

Figure 11.34 The drop method for fast pluviation.

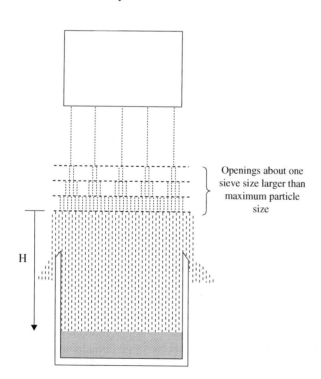

Figure 11.35 Reconstituting specimens using multiple sieve pluviation.

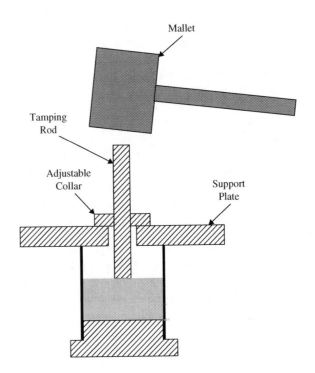

Figure 11.36 Using a mallet to tamp the layers into an under-compaction mold.

The specimen is formed in a rigid mold using the following procedure. A small batch of material is mixed and tempered at the required water content. The total specimen mass is computed based on the mold dimensions. This mass is then divided into equal portions based on the number of layers. Each layer is compacted using a tamper of the required height. Compaction control is achieved by setting the height at the top of each layer, which is calculated using Equation 11.3:

$$h_n = \frac{h}{m}\left[(n-1) + (1+u_n)\right]$$

(11.3)

Where:

h_n = height to top of layer n (mm)
m = total number of layers
h = total height of specimen (mm)
n = layer number
u_n = undercompaction for layer n (decimal)

The percentage of undercompaction for each layer is calculated using Equation 11.4:

$$u_n = u_1 - \left(\frac{u_l}{m-1}\right)(n-1) = u_1\left(\frac{m-n}{m-1}\right)$$

(11.4)

Where:
u_1 = the first layer, which is the bottom layer

The above equations vary the amount of undercompaction from the maximum value at the base to zero at the top. While this is only one possible assumption for the layer variation, experience with the method has shown it to provide suitable control for a range of materials. The procedure is presented schematically as Figure 11.37.

The resedimentation process is used exclusively for fine-grained materials. Resedimentation simulates the one-dimensional consolidation process starting from a thick slurry. The process is intended to replicate the stress state and structure found in nature.

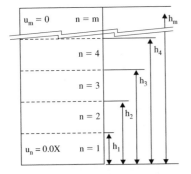

Figure 11.37 Schematic of re-constituting specimens using the undercompaction procedure.

Source material is collected from the field in bulk. It is processed either wet or dry. In the wet method, enough water is added to the batch material to create a thick slurry. The slurry is passed through a 2 mm sieve to remove debris. The salt content is adjusted by adding sea salt or potassium chloride, usually to model in-situ conditions, and then the water content is increased to about twice the liquid limit. The goal is to create a stable slurry having as much water as possible without allowing particle sedimentation. The dry method is essentially the same except the material is dried and ground to a fine powder prior to creating the slurry.

The conditions of the slurry determine the initial soil fabric. The salt content of the pore fluid during formation has considerable effect on the fabric (Lambe and Whitman, 1969). The slurry is evacuated to remove air and tremmied into a rigid-walled one-dimensional consolidation chamber. The material is then incrementally consolidated with a load increment ratio, LIR,[1, 2] of one to the desired stress level. Refer to Figure 11.38 for a photo of the resedimentation process. The consolidation process modifies the soil fabric to mimic geostatic conditions. The resedimentation process takes a considerable amount of time, which increases with the number of stress increments and the square of the batch thickness.

The material is then unloaded to an overconsolidation ratio, OCR[1, 3], that produces a lateral stress ratio, K_0[1, 4], of 1, which eliminates the shear stress. If the material is Boston Blue Clay (BBC), the soil is unloaded to an OCR of 4, since that is the stress level at which the K_0 condition is very close to unity for BBC (Ladd, 1965). The relationship between K_0, OCR, and plasticity index is presented in Brooker and Ireland (1965). The material is then removed from the rigid mold and is essentially an intact block sample. This block can be divided up into individual sealed samples or used directly as a test specimen. This process creates a highly uniform material, ideal for research testing.

TRANSDUCERS

The test complexity increases in Part II of this textbook to a level where making manual data measurements is simply impossible. The test may require simultaneous readings of time, deformation of a specimen, and force applied to the specimen. The measurements could be made through strictly manual devices, such as a stopwatch for the time, a dial gage for the displacement, and a proving ring for the applied force, but as the frequency of the readings or the duration of the test increases, it quickly becomes necessary to convert to an automatic recording system (data acquisition). Converting to a data acquisition system (DAQ) for making test measurements mandates that manual sensors be replaced with electronic devices.

This is not a new concept. In fact, the pervasive use of electronics has had a significant impact on the geotechnical laboratory. Today, the cost of a transducer is generally less than that of an comparable mechanical sensor. The term "transducer" is generally applied to any device that generates an electrical response to a physical stimulus. There are literally hundreds of different types of transducers and many resources available to learn about the options and details of operation. This text is limited to the most common devices used in the geotechnical laboratory, which are direct current voltage-to-voltage devices.

Transducers are a common part of today's laboratory setting. They provide many advantages over manual devices. Most importantly, they have better overall performance than purely mechanical devices, enabling more precision and more variety in

---

[1] Refer to Chapter 17 for more information on the LIR, OCR, and K_0.

[2] The load increment ratio is the increment in load between load n and load n-1, divided by the load at n-1. LIRs of 0.5 to 1.0 are typically used during the loading portion of incremental consolidation.

[3] The overconsolidation ratio, OCR, is the maximum consolidation stress divided by the final axial effective stress.

[4] The lateral stress ratio, K_0, is the horizontal effective stress divided by the vertical effective stress.

Figure 11.38 Resedimentation of fine-grained material.

testing options. Transducers also allow automatic observations, simultaneous sensor recording, much higher recording rates, and far higher measurement resolution, and they eliminate the need to have someone in the laboratory at all times.

The output of transducers can be recorded manually using the readout display on a digital voltmeter or can be recorded by a computer using a DAQ. Transducers are also an integral component for computer-assisted testing.

Transducers generally have much better physical performance characteristics than their mechanical counterparts. Probably the most important factor relative to geotechnical testing is the much higher physical stiffness of transducers. Mechanical sensors generally require conversion of the sensed parameter into motion of some indicator to allow the observation. A good example is a pressure gage in which the pressure causes the rotation of a pointer around a scale of calibrated values. Fluid must flow into the device for the mechanism to function. A pressure transducer has a rigid diaphragm and electrical circuitry to convert a very small strain in the metal to an appreciable voltage output. This dramatically reduces the couple between pressure and volume change (or force and displacement), allowing the device to become much closer to being a passive observer. A classic example of the negative effect due to this couple relates to testing concrete cylinders. When testing concrete cylinders with old load frames, it is common to observe explosive behavior as the cylinder shatters under the peak force. This behavior is caused by the fact that the energy stored in the load frame is released more quickly than it can be absorbed by the concrete. If the frame were infinitely stiff, the cylinder would develop cracks, slowly lose load, and finally crumble.

Backlash is another undesirable performance feature of mechanical systems. Any device that has moving parts suffers from this problem. Backlash is caused by the fact that a gap is required between each pair of moving parts. This is true of slides, gears, bearings, and so on. Each pair must also have a contacting surface to transmit the force necessary for movement. When the direction of motion reverses, all the contact points must switch to a new contact position on the opposite surface, jumping over the gap. Accumulation of all these repositioning displacements within the mechanical device causes errors in displacement measurements and hysteresis in force and pressure measurements. Backlash is greatly reduced in some devices by adding an internal force to the moving parts. This force keeps all the contact points on the same side of the clearance gaps. The internal force is ultimately transmitted externally and must be overcome when using the device. For example, a considerable force is required to move the shaft of a dial gage. Most transducers eliminate moving parts and therefore are not subjected to backlash problems.

All measuring devices have a limited frequency range. The moving parts inside mechanical devices impose severe frequency limits. The inertia of gears and levers in a dial gage provide a perfect example of this limiting factor. Transducers have a much higher dynamic range due to the lack of moving parts. This makes transducers ideal for cyclic loading applications.

Transducers do have some disadvantages. They require knowledge of electronic instrumentation in addition to the traditional knowledge of the experiment. Misuse of a transducer can result in a variety of consequences from incorrect readings (bad data) to damage of equipment. As an example, it is easy to see when a proving ring is distorting excessively, but far more difficult to quickly translate the force transducer output on a voltmeter to the physical danger. The following discussion provides a bit of practical information on the use of electronic devices. As with other topics in this introduction, entire textbooks are devoted to the topic of measurements and the operation of transducers.

A transducer converts some physical quantity to an electrical signal. The exact details of the conversion are dependent on the design of the device. A number of transducer types will be covered below. Common features of all Direct Current (DC) voltage-based transducers are input voltage, output voltage, capacity, and a calibration relationship.[5]

- Input voltage, V_{in}: Each transducer requires an input voltage to power the device. In most cases the voltage will be in the range of 4 to 20 volts.

---

[5] For further information on issues relative to electronics and data acquisition in the geotechnical laboratory, refer to Jamiolkowski et al. (1985), upon which these discussions are based.

- Output voltage, V: The device produces an output voltage in response to the quantity being sensed. The output voltage can be at millivolt (mV) levels or volt (V) levels, depending on the device.
- Capacity: The capacity of the transducer is the physical limit of the device. The limit must always be known for a specific device. Transducers are selected for a particular application based on matching the capacity to the expected observation.
- Calibration relationship: A conversion relationship between the output voltage and the physical quantity must be established to convert the output voltage back into the quantity of interest. If the relationship is linear, the conversion is called a calibration factor.

The input voltage provides power to the transducer and has very specific limits. Too little power will not be sufficient for the device to function. Too much voltage will burn up the electronics within the device. The manufacturer's specification sheet will always provide the acceptable operating voltage range. General-purpose direct current power supplies are used to excite transducers. These devices usually have adjustments for both voltage levels and over current limits. The voltage being generated should always be confirmed before plugging a transducer into the power source. Integrated data acquisition systems usually contain the power supply used with a subset of recommended transducers. Most transducers can easily operate on a 0.25 watt source. The power supply must be stable (i.e., give a constant output and have little noise). Any power irregularity will be amplified by the transducer and appear in the final output signal.

The output voltage will also have a minimum and a maximum limit. For many devices these limits represent physical damage values, such that if the maximum limit is exceeded, the device may be permanently damaged or destroyed. In most cases, the output voltage is altered by the input voltage, thus creating the need to make two measurements for every reading. When using electronic transducers, a measuring device must be present to monitor the output during the test. In many applications, a voltmeter will be set up so that all transducers can be read at any point during the test. The readout box can toggle through the transducers and input voltage using a switch. The voltmeter does not have a means for recording these values. Instead, hand readings are usually recorded as specific points during the test as check values. Examples are "zeroes," seating load values, and so on. The voltmeter will provide values in volts. However, some systems have readout boxes that are adjustable voltmeters, calibrated to display values directly in engineering units. If in volts, the values will need to be converted to engineering values using the calibration factor, CF.

The capacity of a transducer is essentially the operating range of the device. From a practical perspective, this would be the range over which a transducer provides coherent output voltages. The manufacturer must specify the intended operating range, along with an over range limit and a damage limit. The operating range generally accounts for fatigue of the material and should be respected if longevity is expected of the device. Capacity for displacement transducers is more related to the useful operating range of the device and really has no damage consequences. Capacity is used to specify a device for a particular application. The capacity should be matched to the expected test results. Using a device with too much capacity will result in loss of resolution in the measured result.

The calibration relationship for a transducer is the equation that relates the voltage to the physical quantity. For many (even most) devices, this can satisfactorily be expressed as a linear relationship characterized with a slope and intercept. However, most devices exhibit some deviation from linearity. Linear voltage displacement transformers (LVDT) are highly nonlinear devices and must be used with great caution since the relationship is more correctly described by a distorted SIN wave. The LVDT is then used within a limited domain where the function is more nearly linear. An additional complication of transducers comes from the fact that the output voltage is sensitive to the input voltage. This effect is accounted for by dividing the output by the input to obtain the normalized output. Using the normalized output will compensate for some variations in the input voltage, but is not an invitation to use unregulated power supplies. The normalization is good, but not perfect. Normalization requires that the input voltage be recorded with each transducer output measurement.

The relationship between the electronic response and mechanical stimulus (e.g., normalized voltage and pressure, or normalized voltage and displacement) is expressed as a calibration curve. It is possible to derive this relationship based on the physical properties of the device and the elements of the electronic circuit. Such an analysis would be complicated and, when combined with small manufacturing irregularities, would render the relationship approximate. Therefore, the only method to obtain a sufficiently precise calibration relationship is to use an experimental method. The device is calibrated by applying a reference input (displacement, force, pressure, and so on) and then recording the voltage output. A proper calibration requires a number of measurements be taken in both directions of application. For a force transducer, this would be done during both loading and unloading. A typical calibration would include two complete cycles containing 10 to 20 data points over each of the four legs. Each reference point is established by applying a known physical input condition. Ideally, the reference is traceable to the National Institute of Standards and Technology (NIST) or some other certifying agency.

The output voltages from the experimental calibration are normalized to the input voltages and then compared against the reference condition. Typically, the relationship will be linear, but this is not a necessary condition. It is perfectly acceptable to represent the relationship with a higher-order equation. The following discussion assumes a linear relationship, but the same concepts apply equally to other functions.

The results are plotted in a graph of normalized output versus the reference or the range of the transducer (or the range of interest if less), as shown in Figure 11.39. The

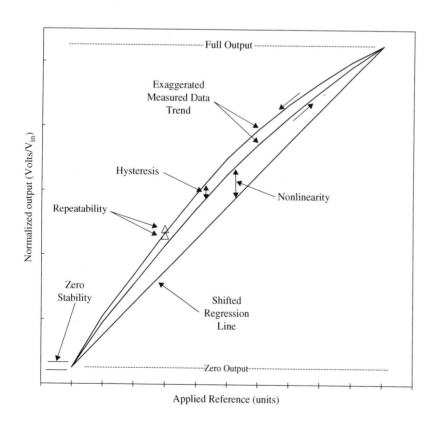

Figure 11.39 The components of a calibration curve. (Adapted from Jamiolkowski et al., 1985)

data set contains enough information to apply a linear regression analysis to obtain the slope and the intercept of the calibration equation. The inverse slope is the calibration factor, CF, and has units of the reference divided by the normalized voltage (V/V_{in}). While the reference has a meaningful zero value, the intercept voltage will be an arbitrary value with no physical meaning. It is important to the calculation and can be tracked as an indication of the wellness of the transducer. A shift in the zero voltage usually indicates the device has been overloaded or damaged in some way.

Once the calibration factor is established, the physical quantity (P in this example) can be calculated for each reading using Equation 11.5:

$$P_m = \left[\frac{V_m}{V_{in,m}} - \frac{V_0}{V_{in,0}} \right] \times CF$$

(11.5)

Where:

P = physical quantity being measured (Engineering Units)
m = reading number
V_m = output voltage for a particular reading (V)
$V_{in,m}$ = input voltage for a particular reading (V)
V_0 = initial output voltage (V)
$V_{in,0}$ = initial input voltage (V)
CF = calibration factor (Engineering Units/V/V_{in})

Several numerical measures are used to evaluate the reasonableness of representing the calibration relationship as a linear function. These measures are the coefficient of determination, nonlinearity, repeatability, hysteresis, and zero stability. These parameters are shown graphically in Figure 11.39. The image is exaggerated to demonstrate the concepts.

The purpose of computing the various parameters is to understand the goodness of fit of an individual calculation. The following paragraphs present a short description of each parameter. The relative importance is a function of the particular application of the measurement and is why several parameters are used in the characterization.

The coefficient of determination, R^2, is the first and most important measure of linearity. R^2 provides a global measure of linearity. R^2 increases from zero to unity as linearity increases. Spreadsheet programs can calculate the coefficient of determination for a data set. When evaluating a calibration equation, the number of nines following the decimal point is the most important measure. If R^2 is 0.98 then the relationship is far from linear and rounding this to 1.0 (mathematically correct) is completely misrepresenting the analysis. The R^2 should be reported to one significant digit more than the last nine in the series and must be greater than 0.9998 for the transducer to be considered linear.

The nonlinearity of the response is computed as the difference between the average measurement at each reference value and the computed value using the *offset* calibration relationship. The offset calibration relationship is obtained by setting the zero value of the linear regression equation to the normalized voltage reading corresponding to the zero reference value. This is essentially how the transducer would be used in an experiment. A plot of the nonlinearity versus the reference will amplify the approximate nature of linearizing the relationship. The maximum nonlinearity is reported for the transducer. Typical values of nonlinearity range between 0.15 to 1.0 percent of the full scale output of the device. Nonlinearity is a good measure of the maximum error one would expect solely due to the linear approximation.

The hysteresis in the response is the difference between the average of the two loading points and the average of two unloading points for every reference value. These differences are plotted versus the reference value to illustrate the trend. Typical values for hysteresis range between 0.05 and 0.25 percent of the full-scale output of the device. Hysteresis would be important when performing cyclic testing.

The repeatability in the response is the difference between any two readings on the same leg of the calibration at each reference value. Repeatability can be plotted versus the reference value to look for a trend. In most situations, repeatability will be represented by a random band of numbers indicating that the calibration process is less repeatable than the transducer itself. Typical values for repeatability are in the range of 0.05 to 0.10 percent of the full scale value. While this value is interesting, it is not very useful because it only reflects a very short-term performance of the device.

The stability of the device is far more important than repeatability. Stability is measured as the variation in the normalized zero value with temperature and time. The temperature stability can be quantified in much the same manner as the calibration curve. The result would be a relationship between the normalized zero and temperature. Temperature sensitivity is typically in the range of 0.2 to 1 percent of the full scale value. In some devices, the temperature will also effect the calibration factor. Stability over time is far less deterministic, tends not to be systematic, and is historically a significant problem in long term experiments. It can be evaluated by observing the normalized zero output over periods of time at matching test conditions. Zero drift is typically quoted as 0.2 to 1 percent of the full scale value, but varies considerably for each individual transducer.

Four physical parameters are most frequently measured in the geotechnical laboratory. They are displacement, force, pressure, and volume. Among these categories and commonly used within geotechnical practice, there are two very different principles of transducer operation: induction-based devices and strain-gage-based devices. The following sections provide an overview of each type of transducer and the particular features that deserve further discussion.

Displacement

Displacement is a fundamental measurement in many professions, resulting in a very large market and a wide variety of commercial devices. Displacement is a *relative* measurement that requires a stable reference location. The device senses the movement of a target relative to this reference position. Errors in measured deformations are very common due to poor reference selection or movements of the displacement support system. The mechanical device used to measure displacement is a dial gage. Dial gages provide a direct read-out of displacement and make use of a very complex set of gears, springs, and levers to amplify the movement and combat the various problems associated with moving parts.

Displacement transducers in the geotechnical laboratory are usually either induction- or strain-gage-based, but many other technologies are available. Transducers have capacities ranging from a few millimeters to several hundred millimeters. Figure 11.40 shows an example of some typical displacement measurement devices.

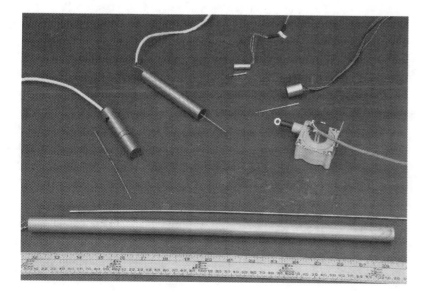

Figure 11.40 Typical displacement measurement devices.

The induction-based displacement device is called a Linear Variable Differential Transformer (LVDT) and carries the misnomer of Direct Current Displacement Transducer (DCDT). This device comes in a very wide selection of ranges and is similar in operation to proximity sensors and Hall Effect devices. LVDT's are actually alternating current (AC) devices but contain electronics within the device to convert the DC power to AC with an oscillator and then the AC signal back to DC with a demodulator for the output. This is important for two reasons. The device only operates when the DC power current is in the correct direction, and the device generates electrical noise, which can have a severe negative impact on some data acquisition systems. On the positive side, LVDT's have infinite resolution. This means the sensitivity of the measurement depends on the stability of the power supply and the data acquisition system rather than the operational details of the device.

An LVDT consists of two completely separate pieces. The body of the device contains the electronics and is normally attached to the reference location. The shaft (or core) is attached to the element to be observed and slides inside the body. The two pieces do not need to touch for the device to function. The body is surrounded by a metal Faraday cage that provides isolation from external electrical interference. The center of the device contains a primary coil, which is excited with an AC current. This generates a symmetrical magnetic field inside the device (faraday cage). On both sides of the primary coil are equal-sized secondary coils. The magnetic field generates a potential in these coils. Electrical components in the far end of the device convert the signals from AC to DC. The shaft is a ferrous metal that can distort the magnetic field. When the body is empty or when the shaft is in the center of the field, the potential in the two secondary coils is balanced and the output voltage is zero. As the shaft is displaced relative to the center of the body it distorts the field, creating a potential difference (the output voltage).

Figure 11.41 presents a conceptual representation of the typical relationship between the DC voltage output of an LVDT as a function of the shaft position, x. When the shaft is outside the body the voltage will be zero (or near zero depending on the symmetry of the secondary coils). Once the shaft approaches the edge of the body, the field will distort and the voltage will increase. The trend will continue to a maximum output voltage and then decrease to zero when the shaft is centered. This is the null position of the shaft and indicates the center of the operating range. The maximum output voltage is nearly equal to the input voltage. As the shaft progresses through the body, the negative signal will be symmetrical to the positive signal. Clearly this is not

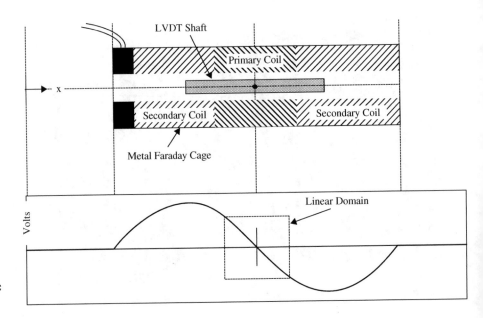

Figure 11.41 Typical schematic and response of an LVDT.

an ideal transducer response. The signal is not unique over the range of the device and actually has opposite slopes depending on the position of the shaft. However, there is a nearly linear range on either side of the null position of the body and this is the operating domain of the transducer. The input voltage for LVDT's varies from 5 V to 20 V. The input voltage strengthens the magnetic field and yields a higher output. This is why the output is normalized to the input voltage. The "linear" range is established by calibration and because the device is not truly linear, the precision will increase as the operating range is decreased. Typical output over the operating range is about 40 percent of the input voltage.

The strain-gage-based displacement device is generally referred to as a Linear Strain Transducer (LST) or a Linear Displacement Transducer (LDT). These devices rely on a more complicated mechanical design and have contacting moving parts. They have a more limited displacement range. They are DC-based devices and do not add electronic noise to the system.

LST's are conceptually very different from LVDT's. Figure 11.42 presents a conceptual schematic diagram of the operating principle. The body contains two cantilevers with slides attached to the free ends. The shaft is formed with a taper. As the shaft is displaced into the body, the slides are forced up the taper, causing the cantilevers to bend. The bending is detected by a strain gage bonded to the cantilevers, resulting in a change in the output voltage. The shaft is spring-loaded and has stops to limit the motion. As such, there is a slight force generated by the device, which changes with position. The physical stops control the operating range to be within the linear operation range of the materials.

The strain gage is essential to the LST operation as well as hundreds of other commercial sensors. It is fair to say that the strain gage has revolutionized the measurements industry. Figure 11.43 provides an illustration of the simplest strain gage design. The gage is made of a backing material and a very long, thin strand of conductive material. The conductive material varies depending on the sensitivity, application, and manufacturer but is essentially a material that changes resistance with changes in dimension. As shown in the figure, the wire is wrapped back and forth in one direction, $l_{0\backslash}$, creating the most sensitive axis of the gage. The wire is bonded onto a backing material to hold everything in place. The gage is then epoxied to the component of interest (the cantilevers in this case) and is assumed to be much softer than the component.

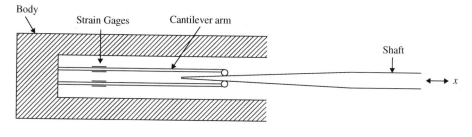

Figure 11.42 Typical schematic of an LST.

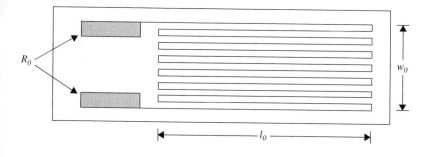

Figure 11.43 Typical schematic of a strain gage.

The wire has an initial resistance, R_0, which will change as the wire deforms. The relationship between the dimensional strain along the axis of the gage and the change in resistance is called the gage factor. The gage factor is presented in Equation 11.6:

$$GF = \frac{\Delta R / R_0}{\Delta l / l_0} \qquad (11.6)$$

Where:

GF = gage factor of the strain gage (dimensionless)
ΔR = Change in resistance of the strain gage wire (ohms)
R_0 = Initial resistance of the strain gage (ohms)
Δl = Change in length of the sensitive axis of the strain gage (mm)
l_0 = Initial length of the sensitive axis of the strain gage (mm)

The denominator ($\Delta l / l_0$) is the axial strain along the major axis of the gage. The strain gage has a linear relationship between dimensional and resistive strain. This is true as long as the material remains elastic. Also note that elongation produces an increase in resistance. The gage factor depends on the strand material. The gage factor of foil gages is about 2 and for semiconductor gages it can be as high as 200.

The strain gage must be integrated into an electric circuit in order to convert the change in resistance to a change in voltage. The simplest useful circuit is a Wheatstone Bridge configuration, which is composed of a minimum of four resistors. Figure 11.44 presents the electrical configuration for this circuit. The input voltage, V_{in}, is DC and constant. The low-voltage side of the circuit is connected to ground, GND, for a stable reference and a drain for stray currents. The current can pass through the circuit along two independent paths; one through R1 and R2 and the other through R3 and R4. The magnitude of the current through each path will depend on Ohms law (voltage = current $\times$ resistance). The output voltage is measured between nodes A and B in the circuit. If the two resistors (R1 and R2) along the leg are equal, the voltage at node A will be half of the input voltage. The same situation applies along the second leg. It is important to recognize the fact that all four resistances do not need to be equal but rather the pair along each path must be equal. At this point, the Wheatstone Bridge is balanced and has zero output.

If R1 is changed from a fixed resistor to a variable resistor, the difference in resistance from the balanced condition, R_0, can be given the symbol ΔR. It is now possible to derive an equation relating the output voltage (difference between A and B) to the variation in resistance of R1. This circuit, where one of the four resistors is variable, is called a one-quarter bridge. The relationship between normalized voltage and the change in resistance for a one-quarter Wheatstone Bridge is provided in Equation 11.7:

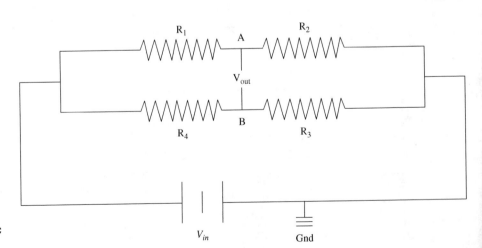

Figure 11.44 Typical schematic of a Wheatstone bridge.

$$\frac{V}{V_{in}} = \frac{-\Delta R / R_0}{\left(4 + 2 \times \Delta R / R_0\right)} \tag{11.7}$$

Where:

ΔR = variation in resistance of resistor R1 (ohms)
R_0 = initial resistance of the resistors R1 and R2 (ohms)
V = output voltage of quarter bridge (volts)
V_{in} = input voltage used to power bridge (volts)

Once again, the output voltage is normalized to the input voltage. The real value of the Wheatstone Bridge circuit is the fact that the output voltage starts near zero and is on a scale proportional to the change in resistance. Consider a situation where the resistance changes by 1 percent. If the multimeter measured 1 part in 1000, then making a direct reading of resistance and subtracting the two large resistance values (initial and final) to determine the change in resistance would yield at best a resolution of 10 percent of the change. On the other hand, measuring the voltage on a scale appropriate to the voltage output would allow a resolution of nearly 0.1 percent. As seen in Equation 11.7, there is a small nonlinearity inherent to the circuit, but this is seldom important in practice.

Combining Equation 11.6 and Equation 11.7 yields the governing equation for a one-quarter bridge circuit using a strain gage as the variable resistor, given by Equation 11.8:

$$\frac{V}{V_{in}} = \frac{-GF\left(\Delta l / l_0\right)}{\left(4 + 2 \times GF\left(\Delta l / l_0\right)\right)} \tag{11.8}$$

Based on Equation 11.8, a gage factor of 2, and a limiting strain of 0.4 percent, the maximum normalized output would be about 0.002 or 2 mV for every volt of input. A reasonable limit for the input voltage would be about 4 volts, leading to a maximum output of around 8 mV. In practice, the Wheatstone Bridge circuit is used with more than one variable resistor placed strategically on the loaded member. When two strain gages are employed it is a half bridge, and four gages results in a full bridge. Making use of locations of tension and compression strain can eliminate bending effects or temperature variations. Addition variable resistors can be added to the circuit to provide additional flexibility.

The output of the LST would be maximized by using a full Wheatstone Bridge configuration with the strain gage corresponding to R1 on top of the upper cantilever. This strain gage will be placed in compression when the cantilever deflects. R2 would be in tension and on the bottom of the upper cantilever, R3 would be on the bottom of the lower cantilever and R4 would be on top. This would create the maximum shift between the voltage at points A and B in the circuit of Figure 11.44. While it is possible to compute the voltage versus displacement relationship based on Equation 11.8 combined with a physical model of the cantilever, this would be too approximate for a practical transducer. Therefore, the transducer must be formally calibrated against a standard reference.

Force

Force measurement is also common in many professions and applications. The interaction of masses results in a force. As such, force is a derived quantity and must be measured indirectly. In mechanical applications, an elastic element is located in series with the element of interest. The force acts on the elastic element and the resulting deformation is measured. The geometry of the element and the modulus of elasticity of the material are then used to compute the force from the deformation. Deformation is again the fundamental measurement. But unlike deformation, force is an *absolute* quantity. Zero force has real physical meaning.

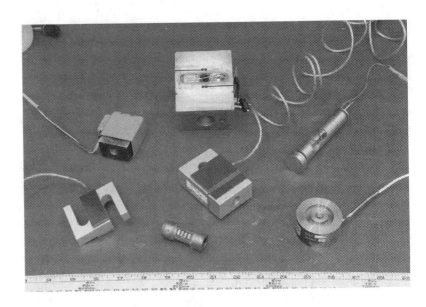

Figure 11.45 Several force-measuring devices.

The mechanical device for measuring force is called a proving ring. This is a metal hoop with a displacement-measuring device in the center. The force deforms the hoop. The geometry of the hoop and the type of material determines the stiffness and capacity. The proving ring must be calibrated to obtain reasonable accuracy. Proving rings are seldom used now due to the many mechanical performance issues.

Electronic force transducers are commonly referred to as load cells. Load cells are available in a wide range of geometries, from solid cylinders to delicate cantilevers. The shear beam design is extremely popular for laboratory scale loads. The geometry influences the stiffness of the device and the sensitivity to off-axis loading. The operating concept of the load cell is still to convert force to deformation and then uses strain gages in a Wheatstone Bridge configuration to generate a voltage. One of the central design elements of the device is to localize the strain to maximize stiffness while maintaining sensitivity. Typical maximum deflections for a load cell are on the order of 0.02 to 0.05 mm. Exceeding the overload limit will cause plastic deformation and rupture the strain gage. Figure 11.45 provides images of several force-measuring devices.

Load cells generally have an input voltage range between 5 to 12 V (depending on manufacturer) but the range will be rather narrow for a specific device. The output voltage is always in the millivolt range and the maximum values are between 30 and 200 mV. The resolution of such devices is 0.01 percent of full scale and can be read to 0.01 mV. Load cells are available with capacities ranging from 45 N to 45,000 N (10 lbf to 10,000 lbf).

Pressure

Pressure is another common measurement. The mechanical device for measuring pressure is a manometer or a pressure gage. The gage device uses a bourdon tube to convert pressure change to deflection. This requires fluid to flow into the device. A gear, lever, and spring are then used to amplify the deflection and convert it to rotation. Pressure gages are now used as visual indicators and for approximate measurements in the geotechnical laboratory, due to imprecision and mechanical performance problems.

Pressure transducers are used exclusively for precision testing. There is tremendous variety in pressure transducer design. Pressure transducers are used in countless applications. Devices are available to make measurements in fluids and gases over a large range of temperatures. In geotechnical applications, it is essential to have a water-compatible interface. Figure 11.46 presents a few of the more common devices.

Pressure is force per unit area and is measured using the same principle as when measuring force. The fluid pressure is applied to a membrane, which causes the membrane to deflect and the deflection is converted to a voltage. As with force, pressure

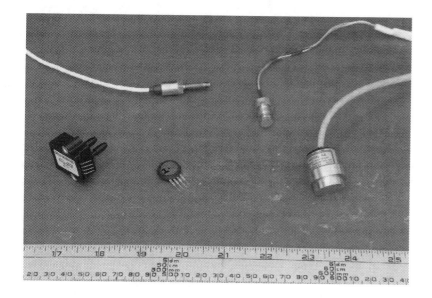

Figure 11.46 Typical pressure transducers.

is an *absolute* quantity. However, there are three types of pressure transducers: absolute, gage, and differential. The back of the membrane is completely sealed in an absolute pressure transducer. This makes the device sensitive to changes in barometric pressure. Gage pressure transducers are open to the atmosphere and do not respond to barometric pressure changes. Differential pressure transducers have a connection on both sides of the membrane allowing more sensitive measures of the difference between two pressures. All of the affordable devices use strain-gage-based technology and function the same as a force transducer in terms of input limits and output characteristics. Pressure transducers can have a sensing area as small as 1.6 mm in diameter and can sense pressure changes due to deflections of the sensing zone as small as 0.08 mm. Sensitivity is on the order of 0.01 percent of full-scale output. Capacities of 40 to 140,000 kPa (6 to 20,000 psi) are available, enabling measurements extremely low and very high.

Volume

Volume measurement is also a very broad application technology. By limiting the flow rate to a few mL per hour and the capacity to a few hundred mL, and requiring the fluid to be under pressure, this eliminates all but a very few devices. Under these conditions, volume measurement is analogous to displacement and is more commonly referred to as volume change. Volume change recognizes the fact that the measurement is *relative* to an initial starting condition. In addition, air/water interfaces are forbidden in many applications.

Mechanical volume measurement is accomplished using a factory-calibrated volumetric burette. The measurement is performed under pressure by locating the burette inside an acrylic tube. A conceptual drawing of this concept is provided in Figure 11.47. The tube is pressurized with air (σ_a) and the position of the air/water interface is monitored to determine the volume change.

Elimination of the air/water interface is accomplished with a more complicated revision of this concept. The outside tube is also filled with water. Oil (usually dyed a color) is floated in the top of the tube and down into the burette. The oil/water interface provides a well defined interface within the burette. Practical devices must include additional components to allow initial deairing, recharging the device, and repositioning the oil interface. These devices all have limited capacity and a resolution that depends on the size of the burette. The burette can be selected to match the experimental requirements. A reading sensitivity of 0.02 mL is reasonable for burette-based devices.

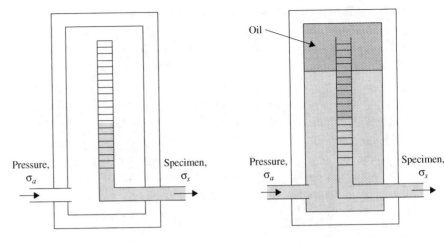

Figure 11.47 Schematic of a volumetric burette.

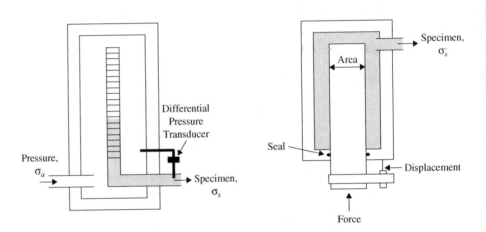

Figure 11.48 Typical setup using a volume-measuring transducer.

Electronic volume change devices are all "mechanical" devices with transducers added to measure either pressure or displacement and the volume change is computed based on a known area. The manual burette devices are modified by adding a differential pressure transducer between the inlet and the outlet to the specimen. (See Figure 11.48.) The density of water is used to convert the pressure to the height of fluid in the burette. This is converted to volume based on the area of the burette. This clearly illustrates a very important side effect of the burette-based system: the outlet pressure varies with reading location. The same measurement is made with the oil-filled device. In this case, the calculation is a bit more involved because the oil-water interfaces move at different rates on the outside and inside of the burette. The specific gravity of the oil is about 0.87, so the pressure difference is much smaller for this device.

The displacement-based volume-change devices supply the pressure and eliminate the air water interface, as well as make the necessary measurement. There are many variations on this concept. Figure 11.48 provides the essential details. The device has a piston located inside a rigid cylinder. The piston is either sealed at the end with the cylinder wall or (as shown) along the shaft at the entrance of the cylinder. The movement of the shaft is measured relative to the cylinder using a displacement transducer. The volume is computed from the displacement times the piston area. The resolution is controlled by the area of the piston and the sensitivity of the transducer. The pressure in the fluid is controlled with the externally applied force. Force capacity and the piston area are selected to achieve the desired pressure capacity. The details of such a device are included in Lade (1988).

In its simplest form, a data acquisition system is a combination of two devices: a measuring device and a recording device. This could, in fact, be a person with a data sheet and voltmeter. The recording device watches a clock until it is time for a measurement. When the time arrives, a reading is taken off the voltmeter and the time and voltage is recorded on the data sheet. The data sheet contains additional test information necessary to associate the data to a test, a device, a project, or the like. This archival information is an essential component of data collection. It must be possible to reconstruct the test results far after the person performing has forgotten the event. The recorder then returns to watching the clock in anticipation of the next reading event.

While this human-based system is essentially complete, it is neither practical nor does it make full use of the transducer capability. Many components should be added to the system, which will be discussed below. The example does, however, completely and accurately describe the first and primary function of the data acquisition system: to obtain and store measurements at specified times or events. If a system provides one hundred optional features but cannot completely fulfill the data recording requirement, then it is of little value for data acquisition.

Computer-based data acquisition systems are usually quite flexible, allowing setup of complicated recording schedules, plotting of the data during the test, and alterations to the reading schedule while the test is still running. For example, readings during a consolidation test are taken on a variable time scale that follows a square root or logarithm of time scale. Plotting during a test is invaluable for quality control purposes, problem detection during a test, decision making during various stages of a test, and general test efficiency. It is important to record hand readings of the transducer output at benchmark stages of tests to provide an independent cross check of the electronic data file.

Many data acquisition system configurations are available. Perhaps the most common is to use a dedicated data acquisition system for each testing station. This is necessary for fast strain rate tests, cyclic tests, and resonant column tests. However, in a large laboratory, it may be desirable to have data from all the electronic transducers from "slow" tests collected using a centralized facility. This provides a single repository for all transducer measurements. It is also very cost effective considering the fact that many tests will require one set of readings every minute or more. The specifications for such a system must account for the needs of all the transducers at the same time. Regardless of the configuration implemented, the system must be designed (both hardware and software) to provide reliability and accountability in the recorded data. The loss of data after a destructive test is complete can be a most depressing experience.

It should be apparent that the data acquisition system combines much more than one measuring device and a computer. In fact, the greatest advantage of the computer-based system is its component flexibility. The devices can easily be changed to fit individual needs. Figure 11.49 schematically depicts the basic components of a general purpose system.

Proper use of the technology requires an understanding of the overall system as well as the details of the data acquisition. The elements are collectively referred to as the data acquisition system and can be viewed as having four basic functional components, plus optional peripherals:

- The laboratory device (e.g., triaxial apparatus), which contains the transducers, power supply for the transducers, and usually a junction box and local voltmeter. Proper grounding of the electronics is essential for electronic stability. The ground is normally established at the low-voltage side of the power supply.
- Since many transducers will be connected to one computer, a switching mechanism is required to allow the computer to select a particular transducer for the measurement. As shown in the figure, this device is located before the analog to digital (A/D) converter and sequentially connects each transducer for the measurement. This configuration causes a time shift between each measurement and limits the speed of operation. A faster alternative is to select a configuration providing one A/D converter for each transducer and then switch the digital outputs.

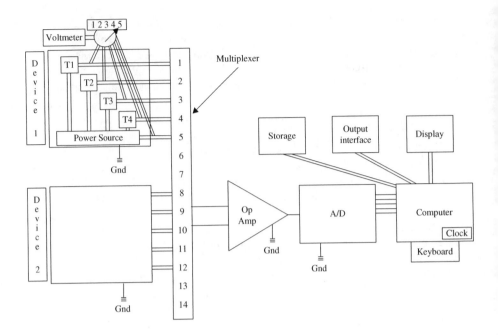

Figure 11.49 Schematic diagram of a centralized data acquisition system.

Table 11.1 Difficulty in attaining various levels of resolution

1 part in	Mechanical	Electrical
1	Unnecessary	Logic
10	Estimation	
100	By eye	
1,000	Easy	
10,000	Difficult	Common
100,000		Difficult
1,000,000		Special

The computer can then instruct all the A/Ds to convert at the same moment in time and then sequentially transfer the results.

- The voltages from each transducer and the power supply are changed to digital code by the analog to digital (A/D) converter. The characteristics of the A/D converter are critical to the precision of the final measurement.
- The computer orchestrates the components and performs the administrative and computational tasks. The data acquisition system can be dedicated to a single test device or centralized to many devices in a room or the entire laboratory.
- The peripheral components include anything from data storage to a network connection. Fewer components are better because the primary function of the system is to collect and archive the information. Data should be stored as received (in volts) with sufficient information about the test to allow complete reanalysis if questions arise in the future. Adding fancy capabilities is fine as long as they do not add risk to data collection.

Measurement Resolution

It is always important to keep perspective on the measurement requirements. One of the key elements to consider is the required (or necessary) resolution on the measurement. In most situations, the challenge imposed in making a measurement is evaluated in terms of the range of possibilities. One can think in terms of dividing the whole into some number of parts. Table 11.1 presents an evaluation of the level of difficulty for both a mechanical measurement and an electrical measurement. Each row adds an additional significant digit to the measurement.

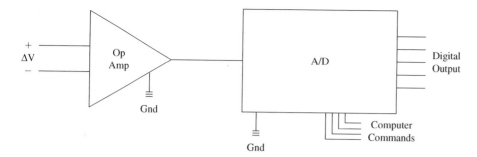

Figure 11.50 Schematic diagram of analog to digital converter with isolation amplifier.

At the top of the table is digital logic, which is a trivial measurement case represented by "on" or "off." This is clearly the most reliable measurement condition and the reason computers function so effectively. The mechanical measurements are far more difficult to make than the electrical measurements, which is essentially due to the fact that some level of physical amplification will be required. Getting to 1 part in 10,000 will require careful control on environmental factors such as vibration, friction, and temperature. This is also approaching the quoted limits of transducer nonlinearity. Electronic measurements are far more sensitive. Making measurements to 1 part in 10,000 is common for A/D converters. Devices are also readily available to measure 1 part in 1,000,000, but it is still a challenge to achieve electrical stability at these levels. Useful measurements require careful attention to the details of the wiring, grounding, and shielding of the circuits.

A/D Converter

The Analog to Digital (A/D) converter transforms analog voltage signal to a digital word. Once converted, the signal is no longer subject to degradation by transmission losses or noise. Therefore, it is advantageous to make the conversion as close to the transducers as practical, and then transmit the digital signal.

Figure 11.50 presents a schematic diagram of the A/D converter. The differential signal from the transducer (or voltage supply) is connected to an operational amplifier. The amplifier has a very high impedance (10^9 ohms), making it a passive observer of the voltage potential produced by the transducer. The amplifier outputs a voltage level referenced to ground. This component can also be used to amplify the signal. Amplification is referred to as "gain" in this context. The voltage is then input into the A/D converter, where it is converted to a digital word according to the characteristics of the converter and the commands from the controlling computer.

The A/D converter establishes the sensitivity of the system. It is the most important element in the data acquisition system. This device is a binary encoder that divides the voltage into a number of increments and outputs the result at discrete times. As a consequence of this operation, the continuous analog signal is transformed into discrete values of voltage and time. The voltage precision (ΔV) of conversion depends on: (1) the number of possible increments, and (2) the range (size) of the voltage scale. The number of increments is determined by the bit precision of the device. The second variable is the voltage range of the converter. This is the absolute limit on the voltage level convertible by the device. Most devices allow the range to be centered about zero, limited to all positive values, or limited to all negative values. The first setting is called bi-polar.

The time increment (Δt) between readings is limited to the sampling time of the converter. Two types of devices are common: the successive approximation and the sigma-delta converter. Successive approximation devices have the smallest sampling time, which is 0.01 msec or less. The sigma-delta converters perform a digital integration process that provides an average voltage over the sampling window. These devices provide higher bit resolution, but sample much more slowly. Typical sampling windows (or the integration time) starts at 1 msec. The sampling window determines

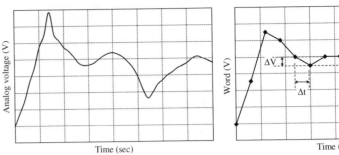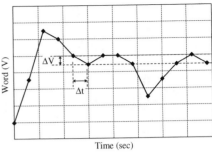

Figure 11.51 Comparison between the input analog signal and the output of the A/D converter.

Table 11.2 Sensitivity achievable for various bit sizes of an A/D converter in a data acquisition system

Bits	Options	S (on ±10V)	Int. Time (ms)
1	Logic	20	n/a
8	256	0.078	0.01
12	4,096	0.0049	0.01
16	65,536	0.00031	1
18	262,144	0.000076	16.7
20	1,048,576	0.000019	100
22	4,194,304	0.0000048	166.7

the maximum reading rate, but readings are most often collected at a much slower rate. It is impractical and even wasteful to collect too much data. The actual sampling rate must be selected to match the process being observed while being careful not to miss important events. Figure 11.51 provides a comparison between the continuous input signal and the resulting output digital sequence.

The voltage sensitivity, S, of an A/D converter is a function of the number of bits and the voltage range of the device. Most commercial devices have a range of 20 volts. Bit resolution varies depending on type of conversion and integration time. The sensitivity is expressed as Equation 11.9:

$$S = \frac{R}{2^B - 1} \tag{11.9}$$

Where:
S = sensitivity (volts)
R = range (volts)
B = number of bits

The number of output variations (options) for a converter is 2 raised to the power of bits. This is essentially the number of parts available over the voltage range of the converter. Table 11.2 shows the number of options for the various bit precisions used for converters. The table also shows the voltage increment associated with one bit and gives an estimate of the required integration time to provide reasonably stable readings at the given bit precision. There is definitely a need to sample longer in order to achieve higher bit precision. This becomes an important consideration when selecting a system to use with fast strain rates or cyclic loading.

Amplification and Range Matching

One of the unavoidable realities of the transducer characteristics used in the geotechnical laboratory is the variety in scale of output voltages. Table 11.3 provides a list of the most common measurements and the magnitude of the output at capacity. This list includes the input voltage because the calibration relationship is based on a normalized measurement.

Device	Full-Scale Output(Volts)	Required Bits	Gain for 18 bits
Power (Input Voltage)	10	15	1
LVDT	2	17	1
Load Cell	0.2	20	3.8
Pressure	0.1	21	7.6
LST	0.04	23	19
Quarter Bridge	0.005	26	152

Note: Values based on an A/D range of 20 volts.

Table 11.3 Comparison of data acquisition system requirements to achieve 1 part in 10,000 for various measurements

The need to connect all these devices to the same data acquisition system clearly presents a challenge, and choices will be necessary. We must first set a measurement resolution target. A reasonable starting point would be to measure each transducer to the resolution of 1 part in 10,000. This would be consistent with the transducer performance parameters and provide measurements to 0.1 percent when performing a test that only uses 10 percent of the device. It also recognizes the fact that one transducer will be used to test a variety of materials. Table 11.3 lists the bit precision for the various transducers that would be required to achieve the target resolution of 1:10,000. Connecting all these devices to one A/D would not be possible under even the best of conditions.

The A/D requirements can be relaxed by creating better matching of the transducer full-scale output with the range of the converter. This range matching is achieved using the operational amplifier placed in front of the A/D converter shown in Figure 11.50. This device increases the signal voltage by a constant factor before it is output to the A/D converter. The digitized output is then divided by the constant to return back to the original value. The gain or amount of amplification required to use an 18 bit A/D converter with each transducer is listed in Table 11.3. This amplification is provided under different names such as auto-ranging (done in factors of 10), preamplification (often customized to the transducer) and range selection (done in factors of 2). Amplification should only be considered when absolutely necessary because it adds another variable to the conversion relationship. Measurement errors must be evaluated prior to selecting the amount of amplification.

The amount of beneficial amplification (improved precision) will depend on the transducer, power supply, and A/D converter. An easy measure of precision can be obtained by collecting 50 readings of a constant signal and computing the standard deviation. The data acquisition system will have an inherent precision level that is obtained by measuring a grounded shunt. This is the best possible performance of the system. The unamplified transducer signal will have a specific precision value. Increasing the amplification will also increase the standard deviation of the measured signal. There will be a particular amplification that will match the transducer and system precision. This is the optimal gain. For example, a transducer having a 35 mV output at full scale with a precision of +/− 0.01 percent connected to a 18-bit A/D converter having +/− 10 V range and +/− 4 bits of noise will have reading precisions of +/- 0.000035 mV and +/− 0.0003 V for the transducer and converter respectively. Therefore, the precision is limited by the converter and the signal can be amplified by as much as 8.5 times before the transducer precision controls the response.

Noise

Direct current transducers are expected to generate a stable output voltage that varies due to the change in physical stimulus. Figure 11.52 is a conceptual diagram of the output of a force transducer during a triaxial compression test. The measurement is made over the time scale of tens of minutes. Electrical interferences will induce a variation in voltage around this "average" voltage output but these fluctuations will generally have a relatively high frequency. These interferences are referred to as noise. Electrical noise

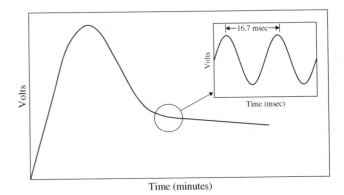

Figure 11.52 Illustration of AC-power-induced noise, superimposed on a transducer voltage signal.

can be due to a variety of causes such as the AC electrical power to the building (60 cycle "hum"), motors, computer monitors, power amplifiers, loose connections, poor grounding, ground loops, or a faulty device. The list is long, and often the cause of noise is never identified. The inset of Figure 11.52 provides an example of the interference caused by AC electrical power. The cyclic noise component has a period of 16.7 msec. Setting the integration time of the A/D converter to 16.7 msec will completely eliminate this noise. The same would be true for any noise source that has a shorter period. In situations where the required reading frequency is slow, integration is the best method to eliminate noise.

Finding and correcting noise problems can range from very difficult to virtually impossible. Provided that all the devices in the system are functioning adequately, there are several best practices that will greatly reduce noise problems:

1. Improper grounding is the most common cause of noise. Each pair of signal wires should be twisted and wrapped with a continuous shield. This protection should pass individually through junction boxes (not one common ground for the box) and continue on with the wire. The grounding network should branch out like a tree with one common sink at the transducer power supply, thus avoiding ground loops.

2. Powering all the components of the measurement system (power supply, voltmeter, computer) with a dedicated power line conditioner. This will provide stable voltage and filter out noise in the line power.

3. Noise is further reduced by having the A/D converter read ground between each and every reading. This provides a sink to drain any residual power.

4. The low voltage output side of LVDTs and similar induction devices should be connected to the low-voltage side of the power supply. This provides a drain for static charge in the secondary coil of the device.

5. The A/D converter should make true differential voltage measurements rather than subtracting two single-ended measurements.

Data Reduction

Data reduction is the process of converting the electronic voltages into physical units (force, pressure, displacement, and so on) and then using the geometry of the experiment to compute the engineering units (stress, strain, gradient, and the like). Data acquisition systems usually collect extremely large data sets, making it impossible to do the calculations by hand. However, the laboratory quality-control program should require hand calculations on a routine schedule to confirm the data reduction process is in working order.

The data reduction calculations can be performed using any of three options: commercial programs, in-house programs, or spreadsheets. These programs use a set of test-specific equations to operate on two input data sources to create one set of results. One input source is the data file. The second source is the test-specific reduction information, which includes calibration factors, specimen information, and project information. The results maybe contain complete or selected tabulated and graphical results. All output must identify the input information and the reduction program in order to maintain accountability of the information. It is best to have a structured and formally approved system of naming tests, data files, reduction files, and results files.

REFERENCES

Brooker, E. B., and H. O. Ireland. 1965. "Earth Pressures at Rest Related to Stress History," *Canadian Geotechnical Journal*, 2(1).

Corps of Engineers. 2001. *Geotechnical Investigations*, EM-1110-1-1804, Washington, DC.

Hvorslev, M. J. 1949. *Subsurface Exploration and Sampling of Soils for Civil Engineering Purposes*, U.S. Army Engineer Waterways Experiment Station, Vicksburg, MS.

Jamiolkowski, M., C. C. Ladd, J. T. Germaine, and R. Lancellotta. 1985. "New Developments in Field and Laboratory Testing of Soils." *Proceedings of the 11th International Conference on Soil Mechanics and Foundation Engineering*, San Francisco.

Ladd, C. C., and D. J. DeGroot. 2003, revised May 9, 2004. "Recommended Practice for Soft Ground Site Characterization: Arthur Casagrande Lecture," *Proceedings of the 12th Panamerican Conference on Soil Mechanics and Geotechnical Engineering*, Cambridge, MA.

Ladd, R. S. 1965. "Use of Electrical Pressure Transducers to Measure Soil Pressure," *Report R65-48*, No. 180, Department of Civil Engineering, Massachusetts Institute of Technology, Cambridge.

Ladd, R. S. 1977. "Specimen Preparation and Cyclic Stability of Sands," *Journal of the Geotechnical Engineering Division*, ASCE, 103(6), June.

Lade, P. V. 1988. "Automatic Volume Change and Pressure Measurement," *Geotechnical Testing Journal*, ASTM, 11(4), December.

Lambe, T. W. and R. V. Whitman. 1969. *Soil Mechanics*, John Wiley and Sons, New York.

Wilson, S. D. 1970. "Suggested Method of Test for Moisture-Density Relations of Soils Using Harvard Compaction Apparatus," *Special Procedures for Testing Soil and Rock for Engineering Purposes*, ASTM STP 479, Philadelphia, PA.

Chapter 12

Compaction Test Using Standard Effort

SCOPE AND SUMMARY

This chapter provides information on determining the compaction characteristics of soil. The compaction test measures the variation in compacted dry density as a function of water content. The test also provides the maximum dry density and the optimal water contact for compaction. Procedures are detailed to perform a laboratory compaction test on a coarse-grained soil using the standard effort dynamic hammer in general accordance with ASTM D698 Laboratory Compaction Characteristics of Soil Using Standard Effort $(12\,400\ \text{ft-lbf/ft}^3\ (600\ \text{kN-m/m}^3))$. The background section discusses other methods of determining compaction characteristics, such as the compaction test using modified effort for simulation of larger field equipment, and the Harvard Miniature test for fine-grained soils.

TYPICAL MATERIALS

The compaction curve, including the maximum dry density and optimal water content, is commonly determined on soil materials ranging from clay to gravel. While larger particles are often included in field construction, these particles are removed (and quantified) prior to applying compactive effort in the laboratory and reintroduced into the results by calculation. The limitation on largest particle size is dependent upon the composition of the material and can vary from 4.75 mm (No. 4) to 19 mm (3/4 in.) in diameter.

Compaction is the reduction in void ratio (or increase in density) due to the application and removal of a static or dynamic force. The process generally happens at constant water content and only air is expelled from the material. As such, the process occurs very quickly for all types of soils. Saturation is always less than 100 percent and the air voids are continuous throughout the specimen. Compaction also imposes a complicated and nonuniform stress increment to the material involving both normal and shear stresses. Compaction should not be confused with consolidation, compression, creep, or shear, which are also volume change processes of soil. These processes will be discussed in subsequent chapters of this text.

The dry density achieved by imparting energy on the soil is dependent upon the initial water content, termed the molding water content. Understanding the relationship between the achievable density and the molding water content has been a research topic for many decades. Starting from the dry condition, an increase in the molding water content will result in a higher compacted dry density. This trend will continue up to the maximum dry density, which occurs at the optimum molding water content. Further increases in water will result in a continuous reduction in the dry density. This relationship is referred to as the compaction curve.

The process of compaction depends on a complicated interaction of interparticle forces and relative particle movements. On the dry side of optimum (water content is implied), surface friction prevents particles from sliding past each other, limiting densification of the material. Addition of water provides better lubrication allowing the particles to move into a denser arrangement under the dynamic force. Counteracting the lubrication mechanism is energy absorption by the water and air in the pore space. Energy lost to the pore fluid reduces particle movement resulting in a lower density. On the wet side of optimum, addition of water causes further reduction in the achievable dry density. The competition of these mechanisms results in a compaction curve that resembles an inverted parabola.

The test method measures the change in density achieved under a given applied energy over a range of water contents. The molding water content and the dry density resulting from applying the same amount of energy to six specimens (points) prepared at different water contents are plotted, as shown in Figure 12.1.

A curve is then constructed connecting the points, and the maximum dry density (MDD) and optimum moisture (ω_{opt}) content are interpreted using some level of judgment. It is unlikely that the peak of the interpreted compaction curve will correspond to a particular compaction point. Refer to Figure 12.2 for an example determination of the maximum dry density and optimum water content.

The water content and energy used to compact a particular soil has a significant impact on the resulting engineering properties. On the dry side of optimum, the material is generally stiffer and stronger, and has a higher hydraulic conductivity. As the dry density approaches the maximum dry density, the engineering properties of that material

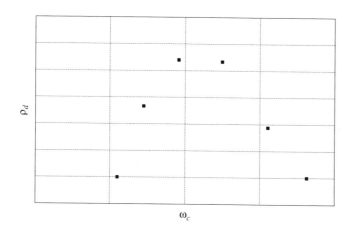

Figure 12.1 Example plotted compaction points.

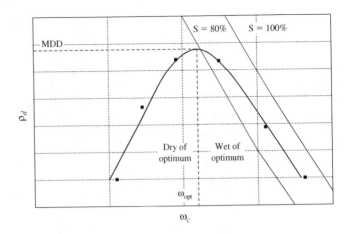

Figure 12.2 Determination of the maximum dry density (MDD) and optimum water content (ω_{opt}).

generally improve. Strength is increased, and hydraulic conductivity, compressibility, and brittleness are decreased. On the wet side of optimum, the material is weaker and more ductile, and has much lower hydraulic conductivity. These principles have applications in highway construction, dam design, hazardous waste containment, and, as seen in the previous chapter, in laboratory specimen reconstitution.

The degree of compaction is the ratio, in percent, of a specimen density to the maximum dry density for that soil as obtained by a specified method. Degree of compaction is the primary quality control tool for field placement of soils. A laboratory test is performed to determine the expected field behavior. The results are adjusted for the presence of large particles, resulting in the maximum dry density and optimum water content for the field compaction. There is usually a maximum limit on the amount of fines acceptable in materials for highway construction because fines cause pumping or freezing problems. During construction, field testing methods are used to verify that the soil has been compacted to the specified criteria.

There are several laboratory methods for determining the compaction characteristics. The two most commonly used methods are the standard and the modified dynamic hammer tests. These methods are standardized as ASTM D698 Laboratory Compaction Characteristics of Soil Using Standard Effort (12 400 ft-lbf/ft³ [600 kN-m/m³]) and D1557 Laboratory Compaction Characteristics of Soil Using Modified Effort (56,000 ft-lbf/ft³ [2,700 kN-m/m³]). The compactive effort (energy per unit volume) applied to the specimen is selected to model the equipment that will compact the soil in the field. Drum and vibratory field compactors compare favorably with compaction achieved in the laboratory using dynamic hammer tests. With the advent of larger, more efficient field compaction equipment, the modified effort energy more typically models the compactive effort that can be achieved in the field. Fine-grained soils, which are typically used for hydraulic conductivity barriers, are often compacted in the field with sheep's foot rollers. This equipment relies on a shearing action to compact the material. Field performance for this type of equipment is better simulated by kneading action using the Harvard Miniature apparatus.

For the dynamic compaction methods of D698 and D1557, the applied energy is determined by the number of drops of a mass from a specified height divided by the volume of the final specimen. For D698, the standard effort test, a 2.5 kg mass (having a static weight of 24.5 N or 5.5 lbf) hammer is used to compact three layers and a drop height of 305 mm (12 in). The equipment used for the test method is shown in Figure 12.3. For D1557, the modified effort test, a 4.5 kg mass (having a static weight of 44.5 N or 10 lbf) hammer with five layers and a drop height of 457.2 mm (18 in) are used. The mold size and the number of drops per layer are varied to accommodate different material gradations.

Each permutation is a different "method" within the test, as shown in Table 12.1. The large particles are removed from the sample prior to compaction. These particles are referred to as the oversize fraction. While compaction is generally characterized by

Figure 12.3 24.5 N (5.5 lbf) compaction hammer, 101.6 mm (4 in.) and 152.4 mm (6 in.) diameter molds with collars used to perform ASTM D698. Also pictured is a straight edge used to square off the top of the specimen.

energy per unit volume, it is recognized that the results do vary slightly with specimen size and geometry of hammer.

A compaction test is sometimes referred to as a "Proctor," as in Standard Proctor or Modified Proctor. The name is in recognition of R. R. Proctor (1933), who performed a tremendous amount of work to determine compaction properties.

The procedure for performing a compaction test according to D698 and D1557 requires a measure of the oversize fraction in order to select the appropriate mold size and number of drops per layer. The first step is to estimate which method would be most likely and then to separate the material on the appropriate sieve (4.75 mm, 9.5 mm, or 19 mm). While a material meeting the limitation of Method A will also meet limitations of the other two methods, it is not permissible to use Method B or C unless so specified by the requesting agency.

In most cases, a grain size analysis will be performed along with the compaction test. It is most efficient to process and separate the material for both tests at the same time using one of these sizes as the Designated A sieve as discussed in Chapter 8 for the composite sieve analysis.

As with previous tests, if fines are present the material should not be oven-dried prior to compaction. The total mass of the sample is obtained along with the portion of coarser and finer fractions in the moist condition. The oversize percentage can be estimated immediately using these mass measurements (and then finalized after the grain size analysis has been performed), eliminating any delay in proceeding with the compaction test. Blend the finer fraction and estimate the initial water content and the optimum water content. Experience with compaction and with the specific material is

Method	Mold Diameter	Material used in Specimen	Number of Blows per Layer	Maximum Particle Size Limitations
Method A	101.6 mm (4 in)	Fraction passing the 4.75 mm (No. 4) sieve	25	25% or less retained on the 4.75 mm (No. 4) sieve
Method B	101.6 mm (4 in)	Fraction passing the 9.5 mm (3/8 in) sieve	25	25% or less retained on the 9.5 mm (3/8 in) sieve
Method C	152.4 mm (6 in)	Fraction passing the 19 mm (3/4 in) sieve	56	30% or less retained on the 19 mm (3/4 in) sieve

Table 12.1 Table of mold size, number of blows per layer, material used, and material limitations for the various methods in ASTM 698 and ASTM D1557. (Adapted from ASTM D1557-07)

extremely helpful in making these two estimates. Split the material into six portions. Use one portion for a water content determination. Using distilled water, prepare the remaining portions at five different water contents, bracketing what is expected as the optimum moisture content. Prepare the middle specimen at the estimated optimum water content, along with two higher water contents and two lower water contents. The water content increments will increase as the gradation becomes finer and as the plasticity of the material increases.

Seal each portion in a container and temper the soil if necessary. Refer to Table 9.2 for the minimum tempering time according to soil classification. Compact the specimen in the required number of layers and blows per layer as specified for each method. To prevent arching, be sure to use the edge of a frosting spatula or knife to make grooves in (i.e., scarify) the top of each layer before adding material for the overlying layer. Trim the top of the specimen using the mold as a guide. Obtain the mass of the total specimen. Extrude the specimen from the mold and obtain a water content specimen from a pie-shaped slice cut throughout the entire depth of the specimen. Repeat the process for the other points of the test.

The total mass density (ρ_t) of the compacted specimen is calculated using Equation 12.1:

$$\rho_t = \frac{M_t}{V_t} \tag{12.1}$$

Where:

ρ_t = total mass density of the compaction specimen (g/cm^3)
M_t = total mass of the compaction specimen (g)
V_t = volume of the compaction mold (cm^3)

The water content is obtained from a pie-shaped section of the compacted specimen. The value of the dry mass density of the compacted specimen is calculated using Equation 12.2:

$$\rho_d = \frac{\rho_t}{1 + \dfrac{\omega_C}{100}} \tag{12.2}$$

Where:

ρ_d = dry mass density of the compaction specimen (g/cm^3)
ω_C = water content from pie slice shaped section of the compacted specimen (%)

As shown on Figure 12.2, lines of constant saturation are also included on the density versus moisture curve for reference. Sometimes the 100 percent saturation line is referred to as the zero air voids (ZAV) line. The lines of constant saturation are calculated using $G_s\omega_C = Se$, which can be rearranged as shown in Equation 12.3:

$$S = \frac{G_s\omega_C}{\left(\dfrac{\rho_w}{\rho_d}\right)G_s - 1} \tag{12.3}$$

Where:

G_s = specific gravity (dimensionless)
ρ_w = mass density of water (g/cm^3)

As can be seen in Table 12.1, there is a limit to the maximum particle size that can be accommodated in the specimen. These limits are similar to other mold to particle diameter specifications and in this case are set at between 8 and 12. The limit is intended to provide adequate mobility of particles during densification and to prevent excessive particle crushing due to impact with the hammer.

The Proctor mold is relatively small (4 in. or 6 in. diameter) compared to the maximum size of particles used in construction. Therefore, the largest of the particles are scalped off the material prior to the compaction test. The dry density achieved for the modified test specimen and the molding water content are then adjusted to account for the removed oversize particles. The concept behind the corrections is that the oversize particles float in a matrix of the finer material (i.e. there is no particle to particle contact of these oversize particles). Figure 12.4 is a schematic of the coarser materials in the finer matrix. The numerical limit of this assumption is clearly approximate and has been set based on engineering judgment. Considerable research has been conducted over the years to better quantify the effect of removing oversize particles from the test specimen, but this simplification remains an important approximation in the application of compaction data to the field.

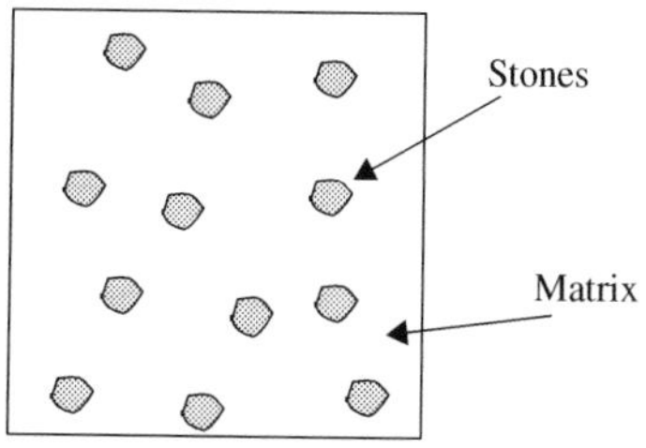

Figure 12.4 Example of coarse particles sitting in a finer matrix.

The method of applying the oversize correction is covered in ASTM D4718 Correction of Unit Weight and Water Content for Soils Containing Oversize Particles. The percent finer fraction (F) is calculated using Equation 12.4. A companion sample (or the results of the grain size analysis) is used to determine this fraction rather than drying the test specimens prior to compaction.

$$F = \frac{M_m}{M_m + M_s} \tag{12.4}$$

Where:

F = fraction of dry soil mass less than the separation sieve used to prepare the compaction specimens (decimal)

M_m = mass of the dry soil comprising the fine matrix (g)

M_s = mass of the dry soil comprising the coarse particles (g)

The dry density that would be achieved in the field when both the finer fraction and coarser fraction are compacted is then calculated using Equation 12.5. The equation is applied to each compaction point obtained in the laboratory.

$$\rho_{df} = \frac{\rho_d G_{s,C} \rho_w}{\rho_d (1 - F) + G_{s,C} \rho_w F} \tag{12.5}$$

Where:

ρ_{df} = dry density achievable in the field (g/cm^3)

ρ_d = dry density achieved in the lab (g/cm^3)

$G_{s,C}$ = specific gravity of the coarse materials (dimensionless)

ρ_w = mass density of water (g/cm^3)

If not assumed, the bulk specific gravity of the coarse materials can be determined using ASTM C127 Density, Relative Density (Specific Gravity), and Absorption of Coarse Aggregate. This test method describes three measures of bulk specific gravity, namely the oven-dry, saturated surface-dry, and apparent specific gravity. The value for apparent specific gravity used in C127 is the actual specific gravity of the minerals contained in the coarse materials, that is, the same specific gravity as measured in Chapter 3 of this text.

The water content must also be corrected to account for the presence of the coarse particles. Calculate the water content of the coarse particles (ω_{cs}) and use Equation 12.6 to determine the corrected field water content (ω_{cf}):

$$\omega_{cf} = (1 - F)\omega_{cs} + F(\omega_{cm}) \tag{12.6}$$

Where:

ω_{cf} = corrected water content of the material, as compacted in the field (%)

ω_{cs} = water content of the oversize material, as received (%)

ω_{cm} = water content of compaction test specimen (%)

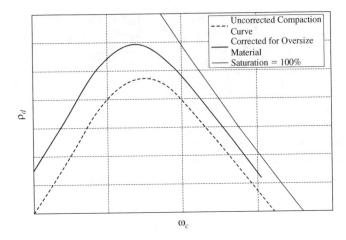

Figure 12.5 Example compaction curves demonstrating the effect of correcting for oversize particles.

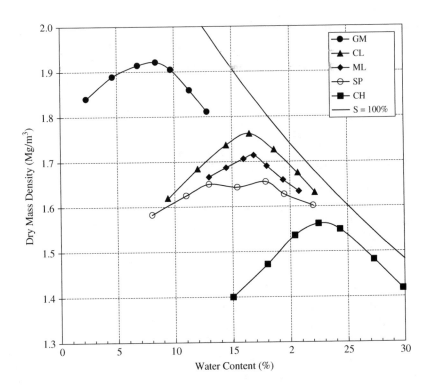

Figure 12.6 Typical compaction curves for various materials.

Since the density of the coarse particles will be greater than the density achieved by the finer fraction for that volume, the calculated dry density achievable in the field will be higher than that for the specimen with the material scalped. Likewise, after correction, the water content will be less since the largest particles will usually have a much lower water content than the finer matrix. The result is a corrected compaction curve as shown in Figure 12.5.

There are numerous other factors to consider on the topic of compaction, the first of which is the impact of the soil type on the position and general shape of the compaction curve. The soil type will have dramatic effects on the shape of a compaction curve achieved. In general, coarser materials will have a higher maximum dry density and lower optimum water content than soils with fine-grained components. In addition, well-graded coarse materials will have well defined peaks, while uniform coarse-grained materials will be flatter, and may not even have a peak. Figure 12.6 shows typical compaction curves for a range of soil types.

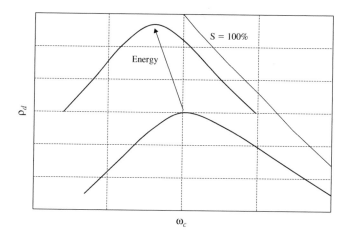

Figure 12.7 Effect of energy level on the position of the compaction curve.

The amount of energy imparted on a specimen causes a shift in the curve as well. In general, as the energy increases, the compaction curve moves up and to the left, as shown in Figure 12.7. This is why the energy level used to generate the compaction curve must be matched to that of the field equipment placing the material. Using too low an energy level in the laboratory can mislead the contractor to use too much water and overcompact the soil.

Compaction can be performed using various forms of input energy. Dynamic compaction, sometimes referred to as impact compaction, is the method described thus far in this chapter: dropping a mass repeatedly onto soil. Static compaction involves applying a constant compressive force. Vibratory compaction is a force that rapidly and repeatedly increases and decreases on the soil surface. Kneading action is when soil particles are moved past one another. Each method creates a different interparticle structure, resulting in different mechanical properties. As an example, Figure 12.8 (After Seed and Chan, 1959) shows the relative strength of a particular soil achieved with various compaction methods versus molding water content as compared to kneading compaction. Other

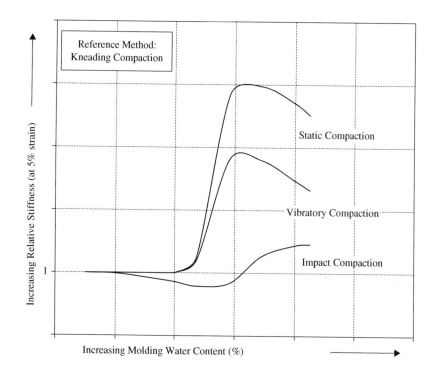

Figure 12.8 Effect of method of compaction on the strength of soil as referenced to kneading compaction. (Adapted from Seed and Chan, 1959)

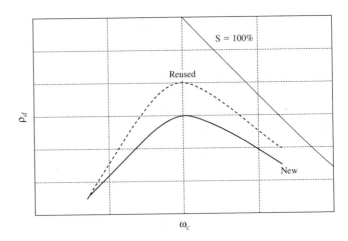

Figure 12.9 Effect of reusing material on the position of the compaction curve.

soils will have different scales, but the relative positions of the curves with increasing water content should remain about the same.

Reusing material causes a considerable increase in the dry density achieved at a particular water content. Previously compacted fine-grained materials will clump. Coarse-grained materials will experience particle crushing during compaction. Reusing material will affect the shape of the curve and the resulting interpretation of the maximum dry density. A schematic comparison between using completely new material versus reusing the same material for each point is shown in Figure 12.9. For the plot of the reused material, the driest point is compacted first, and then the same material is reused for each successively wetter point.

Other important considerations for determining compaction characteristics include:

- The compaction properties of some soils are sensitive to drying and therefore soils should not be dried prior to performing a compaction test unless it can be shown that the results are independent of drying. The fine fraction of soils should not be oven-dried prior to compaction. It is important to note that air-drying is not precluded in test methods D698 and D1557; however, the moist method is the preferred specimen preparation method and a warning is given concerning possible altering soil properties by air-drying.
- Make sure the material is homogeneous across the portions. Use the blending and splitting procedures described in Chapter 1 to obtain the testing specimens.
- Temper soils that contain fine-grained materials. Refer to Table 9.2 in Chapter 9, "Atterberg Limits" for suggested tempering times according to USCS group classification.
- Secure the mold to a solid base prior to imparting energy. Make sure the solid base has a mass of at least 100 kg, and the mold is securely attached using devices such as clamps. The surface of the base shall be level, and flat such that the compaction mold is fully supported and does not tilt or translate during application of compaction energy.
- Make sure the layers have approximately the same volume to distribute the energy, and that each layer is spread out evenly across the mold.
- Compact each layer using the sequence of hammer drops shown in Figure 12.10 to ensure uniform energy for the 101.6 mm (4 in.) diameter mold or Figure 12.11 for the 152.4 mm (6 in.) diameter mold.
- After compaction, the surface of each layer should be relatively flat. If significant unevenness occurs, adjust the tamping pattern to obtain a flat surface. Note that pumping may occur at water contents above optimum, particularly for fine-grained soils.

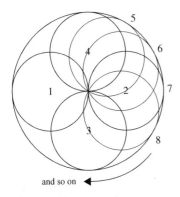

Figure 12.10 Hammer drop pattern when using the 101.6 mm (4 in.) diameter mold.

- After compaction, verify that the mold is not overfilled. Specifically, the surface of the final compacted layer must not extend more than 6 mm (0.25 in.) above the top of the mold.
- Obtain a representative water content of the extruded specimen by cutting vertically down through the material and obtaining a pie-shaped slice through the center.
- Check to see if water is seeping out of the bottom of the mold of any compaction point. This is referred to as bleeding and indicates that the water content obtained will not be representative of the molding water content. This is more likely to occur for poorly graded coarse-grained soils and for points very wet of optimum. Such points should be noted on the data sheet.
- Do not reuse material compacted in the laboratory, because the resulting dry density will be erroneously high.
- Be sure to choose a reasonable scale to plot compaction curve. Interpretation of the curves is very sensitive to the scale of the plot. The ASTM standards provide examples of reasonable scale dimensions. The suggested plot has a range in water content of 14 percent and a range in dry density of 5.6 kN/m^3 (28 lbf/ft^3). If the aspect ratio is distorted or the scale is overexpanded, it will be very difficult to define the curve.

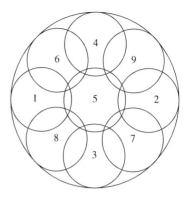

Figure 12.11 Hammer drop pattern when using the 152.4 mm (6 in.) diameter mold.

TYPICAL VALUES

Typical values of the maximum dry density and optimum moisture content as determined by the method specified are listed in Table 12.2.

Equipment Requirements

1. Scale with a capacity of at least 10 kg and readable to 1 g for determination of the mass of the compacted specimens in the mold
2. Mold: Either a 101.6 mm (4 in.) or 152.4 mm (6 in.) diameter mold, depending on grain size distribution characteristics of the material being tested. The actual dimensions of these molds are 101.6 +/− 0.4 mm in diameter by 116.4 +/− 0.5 mm in height or 152.4 +/− 0.7 mm in diameter by 116.4 +/− 0.5 mm in height. Include the base plate and the collar assembly for the mold used.
3. Base: A solid base with a mass of at least 100 kg or a weight of about 200 lbf, and clamps to secure the mold to the base
4. Hammer: A 24.47 +/− 0.09 N (5.50 +/− 0.02 lbf) dynamic hammer having a drop height of 304.8 +/− 1 mm (12 +/− 0.05 in.). The mass of this hammer at sea level is 2.495 +/− 0.023 kg. An automated mechanical rammer meeting the specifications in D698 may be used instead, if preferred.
5. Calipers readable to 0.02 mm
6. Sieve for separation of particles, if using oversize material
7. Frosting spatula or knife
8. Straight edge
9. Extruding jack for removal of specimens from the mold, if necessary
10. Mixing and splitting equipment
11. Equipment necessary for determination of water contents: forced draft oven, desiccator, scale, water content tares, and the like

Table 12.2 Typical values of the maximum dry density and optimum moisture content.

Soil	ρ_{max} (Mg/m^3)	ω_{opt} (%)	Method
Maine Clay*	1.80	17.7	Harvard Miniature
Vicksburg Buckshot Clay**	1.56	22.8	D698
Annapolis Clay**	1.75	16.6	D698
Vicksburg Silt**	1.70	17.1	D698
Well-graded clean gravels, gravel-sand mixtures***	2.0 to 2.2	8 to 11	D698
Poorly graded clean gravels, gravel-sand mix***	1.8 to 2.0	11 to 14	D698
Silty gravels, poorly graded gravel-sand-silt***	1.9 to 2.2	8 to 12	D698
Clayey gravels, poorly graded gravel-sand-clay***	1.8 to 2.1	9 to 14	D698
Well-graded clean sands, gravelly sands***	1.8 to 2.1	9 to 16	D698
Poorly graded clean sands, sand-gravel mix***	1.6 to 1.9	12 to 21	D698
Silty sands, poorly graded sand-silt mix***	1.8 to 2.0	11 to 16	D698

*Unpublished laboratory data.
**After ASTM D698; values converted from pounds per cubic foot (pcf) to megagrams per cubic meter (Mg/m^3).
***After Naval Facilities Engineering Command (1986), *Design Manual: 7.02*; values converted from pcf to Mg/m^3.

CALIBRATION

1. Determine the mass of the sample mold and mold base to the nearest gram. Do not include the collar.
2. Measure the mold depth (3 places) and mold diameter (6 places) to $\pm$ 0.02 mm.
3. Check hammer for damage, square edges, proper mass, and proper drop height.

SPECIMEN PREPARATION

Using a well-graded sand with less than 5 percent fines eases the process of laboratory instruction. These procedures assume such a material is being used. Select material with oversize particles only if it will be desired to demonstrate those aspects of the test.

1. Select enough of the moist material to provide about 12 kg of dry soil.
2. Obtain the natural water content.
3. Estimate how much material is required for each compaction point and prepare five samples having water contents separated by about 1.5 percent. Adjust the water contents such that they bracket the optimum value.
4. Temper the soil overnight. For laboratory instruction purposes, it may be necessary to skip this step; however, not tempering soils may increase the scatter in the data and make it more difficult to define the compaction curve.

PROCEDURE

The compaction test will be performed in general accordance with ASTM Standard Test Method D698, although D1557 can be used instead with minor modifications.

1. Assemble the mold and clamp to the solid base or floor.
2. Compact each specimen in three equal layers using 25 blows per layer.
3. Before compacting the second and third layers, scarify the top surface of the under-lying layer with a knife to about 3 mm depth.

4. The last layer must completely fill the mold but should not exceed the mold by more than 6 mm. If this condition is not met, the point is not valid.

5. Compact each layer using the recommended sequence of hammer drops shown in Figure 12.10 or Figure 12.11 (whichever is appropriate) to ensure uniform energy.

6. Remove compaction collar and check that the surface of the soil is not too high. The soil surface must extend above the top of the mold, but less than or equal to 6 mm (0.25 in.).

7. Scrape the surface flat with a knife, using the top of the mold as a guide. Be sure to check the straightness of the knife edge. A slight curvature of the top surface can result in significant errors in the dry density.

8. Determine the mass of the specimen and mold.

9. Extrude the specimen, using a hydraulic jack if necessary.

10. Obtain a representative water content specimen. Be sure to cut a pie-shaped slice through the entire specimen.

Calculations

1. Calculate the total density at each point using Equation 12.1.

2. Calculate the molding water content at each point from the post-compacted slice. Do not use the prepared value. If oversize material has been included, calculate the corrected water content at each point using Equation 12.6.

3. Calculate the dry density at each point using Equation 12.2. If oversize material has been included, obtain the water content and specific gravity of the oversize fraction, and calculate the corrected dry density at each point using Equation 12.5.

4. Plot the dry density and corresponding molding water content of each point on a compaction curve. If oversize material has been included, plot the corrected compaction points as well, and distinguish between the two curves.

5. Calculate the 80 percent, 90 percent, and 100 percent saturation lines using Equation 12.3, and plot on the compaction curve.

6. Draw a best fit line through the compaction points, and determine the maximum dry density and optimum water content.

Report

Report the uncorrected and corrected maximum dry density and optimum water content for the material, along with the method used to determine the values. Include the raw data sheets, a plot of the compaction curve (ρ_d versus ω_C) with 80, 90, and 100 percent saturation lines indicated.

PRECISION

Criteria for judging the acceptability of test results obtained by test method D698 (Standard Effort Compaction) are given as follows for three soil types used in the inter-laboratory study (ILS) conducted by the ASTM Reference Soils and Testing Program. Note that Method A and the dry preparation method were used on three soil types (CH, CL, and ML) to produce these results. The precision was found to vary with soil type, and it is expected that the precision will also change with the methods and mold size used. The study was performed using units of pounds per cubic foot (pcf). The values have been converted to megagrams per cubic meter (Mg/m^3).

- *Within Laboratory Repeatability:* Expect the standard deviation of your results on the same soil to be on the order of 0.01 Mg/m^3 for the maximum dry density and 0.3 percent for the optimum water content.
- *Between Laboratory Reproducibility:* Expect the standard deviation of your results as compared to others on the same soil to be on the order of 0.02 Mg/m^3 for the maximum dry density and 0.7 percent for the optimum water content.

DETECTING PROBLEMS WITH RESULTS

For laboratory instructional purposes, it is recommended that a minimum of five data points are produced to define the compaction curve. If the data are highly scattered, preventing a typical shaped curve to be created, the experiment should be repeated on previously uncompacted soil. If the results are still varied, probable causes of error are poor blending and splitting techniques, sloppy control of the energy imparted to the specimen, improper clamping of the mold to the base, insufficient tempering time, or lack of a control volume.

If the results fall outside of typical ranges, likely sources of error are in the calculation of the oversize correction, poor establishment of the percent of finer and coarser fractions, incorrect mold volume, or equipment calibration problems.

REFERENCE PROCEDURES

ASTM D698 Laboratory Compaction Characteristics of Soil Using Standard Effort (12 400 ft-lbf/ft^3 [600 kN-m/m^3])

REFERENCES

Refer to this textbook's ancillary web site, www.wiley.com/college/germaine, for data sheets, spreadsheets, and example data sets.

Naval Facilities Engineering Command. 1986. *Foundations and Earth Structures, Design Manual 7.02,* Alexandria, VA.

Proctor, R. R.1933. "Fundamental Principles of Soil Compaction."*Engineering News Record,* vol. *111,* no. 9.

Seed, H. B. and C. K. Chan. 1959. "Structure and Strength Characteristics of Compacted Clays," *Journal of the Soil Mechanics and Foundations Division, ASCE,* Vol. 85, SM5, October.

Calculate the dry density at each point using Equation 12.2. If oversize material has been included, obtain the water content and specific gravity of the oversize fraction, and calculate the corrected dry density at each point using Equation 12.5.

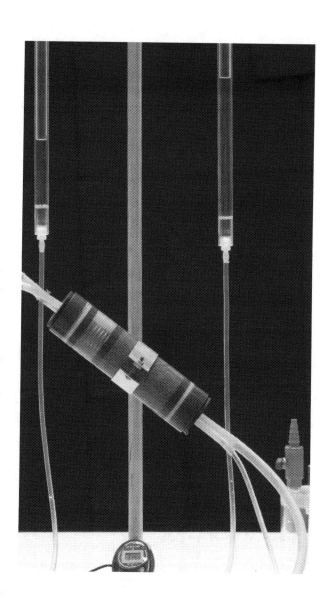

Chapter 13

Hydraulic Conductivity: Cohesionless Materials

This chapter describes general background on the determination of hydraulic conductivity. Specific guidance is presented for performing a constant head and a falling head hydraulic conductivity test using a rigid walled mold on a laboratory-prepared coarse-grained soil specimen with relatively high hydraulic conductivity (i.e., greater than 10^{-7} m/s).

SCOPE AND SUMMARY

This test method is appropriate for determining the hydraulic conductivity of a coarse-grained material with less than ten percent of the mass finer than 75 µm (No. 200 sieve). Additionally, the material must not undergo volume change during the test or be sensitive to changes in effective stress, thus allowing use of a rigid walled mold. Laboratory specimen fabrication methods are presented in Chapter 11, "Background Information for Part II."

TYPICAL MATERIALS

Hydraulic conductivity (k) is defined as the rate of flow through a cross section of soil. The units are length per unit time, which is a velocity. This textbook uses units of meters

BACKGROUND

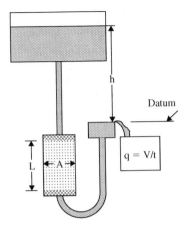

Figure 13.1 Schematic of the components of total head in a system.

per second (m/s) for hydraulic conductivity. Hydraulic conductivity is given numerically by the empirical expression known as Darcy's Law, shown as Equation 13.1:

$$k = \frac{q}{iA} \tag{13.1}$$

Where:

k = hydraulic conductivity (m/s)
q = volume of flow per unit time (m³/s)
i = hydraulic gradient (dimensionless)
A = cross-sectional area (m²)

Hydraulic gradient is a dimensionless measure of the driving energy in the flow system, and is given by Equation 13.2:

$$i = \frac{h}{L} \tag{13.2}$$

Where:

h = total head[1] dissipated (m)
L = external length of flow path (m)

The components of total fluid energy, expressed as total head in dimensions of length, are demonstrated schematically in Figure 13.1.

The empirical law is based on Darcy's Parametric Black Box experiments in which he found that the numerical relationship in Equation 13.1 held for constant head conditions. Darcy's Law is valid only when flow is laminar. During laminar flow, water molecules follow stable flow paths called stream lines. The discipline of fluid mechanics describes the transition between laminar and turbulent flow as gradual and dependent upon the media within which fluid is flowing. Reynolds number (Re) is usually used to establish approximate boundaries between laminar, turbulent, and transitional flows. Reynolds number is defined for pipes using Equation 13.3:

$$\text{Re} = \frac{v_c D_a \rho_f}{\mu} \tag{13.3}$$

Where:

Re = Reynolds number (dimensionless)
v_c = critical velocity (m/s)
D_a = diameter of the pipe[2] (m)
ρ_f = mass density of fluid (Mg/m³)
μ = viscosity of fluid (Pa-s)

Laminar flow in pipes occurs when the Reynolds number is less than about 2000. In porous media, the pore space has highly irregular geometry and the transition between laminar flow and turbulent flow is even less defined than in pipe flow. A wide range on the maximum Reynolds number for laminar flow is quoted in the literature. Frequently, a maximum Reynolds number of 1 is used to ensure that flow within soil is laminar. Laminar flow conditions are easily met for typical laboratory conditions when testing

---

[1] Total head is related to the total energy in a system. Total energy is expressed by Bernoulli's equation and consists of the sum of potential energy, pressure energy, and kinetic energy. Kinetic energy is typically assumed to be zero within groundwater flow studies because flow velocities are very small. In the laboratory, energy is measured at points where the pressure energy is zero. Total energy divided by gravity gives a value referred to as total head, which has units of length. The length is the height of a column of water.

[2] In soils, the diameter of the flow channel is sometimes taken as the effective diameter of the particles of media.

clays, silts, and most sands. However, turbulent flow conditions occur in materials with larger, gravel-sized particles.

A second but no less important condition for Darcy's Law to be valid is that there must be no volume change during the test. Volume change is generally not a concern for clean, coarse-grained materials; however, it is for soils with a fine-grained component. Volume change during a test is indicated by a change in height of the specimen. Particles flowing out of the specimen will also make the outflow water cloudy.

The concept of flow velocity is important to the understanding of hydraulic conductivity. The approach velocity (v) is the velocity of the permeant as it approaches the specimen. Darcy's Law can be rewritten as Equation 13.4:

$$\frac{q}{A} = ki \qquad (13.4)$$

The left side of the equation is the approach velocity or the Darcy velocity, leading to Equation 13.5:

$$v = ki \qquad (13.5)$$

However, porous media consists of both solids and voids. The area available for flow isn't the full cross section of the permeameter, but rather the area of the void space (A_v) only. Porosity is defined as the volume of voids divided by the total volume, which can be rewritten as Equation 13.6:

$$n = \frac{V_v}{V} = \frac{A_v}{A} \qquad (13.6)$$

Where:

n = porosity (dimensionless)
V_v = volume of voids (m^3)
V = total volume (m^3)
A_v = area of voids (m^2)

The seepage velocity (v_s) is the velocity of the permeant as it flows through the pores. Considering the above equations along with continuity of flow, Equation 13.7 can be written.

$$Av = A_v v_s = nAv_s \qquad (13.7)$$

Where:

v_s = seepage velocity (m/s)

Equation 13.8 and Equation 13.9 result from Equation 13.7.

$$v = nv_s \qquad (13.8)$$

or

$$v_s = \frac{v}{n} \qquad (13.9)$$

The seepage velocity is greater than the approach velocity. The seepage velocity determines when a particle will arrive on the other end of the specimen. Note that in actuality, a molecule of water must travel an irregular path through porous media around individual particles and has an actual velocity greater than either the seepage velocity or approach velocity. A dimensionless factor called tortuousity (T) is used to quantify the actual path around solid particles compared to the distance between two particles. Tortuousity is given by Equation 13.10:

$$T = \frac{l'}{l}$$

(13.10)

Where:

T = tortuousity (dimensionless)
l' = the length of the path around soil grains followed by a molecule of fluid (m)
l = the length of a specimen (m)

The term "hydraulic conductivity" has also been referred to as "permeability" in the recent past. Hydraulic conductivity (with units of velocity) is the preferred terminology in this book when referring to the relationship between flow volumes and gradients within porous media. The term "permeability" is reserved for discussions involving intrinsic properties of porous media. Intrinsic (or absolute) permeability (K) is a property of porous media that is independent of fluid, and is in units of area. Equation 13.11 relates the hydraulic conductivity to absolute permeability.

$$K = \frac{k\mu}{\rho_f g}$$

(13.11)

Where:

K = intrinsic permeability (m^2)
g = acceleration due to gravity (m/s^2)

There are many uses for the property of hydraulic conductivity, including the design of filters, drains, and hazardous waste liners; analysis of flow under dams; and consolidation and seepage problems. Hydraulic conductivity is one of the most extensively used engineering parameters with the widest range of values (10^2 to 10^{-11} cm/s) and the largest uncertainty (on the order of a factor of two to five). Hydraulic conductivity can be analyzed with theoretical relationships, estimated through empirical relationships, or determined experimentally.

Theoretical Relationship

The geometry of flow through porous media is extremely complicated, since the flow paths consist of highly variable-sized openings. The flow channels must be simplified to geometries that lend themselves to rigorous analysis. Historically, this has been done by simulating flow within pipes.

Flow through pipes has been analyzed with a number of theoretical relationships. The most common are Poiseuille's Law and the modification to this relationship presented as the Kozeny-Carman equation. These relationships result in a general equation that is insightful but not very practical. One of many forms of the Kozeny-Carman relationship follows as Equation 13.12 for fully saturated materials.

$$k = \frac{\rho_f g}{\mu C_o (G_s \rho_w)^2 SS^2 T^2} \cdot \frac{n^3}{(1-n)^2}$$

(13.12)

Where:

ρ_f = mass density of fluid (Mg/m^3)
g = acceleration due to gravity (m/s^2)
n = porosity (dimensionless)
C_o = shape factor (dimensionless)
G_s = specific gravity of soil (dimensionless)
SS = specific surface of solids (m^2/g)

From the above equation, it can be seen that hydraulic conductivity is theoretically a function of many variables. The most important factors are:

- *Fluid*: The effect of fluid on hydraulic conductivity is easy to account for by determining changes in viscosity with temperature. The typical variation of the fluid properties of the most common permeant (water) lead to changes of about 10 percent in the value of k.
- *Porosity, n:* The porosity (or void ratio) accounts for the amount of area available for flow. Porosity typically ranges from 0.25 to 0.75. Given the operations performed with the parameter in Equation 13.12, this range can impact the hydraulic conductivity by two orders of magnitude.
- *Shape and grading of particles, SS*: The specific surface of solids ranges from $10 \text{ cm}^2/\text{g}$ to $100 \text{ m}^2/\text{g}$. Since this property is squared, the effect of the change in specific surface of solids is ten orders of magnitude. This captures the behavior due to particle size.
- *Tortuosity, T:* Soil fabric affects the flow path and is quantified by the tortuosity. The fabric is difficult to quantify in practice with a significant level of certainty. Tortuosity ranges from 1 to 100, leading to an impact of four orders of magnitude on the interpretation of hydraulic conductivity.

Some of these parameters can be controlled effectively to provide reproducible, limiting values. However, the difficulty lies in estimating the parameters to model the field conditions appropriately. In engineering practice, theoretical relationships are rarely used to estimate hydraulic conductivity because some of these terms are too difficult to quantify with sufficient accuracy. Hydraulic conductivity is more reliably estimated using simple empirical relationships or with experimental methods.

Empirical Relationship

Many correlations have been developed to estimate the hydraulic conductivity from grain size characteristics. This is justified by the fact that the specific surface is the dominant variable in Equation 13.12. Although there are several more complicated relationships based on multiple fractions of the grain size distribution of material, one equation in particular, Hazen's equation, has experienced widespread use in practice.

Hazen's Equation

Hazen's equation was developed by his experimental research with clean sands. Hazen's equation is given by Equation 13.13:

$$k = C_1 D_{10}^2 \qquad (13.13)$$

Where:
k = hydraulic conductivity (cm/s)
D_{10} = particle diameter corresponding to 10 percent finer by dry mass on the grain size distribution curve (cm)
C_1 = Hazen's factor ($\text{cm}^{-1}\text{sec}^{-1}$)

Hazen proposed that a value of about 100 be used for C_1. (Terzaghi and Peck, 1948).

As can be seen from the theoretical Equation 13.12 in the previous section, the factor that has the most impact on hydraulic conductivity is the square of the specific surface of the material, which is directly dependent upon the diameter of the contained particles. Thus, it makes sense that Hazen's equation is based on this factor to the same power. Of course, the equation does not provide for any changes in density, tortuousity, or other factors that impact the hydraulic conductivity. However, these limitations are well known and taken into consideration when using the simple relationship.

Experimental Methods

Numerous measurement techniques exist for determining hydraulic conductivity. The method most applicable to a problem depends on the soil type and purpose. Some of those techniques are applied in the field, and are beyond the scope of this textbook. The laboratory methods are variations on fluid boundary conditions. The most common methods involving flowing water through the material include constant head, falling head, constant flow, constant volume-falling head, and constant volume-constant head. Two other methods are the determination of hydraulic conductivity during the oedometer consolidation test and during the CRS consolidation test. Two of the laboratory methods, the constant head test and the falling head test, are most applicable for coarse-grained soils and those procedures are described in this chapter. Methods for fine-grained soils are addressed in ASTM D5084 Hydraulic Conductivity of Saturated Porous Materials Using a Flexible Wall Permeameter.

An important practical detail is the selection of the container for the specimen, termed the "permeameter." The two basic types of containment are the rigid-walled mold and the flexible-walled mold. The rigid mold is much simpler to use but does not allow for expansion or contraction of the specimen during the test. Coarse-grained soils will not change volume during the test and are most commonly tested using a rigid-mold permeameter. Flexible-mold permeameters are normally used with fine-grained soils. They require explicit control of the stress state.

The diameter of the mold must be bigger than the largest particle in the specimen by a certain multiple. ASTM D2434 requires that the diameter of the mold must be at least eight to twelve times the diameter of the largest particle. The use of smaller diameter molds will limit the flow paths available.

Either the head drop across a specimen can be measured by the use of manometers installed on the permeameter, or the head can be determined by the difference in elevation head between the headwater and the tailwater. In the latter method, head losses within the equipment must be accounted for by calibration.

The hydraulic conductivity is typically reported at 20°C. To correct the experimentally determined hydraulic conductivity at test temperature to the value at 20°C, the relationship in Equation 13.14 is used:

$$k_{20°C} = k_T \cdot \frac{\mu_T}{\mu_{20°C}} \qquad (13.14)$$

Where:

$k_{20°C}$ = hydraulic conductivity at 20°C (m/s)
k_T = hydraulic conductivity at test temperature (m/s)

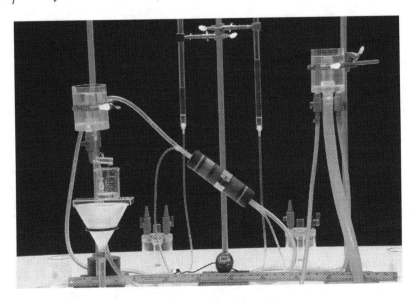

Figure 13.2 Typical equipment setup to measure hydraulic conductivity of coarse-grained materials using a rigid wall cell.

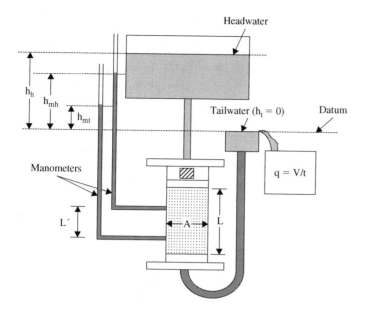

Headwater

Tailwater ($h_t = 0$)   Datum

Manometers

$q = V/t$

L' A   L

Figure 13.3 Schematic constant head hydraulic conductivity test setup including sideport manometers for downward flow through the specimen.

$k_{20^\circ C}$ = absolute viscosity of permeant at 20°C (Pa-s)
μ_T = absolute viscosity of permeant at test temperature (Pa-s)

The equipment for testing granular soils for hydraulic conductivity is fairly simple and versatile. A specimen is prepared within a rigid-wall hydraulic conductivity cell. The aspect ratio, defined as the length of the specimen divided by its diameter, can be altered to control the flow rate and head loss across the specimen. This technique is used with high-hydraulic-conductivity materials. A typical setup for measuring hydraulic conductivity in a rigid-wall cell is shown in Figure 13.2.

The soil can be compacted in almost any state desired. Generally, the laboratory conditions should approximate field conditions. The specimen can be saturated using a back pressure; however, this adds considerable complication to the experimental setup and is not normally done when testing coarse-grained soils. A very large range in gradient can be applied. Springs are used on top of the specimen to prevent fluidization and to maintain the initial specimen density. ASTM Test Method D2434 Permeability of Granular Soils (Constant Head) recommends using a spring or other device capable of applying a force of 22 to 45 newtons (N) (5 to 10 pounds force (lbf)). However, the upward pressure of flow during the test must be known so it can also be resisted by the spring. Manometers can be used to measure head loss across a specimen or portion of specimen, as shown in Figure 13.3.

The few limitations are that undisturbed specimens can not be set up with the device, and there is also no way to control the stress state or prevent volume change. Volume change is controlled by limiting the percent passing the 75 km (No. 200) sieve to less than 10 percent to prevent the fines washing through the sample.

Constant Head Flow Conditions

The constant head test for granular soils is covered by ASTM D2434. By rewriting Darcy's Law into parameters determined during the test, the relationship for hydraulic conductivity becomes Equation 13.15:

$$k = \frac{\Delta V}{\Delta t} * \frac{L}{A} \left\{ \frac{1}{h_h - h_t - h_e} \right\} \qquad (13.15)$$

Where:
ΔV = quantity of permeant passed through the specimen in a measured time interval (m^3)
Δt = interval in time (s)

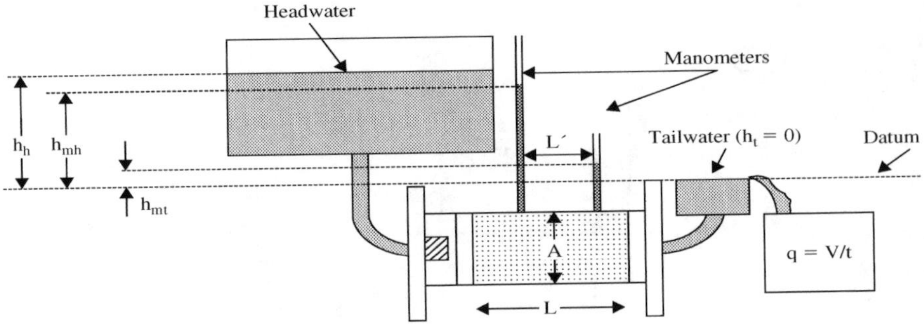

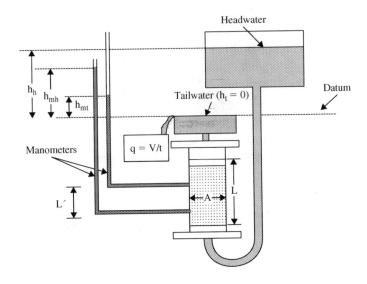

Figure 13.4 Schematic constant head hydraulic conductivity test setup for horizontal flow through the specimen.

Figure 13.5 Schematic constant head hydraulic conductivity test setup for upward flow through the specimen.

L = length of specimen (m)
A = cross sectional area of specimen (m^2)
h_h = total head at headwater (m)
h_t = total head at tailwater (m)
h_e = total head loss in equipment as a function of flow rate (m)

Head loss in the equipment is a function of flow rate, and must be determined during equipment calibration if not using manometers. When manometers are used, the head loss in the equipment is zero, and the portion of the denominator in brackets becomes the total head at the headwater side manometer, h_{mh}, minus the total head at the tailwater side manometer, h_{mt}.

Flow can occur either upward, downward, or horizontally. The direction of flow is controlled by the way in which the equipment is set up. Upward flow helps displace air out of the soil. Downward flow eliminates fluidization. Horizontal flow appears to be an attractive method to avoid these two previous issues, but is prone to developing channels along the top of the specimen as particles rearrange. Figure 13.3 shows the setup for downward flow, while Figures 13.4 and 13.5 show horizontal flow and upward flow, respectively. In all these figures, the datum is shown at the elevation of the tailwater. For the constant head test, the datum can be arbitrarily selected and it is often most convenient to use the top of the bench as the reference point.

During a measurement, the total head difference $(h_h - h_t)$ is kept constant. ASTM D2434 suggests a gradient of 0.2 to 0.3 for loose samples and a slightly higher gradient (0.3 to 0.5) for dense samples. The low end of the range applies to coarser soils, while the high end of the range applies to finer soils. Typically, the equipment is dimensioned for the anticipated value of hydraulic conductivity, while also limiting head loss in the equipment

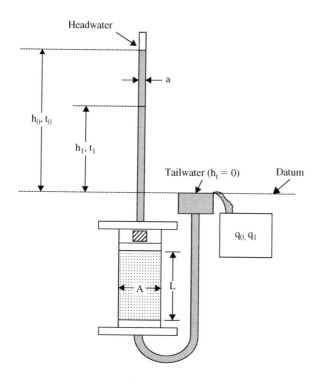

Figure 13.6 Schematic falling head hydraulic conductivity test setup.

to less than 10 percent. Measurements should be collected at 3 to 5 different gradients. Certain testing problems will be masked if the trend in gradients is correlated with time so the gradients should be applied in a random order.

The equipment configuration lends itself to simple quality control checks. This can be done by varying the applied gradient. If the flow is not a constant proportion to the applied gradient, it suggests that turbulence is occurring in the specimen and therefore the laminar flow criteria are not met. If the value of hydraulic conductivity decreases over time, independent of the applied gradient, it suggests that segregation of fines or desaturation is occurring within the specimen.

ASTM does not have a standard test method covering the falling head test within a rigid mold. General procedures are as follows. The applied head decreases over time during the test due to the falling water level on the inflow side. The rate of decreasing head is controlled by the diameter of the inflow tube (i.e., a smaller diameter inflow tube is going to cause the water level to fall more rapidly for a soil with given hydraulic conductivity). A schematic of the falling head hydraulic conductivity equipment setup is shown in Figure 13.6. For the falling head test, the datum location is not arbitrary and must be at the tailwater elevation, as shown in this figure.

Analysis of the falling head condition is also based on Darcy's law. The flow rate through one section of the equipment (such as the inflow tube) must equal the flow rate through other sections of the equipment (such as through the specimen). The governing equations are shown as Equation 13.16 through Equation 13.20:

Falling Head Boundary Conditions

$$q = \frac{dV}{dt} \tag{13.16}$$

$$q_1 = a\frac{dh}{dt} \tag{13.17}$$

$$q_2 = kA\frac{h}{L} \tag{13.18}$$

$$q_1 = q_2 \rightarrow \frac{-dh}{h} = \frac{kA}{La}\,dt \tag{13.19}$$

$$k = \frac{\ln\left(\frac{h_0}{h_1}\right)aL}{A(t_1 - t_0)} \tag{13.20}$$

Where:

q_1, q_2 = flow rate in the inflow tube and in the specimen, respectively (m^3/s)
a = cross-sectional area of the inflow tube (m^2)
L = length of specimen (m)
h_0, h_1 = head levels measured in the flow tube (m)
t_0, t_1 = time when water is at head levels of h_0 and h_1, respectively (s)

The falling head hydraulic conductivity has numerous positive aspects. The test is simple to run, and provides a very fast method of determining hydraulic conductivity. The method is appropriate for all ranges of hydraulic conductivity, and is especially well suited for testing specimens with medium to low values. The gradient that can be applied is limited by the physical height of the ceiling.

There are a few negative points with the test. There is no steady state flow in the system, and no measures of quality control. In addition, capillary rise in the tube must be considered. The temptation is to reduce the cross-sectional area of the inflow tube and therefore test materials with low values of hydraulic conductivity. However, capillary rise in small diameter tubes can cause large errors.

Other important considerations relative to hydraulic conductivity measurements of coarse-grained materials include:

- Segregation of fines during flow (and therefore loss of solids) can occur when the soil material is not well-graded (uniformity). ASTM D2434 limits the amount fines to 10 percent for materials testable with the constant head method.
- The interior diameter of the permeameter must be greater than 8 to 12 times the diameter of the largest soil particles and larger for uniform soils. Otherwise, adverse surface effects and limited flow paths occur. Flow paths along rigid walls will be large and important.
- Soil fabric has a significant impact on the hydraulic conductivity and soil placement should mimic field conditions as closely as practical. Fabric is particularly important for compacted materials.
- The pressure head must always be positive in the soil. To accomplish this, make sure the tailwater level is above the specimen at all times.
- The fluid must be equilibrated to standard conditions and the properties with temperature known. Allow water drawn from the tap to stand for at least 12 hours prior to using in a hydraulic conductivity test. Always use equilibrated water if using water as the permeant. Using water containing air will cause air bubbles to be entrapped within the soil matrix, decreasing saturation over time, and therefore a decrease in hydraulic conductivity will be observed.
- Saturation level can have a significant effect on the measured hydraulic conductivity. In general, a soil with lower saturation levels will have a lower hydraulic conductivity as the air pockets impede flow. Quantification of the effect of saturation on hydraulic conductivity is still in the research phase. Mitchell and Soga (2005) account for the effect of saturation using a cubic term in the numerator of a relationship similar to Equation 13.12. If the saturation varied from 100 percent to 50 percent, the effect on hydraulic conductivity would be an order of magnitude.
- During specimen saturation, introduce equilibrated water under controlled conditions such that the permeant front gradually travels through the specimen from bottom to top so air is not trapped.

- There are many components of this test and the elapsed times of concern must be documented. There are two time scales of importance in the falling head test and three times in the constant head test. In both tests, the elapsed time since flow was initiated through the specimen must be tracked to evaluate whether there are issues with hydraulic conductivity over time, such as due to bacterial growth within the medium, desaturation, or loss of fines. The elapsed time during a falling head trial is tracked to make the necessary measurements to calculate the hydraulic conductivity. In the constant head test, the elapsed time at each gradient is tracked to ensure that steady-state flow conditions have been met. Finally, the time increment over which a quantity of water is collected must be recorded in order to calculate the hydraulic conductivity at that gradient.
- Darcy's Law must be valid (i.e., flow is laminar, no volume change during the test, and h is the total head drop across the specimen). Experimental verification of Darcy's Law consists of obtaining a linear relationship of gradient with hydraulic conductivity. Verification should be performed for materials with a high value of hydraulic conductivity.
- The head loss of the equipment can have significant effects on the test when the hydraulic conductivity of the specimen is greater than 1×10^{-6} m/s. The flow velocity in connecting tubes above this value is so high that significant head losses occur within the equipment.
- ASTM D2434 gives a range of 0.1 to 0.5 for applied gradient, depending on soil type. Field conditions can be much lower or higher than these ranges. Ideally, the gradient applied in the laboratory as well as the material density should match the field conditions. If the laboratory data must be extrapolated significantly further than the data measured in the laboratory, the laboratory measurements should be performed at the largest range of gradients possible to provide confidence in the extrapolation. Extrapolation does have its limits, however, since a high gradient could induce turbulent flow in gravel and other types of coarse-grained materials.
- Organic growth can occur in water, and will accumulate within the tested material, lowering the hydraulic conductivity. To avoid this, keep the hydraulic conductivity setup and circulated water out of the light if a test is to last several days.

Typical values of hydraulic conductivity are listed in Table 13.1.

TYPICAL VALUES

Equipment Requirements

General

1. Commercially available permeameter, with manometer ports
2. Two filter screens (monofilament nylon)
3. Two noncorroding metal filter screens openings of 0.425 mm (No. 40)
4. Flow distribution plates
5. Spring
6. Top and bottom caps with tubing connections
7. Water storage tank, 20 L (nominal) capacity
8. Two adjustable elevation constant head reservoirs
9. Circulation pump
10. Balance readable to 0.01 g
11. Calipers readable to 0.01 mm
12. Two stopwatches readable to 1 second

13. Equilibrated water, at least 12 hours old
14. Water collection bowl

Constant Head Test

1. Plumbing and valves
2. Manometers
3. Two beakers or similar effluent collectors
4. Meter stick

Falling Head Test

1. Inflow tube with known area. The inflow tube diameter should be selected so the experiment can be completed in about 15 minutes. Use this time estimate, an estimated hydraulic conductivity and a ratio of 3 for h_0/h_1 in Equation 13.20 to calculate this area.
2. Inflow reservoir for use during saturation and filling inflow tube
3. Mounted meter stick with millimeter resolution to read fluid elevation of inflow tube
4. Plumbing and valves

Table 13.1 Typical values of hydraulic conductivity.

Soil Type	Void Ratio	k (m/s)
Clean gravel*	N/A	10^{-2} to 1
Clean sand, clean sand and gravel mixtures*	N/A	10^{-5} to 10^{-2}
Very fine sands, silts, sand/silt/clay mixtures, glacial till*	N/A	10^{-9} to 10^{-5}
Silty Sand**	0.25 to 0.35	7×10^{-11} to 3×10^{-10}
Fort Peck Sand**	0.55 to 0.6	2×10^{-5} to 3×10^{-5}
Ottawa Sand**	0.6 to 0.65	6×10^{-5} to 7×10^{-5}
Union Falls Sand**	0.4 to 0.65	4×10^{-4} to 1×10^{-3}

*After Casagrande, 1938 (as in Holtz and Kovacs, 1981).
**After Lambe and Whitman, 1969.

CALIBRATION

Calibration is not necessary for a constant head test when using manometers to measure the head loss across a portion of the specimen. For all other situations, confirm that the head loss of the equipment is less than 1 percent of the head loss across the soil specimen. This is accomplished by following the procedures:

1. Set up the equipment with filters, but without any soil. Be sure filters are sealed against flow around edges of cylinder.
2. Adjust the head to achieve a flow rate within the range of expected rates for the test. Allow the flow to stabilize before proceeding.
3. At each flow rate, record the head difference and the flow rate.
4. Increment the head and repeat.
5. Use these values to establish a calibration curve between equipment head loss, h_e and flow rate. A schematic equipment head loss calibration curve is shown in Figure 13.7.

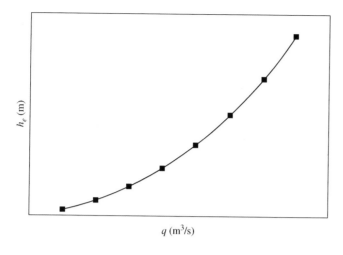

q (m^3/s)

Figure 13.7 Schematic equipment head loss calibration curve.

The specimen preparation guidance and the procedures assume that an initially dry, coarse-grained material with well-rounded grains, such as concrete sand or sand filter pack, is used for the test.

SPECIMEN PREPARATION

The test method will be performed in general accordance with ASTM D2434.

PROCEDURE

Apparatus Preparation

1. Remove and grease all O-rings.
2. Measure and record the inside diameter and mass of the mold, and the distance between manometer ports.
3. Cut (or locate) two pieces of filter fabric to cover top and bottom of specimen. The diameter should be such that the filter covers the soil but does not breech the O-ring seal.

Apparatus Assembly

1. Assemble permeameter with flow distribution plate and filter in base.
2. Deposit dry sand using one of the methods described in Chapter 4 or Chapter 11.
3. Level the top surface.
4. Measure the distance to the top soil surface.
5. Cover with a filter and a flow distribution plate.
6. Place the spring, and secure the top to the permeameter.
7. Connect the headwater, tailwater, and manometer tubes to the permeameter.

Saturate Specimen

1. Saturation and testing should be performed with old water (12 hours at standard conditions). Independent of the specimen orientation during conductivity tests, the flow should be upward for saturation.
2. Start test time. Set the headwater for a gradient of about 0.1 and slowly fill specimen from the base. The goal is to achieve a plug-flow-type water front that slowly displaces all the air in the soil voids. With many specimens this is impractical if not impossible to create. ASTM requires the saturation process to be performed under a partial vacuum. This will certainly improve the degree of saturation, but is an added complication for class instruction.
3. Record the starting time of the experiment. You will be using three different times during the test. The total time is the duration water has been flowing through the specimen. The gradient time is the duration of flow during one specific gradient.

The incremental time is the change in time during the collection of one water sample and is used for the hydraulic conductivity calculation.

4. Record event time. When water flows from the top drain tube, the specimen is ready to be tested. Be sure that the exit water levels are always above the specimen elevation. This keeps the water pressure inside the specimen above atmospheric pressure.

5. Set up a water collection system so the flow can continue for the total duration of the experiment. Remember that you need to measure the total volume of water flushed through the specimen. You can recycle the water.

Constant Head Hydraulic Conductivity Test

1. Perform hydraulic conductivity test using the constant head method at five or more gradients. Vary the gradients between 0.1 and 20.

2. Set the gradient and record the event time.

3. At each gradient adjust the head difference to the desired level and obtain sufficient quantities of the outflow water to make reasonable measurements. Use the mass difference method and assume the mass density of water is 1 g/cm^3. In order to perform calculations to 1 percent, at least 1 mL volume of water must be collected over a minimum of 100 seconds.

4. Record time interval over which water was collected. Record both the incremental time for the conductivity measurements and the total elapsed time of the experiment to evaluate long-term time effects.

5. Make several measurements (more than 6) at each gradient to check for time stability (keep track of flow history, meaning when gradients are changed and so on).

Falling Head Hydraulic Conductivity Test

1. Perform hydraulic conductivity test using the falling head method according to the following steps.

2. Two different times will be tracked during the falling head test. The total time is the duration water has been flowing through the specimen. The incremental time is the duration of flow during one individual falling head measurement.

3. Close the valve to the inflow reservoir and use the valve to the circulation pump to charge the inflow tube to the desired elevation. It is convenient to have a starting mark on the inflow tube. If this is the case, then charge the tube to several cm about the starting mark. Starting elevations should result in a gradient of about 10.

4. Close the valve and allow the water elevation in the tube to drop about 2 cm.

5. Start the timer as the interface reaches the starting mark.

6. Take readings of the water elevation as a function of time. Typical readings would be at 5, 10, 20, 40 sec, and so on.

7. Continue taking measurements until the gradient reduces to about 0.5.

8. Repeat the process to obtain at least two sets of measurements.

Disassemble Equipment

1. Stop flow.
2. Remove water lines.
3. Unscrew permeameter top.
4. Measure the distance to the top soil surface.
5. Remove all the soil and place it in a container for oven-drying.
6. Obtain final dry mass.

Calculations

Calculate the initial values of the specimen condition, including dry density and void ratio or porosity.

Calculate the gradient for each point using Equation 13.21 if not using manometers and Equation 13.22 if using manometers:

$$i = \frac{h_h - h_t - h_e}{L} \qquad (13.21)$$

$$i = \frac{h_{mh} - h_{mt}}{L'} \qquad (13.22)$$

Where:
i = applied gradient (dimensionless)
L' = length between manometers (m)
h_{mh} = total head at the headwater side manometer (m)
h_{mt} = total head at the tailwater side manometer (m)

Calculate the hydraulic conductivity at each gradient using Equation 13.23:

$$k = \frac{\Delta V}{\Delta t} * \frac{1}{iA} \qquad (13.23)$$

Correct the hydraulic conductivity to 20°C using Equation 13.14.

Using Equation 13.20, calculate the hydraulic conductivity at the various time increments during the trial. Calculate the corresponding average gradient during the time increment using Equation 13.24:

$$i = \frac{h_0 + h_1}{2L} \qquad (13.24)$$

Correct the hydraulic conductivity to 20°C using Equation 13.14.

Include the following in a report:

1. A summary table of flow history with columns clearly identified, initial dry density and void ratio of the specimen, and the average hydraulic conductivity for each method.

2. A graph of flow volume (or flow rate) versus time for each gradient. Define zero time as the time when the gradient is set for the increment.

3. A summary plot of the hydraulic conductivity versus gradient.

4. A summary plot of the hydraulic conductivity versus time.

5. A summary table of individual readings including date, time, volume (mass), Δt, k_T, $k_{20\,C}$, etc.

6. A calculation of the number of pore volumes passed through the specimen.

7. A graph of hydraulic conductivity versus gradient.

Criteria for judging the acceptability of test results obtained by this test method have not been determined by ASTM International. However, a typical user should reasonably expect results from other technicians in the same laboratory within a factor of 2 to 5.

DETECTING PROBLEMS WITH RESULTS

If the range of hydraulic conductivity results exceeds the estimates provided above, individual techniques should be evaluated and a second set of measurements should be collected to check repeatability. Erratic results without a time or gradient trend would suggest problems with experimental technique. Increase the quantity of water collected for each constant head calculation and the initial gradient in the falling head test.

Trends in the hydraulic conductivity results indicate a systematic problem. Decreasing hydraulic conductivity with increasing gradient would indicate nonlaminar flow conditions. Check that the trend is independent of time and then make measurements at lower gradients. Continued difficulty would suggest a calibration problem for the equipment head losses. Increasing hydraulic conductivity with increasing gradient is very unlikely. If the trend is independent of time, then it would indicate loosening of the specimen, side wall separation with flow, or exceeding the spring capacity. Check experimental setup and test at lower gradients. Hydraulic conductivity decreasing with time suggests systematic desaturation. Lower the specimen relative to the tailwater reservoir to increase the pressure head. This should cause an increase in hydraulic conductivity. Other possibilities are plugging of the filters and organic growth. If the equipment is nonreactive, bleach can be used to control organics. If hydraulic conductivity increases with time, the degree of saturation may be increasing or fines may be flushing from the specimen. Increase the pressure head to check for saturation effects. If loss of fines is suspected, they should be detectable in the tailwater reservoir.

REFERENCE PROCEDURES

ASTM D2434 Permeability of Granular Soils (Constant Head).

REFERENCES

Refer to this textbook's ancillary web site, www.wiley.com/college/germaine, for data sheets, spreadsheets, and example data sets.

Casagrande, A. 1938. "Notes on Soil Mechanics, First Semester," Harvard University, Cambridge, MA (unpublished).

Holtz, R. D., and W. D. Kovacs. 1981. *An Introduction to Geotechnical Engineering*, Prentice-Hall, Englewood Cliffs, NJ.

Lambe, T. W., and R. V. Whitman, 1969. *Soil Mechanics*, John Wiley and Sons, New York.

Mitchell, J. K., and K. Soga. 2005. *Fundamentals of Soil Behavior*, John Wiley and Sons, Hoboken, NJ.

Chapter 14

Direct Shear

SCOPE AND SUMMARY

This chapter provides background information on the direct shear test and detailed procedures to perform a direct shear test on dry, coarse-grained soil. The effects of particle packing along with the cohesion and friction angle components of frictional resistance are explored. The direct shear test is typically a drained test. Techniques previously presented in this book are used for the preparation of sand at different relative densities.

In addition to direct shear testing, the concepts of stress-strain and Mohr's circle representations of results are introduced.

TYPICAL MATERIALS

The direct shear test can be performed on soil materials ranging from sands and gravels to clays, although the size of commercially available devices limit the maximum particle size that can be present in the tested material. The test is also performed on many processed materials such as glass, corn flakes, and coal. The direct shear test can be used to determine the interface frictional properties of soil with materials such as geosynthetics, or along rock fracture surfaces.

BACKGROUND

The measurement of shearing properties is fundamental to engineering practice. A shear test causes distortion of the specimen while measuring the resulting force and deformation to determine the response of the material. In addition to the resistance to the imposed shear distortion, soils generally have a tendency to change in volume during shearing. For this reason, shear tests can be either drained or undrained, depending on the imposed boundary conditions. A drained test must be performed slowly enough to allow water to flow from the pore space and maintain constant pore pressure. The amount of volume change is measured in a drained test. An undrained test is performed in a manner that prevents volume change and the change in effective stress is measured as the "volumetric" response to the material. Undrained tests can be performed more quickly than drained tests unless pore pressures require time to equilibrate. The drained and the undrained tests provide the two limits that bound the scale of partially drained conditions that may occur in the field.

Soil response to shear is dependant on the rate of distortion, the initial stress conditions, and the direction of loading. Variations in response due to changes in the rate of distortion will occur in any soils test. Rate sensitivity often requires explicit evaluation when applying laboratory measurements to field expectations. Variation in material response due to the stress state or the direction of loading is termed "anisotropy." Soils generally exhibit significant anisotropy. Anisotropy can be caused by the stress state (induced) or the soil fabric (inherent). This reality has fostered the development of many different shearing devices. Each type of device simulates a particular aspect of engineering interest. None of these devices are perfect, and it is important to understand the advantages and limitations of each. The profession does not have a generalized testing apparatus capable of simulating all aspects of shear behavior. It is safe to predict that such a device will never exist. It becomes necessary to match an engineering application to the device that simulates the most important aspect of the problem, and then use soil behavior concepts to apply the measurements.

Shear tests are designed to measure engineering parameters. Unlike routine index tests, more scrutiny is applied to engineering tests both from the perspective of apparatus details as well as evaluation of the test results. Every engineering test should have a companion set of index tests to properly characterize the material and to allow generalization for the measurements. Remember, a project will have many more index tests to broadly understand the site and relatively few engineering tests to obtain specific parameters. Integration of all this information makes for a more cost effective design.

This chapter addresses the condition of applying a force in opposite directions to generate relative displacement along a predefined failure plane of the specimen, called direct shear. The direct shear test is often credited as being the oldest method to evaluate the strength of soils. The direct shear device is certainly one of the simpler devices used by the profession. The test is applicable to measuring the effective stress strength envelope of a wide range of materials under drained conditions. The test provides an indication of volumetric behavior. It does not provide a measure of the stress-strain relationship or the modulus. ASTM D3080 the Direct Shear Test of Soils Under Consolidated Drained Conditions provides the standard test method of the translatory direct shear test.

The test is performed slowly enough to allow complete drainage of excess pore pressure with the specimen loaded by a constant normal force. The results for typical direct shear tests on dense and loose specimens of a cohesionless soil are presented as the shear force, S, and the normal displacement, δ_N, both versus shear displacement, δ_H, in Figure 14.1. The maximum shear force points are marked as peak and the conditions at large deformation are marked as residual. Depending on the material density and normal stress level, the peak may be well defined or it may coincide with the residual condition. In all cases, the normal displacement should be constant at the residual condition.

Multiple tests must be performed on companion specimens in order to evaluate the strength envelope. The specimens are loaded to a range of stress levels that bracket the expected range in the field. Figure 14.2 presents a typical result derived from three

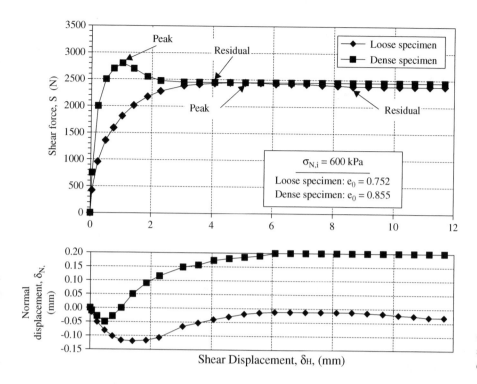

Figure 14.1 Typical force-deformation results from a direct shear test on loose and dense cohesionless soil specimens.

tests. Two envelopes are normally provided: one corresponding to the peak condition and one for the residual condition. The slope of the envelope is the effective friction angle (ϕ') and the intercept is the effective cohesion (c'). All specimens in a series must be prepared by the same method for proper interpretation. The density at the time of shearing will vary with stress level due to the material being compressed by the applied normal force.

When presenting results in any stress space, it is important that the scale of the x and y axes are dimensionally the same. For example, the physical distance between two points of stress on the x-axis must be the same as the physical distance between an increment of the same value on the y-axis. Otherwise, the scale of the plot will be distorted, thereby causing an incorrect presentation of a circle and the wrong interpretation of the friction angle.

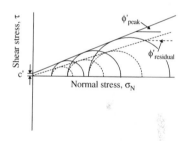

Figure 14.2 Results from a direct shear test, presented as shear stress versus normal stress.

There are many device variations used to perform the direct shear test. These variations all have common features specific to the direct shear configuration. The translatory shear box equipment will be used in the following discussion to illustrate the essential concepts of the direct shear experiment. The direct shear test is very similar in concept to the "block on plane" physics experiment used to measure interface friction properties.

In the soils experiment, the specimen starts out as a continuum of soil particles and the deformation must evolve into the large deformation geometry that mimics the sliding block experiment. The idealized deformed specimen is shown in Figure 14.3. The top half of the specimen is displaced relative to the bottom. As shown, the displacement causes a reduction in the contact area of the two halves of the specimen. This is a significant and important source of uncertainty in the test, and is one of the reasons that the stress-strain results should be reported as force rather than stress. While the contact area may decrease, the effective area is the same for both the normal and shear force.

Considerable effort has been devoted to investigating the deformation pattern inside the specimen. A shear zone develops rather than a single plane. This shear zone is larger in the center of the specimen and tapers toward the ends, as indicated by the shaded area

Evaluation of the State of Strain and Stress

Figure 14.3 Idealized deformed direct shear specimen.

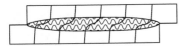

Figure 14.4 Exaggerated deformed direct shear specimen showing larger shear zone in the center of the specimen than at the edges.

in Figure 14.4. It is clear that the shear state is very non-uniform, and that most of the material is not sheared during the test. These observations support the practice to report shear deformation rather than shear strain and to report normal displacement rather than convert this to volumetric strain.

The stress state is one of the most important features of any shear test. In fact, devices are designed to impose specific stress states. The stress state probes particular aspects of the soil anisotropy. For this reason, it is necessary to look carefully at the stress conditions in the direct shear test. As shown in Figure 14.5b, the specimen is contained on all sides by rigid boundary conditions. The normal force is applied to the top cap, resulting in a known normal stress. The sides are sufficiently rigid as to prevent lateral deformation. This constant area condition (or no lateral strain) is referred to as the geostatic condition and is characterized by the stress ratio, K_0. K_0 is the ratio of the lateral stress to the normal stress in this case. The K_0 value depends on the properties of a particular soil. A reasonable estimate for discussion is 0.5. Figure 14.5a provides the well-defined Mohr's circle for the preshear stress state. Satisfying the condition of this Mohr's circle requires that there is no shear on the sides of the specimen. Although this is not strictly true, the specimens are thin enough that the approximation is sufficient given the other limitations of the test.

Application of the shear deformation (or force) increases the stress level and causes a rotation of the principal stress direction. At the start of shearing, the specimen will experience progressive distortion starting from the line of separation of the box halves, which is a location of stress concentration in the apparatus. The shear plane (zone) expands from the two sides and finally develops into a fully mobilized plastic zone. Unfortunately, the state of stress is not sufficiently defined during shear. The normal stress and the shear stress are known on the shear surface but three independent pieces of information are required to draw the Mohr's circle. Assuming that the measured shear stress is the shear stress at failure on the failure surface (τ_{ff}) provides the last piece of required information, and allows construction of the circle and location of the origin of planes, O_p. For this assumption to be valid, sufficient movement must occur along the sliding plane to result in a fully plastic condition. This is reasonable for the residual condition, but may not be strictly correct for the peak strength. Based on the stress conditions portrayed in Figure 14.6a, it is possible to determine the orientation of the major principle stress. Figure 14.6 shows the interpreted stress state (solid black dot) during the shearing process.

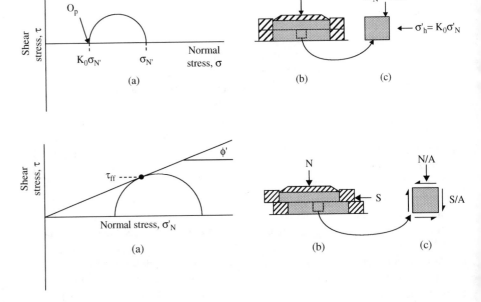

Figure 14.5 State of stress in a direct shear specimen after consolidation, but prior to shear.

Figure 14.6 State of stress in a direct shear specimen at failure.

Figure 14.7 Photograph of typical translatory shear box equipment.

The mechanical behavior of particulate materials is controlled by effective stress, σ'. Effective stress is defined as the difference between the total stress, σ, and the pore pressure, u, as shown in Equation 14.1:

$$\sigma' = \sigma - u$$

(14.1)

Where

σ' = effective stress (kPa)
σ = total stress (kPa)
u = pore pressure (kPa)

The effective stress equation is applicable to fully saturated or completely dry particulate materials, provided the area of contact between grains is insignificant.

The prime notation is used to denote effective stress, and is applicable to any of the normal stress components. While normal stress can either be a total stress or an effective stress, shear stress is generated by total stress. It is inappropriate to denote a shear stress as an effective stress. Total stresses will always act immediately on a specimen. Effective stress requires time for the pore pressure to dissipate. The rate of dissipation is a function of the stiffness and hydraulic conductivity of the material. In the case of fine-grained materials, dissipation of pore pressure can take hours, days, or even longer in a test specimen. One of the major challenges when testing fine-grained materials is to evaluate the effective stress state inside the specimen.

The direct shear test is commonly performed on fine-grained soils. The pore pressure is not measured in the direct shear test. It is necessary to use rate of consolidation measurements in order to properly limit the rate of shearing such that pore pressures are insignificant. Under these conditions, the effective stress will be equal to the total stress.

The translatory direct shear device is by far the most commonly used in practice, and is standardized in ASTM D3080. The translatory direct shear devices are square or circular in plan view. The split rigid box fits into a water bath and the assembly fits into a support frame. Weights are used to apply the normal force and a motor-driven gearbox applies the shear deformation. A typical translatory device is shown in Figure 14.7.

A typical cross section is shown in Figure 14.8. The diameter (or width) to height ratio must be greater than 2, the height must be at least 6 times the maximum particle size, and the diameter must be at least 10 times the maximum particle size. The rigid box is constructed of two equal thickness sections. The box is designed to keep the two halves stationary during preparation and consolidation of the specimen, and then create

Translatory Direct Shear Devices

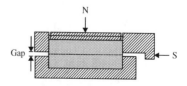

Figure 14.8 Translatory square or circular direct shear device.

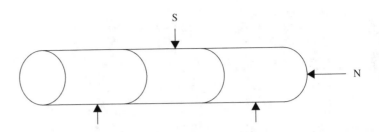

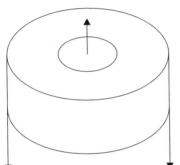

Figure 14.9 Translatory ring direct shear device.

Figure 14.10 Translatory annular direct shear device.

a gap between the two halves in preparation for shearing. One section translates with the respect to the other during shearing. The loading yoke is designed such that the shear force acts along the line of the shear surface.

Other types of translatory devices have been designed to improve loading symmetry. These include the translatory ring direct shear device, which causes a double failure plane, shown schematically in Figure 14.9, and the annular direct shear devices, shown schematically in Figure 14.10.

The translatory devices are well suited to test either intact or laboratory fabricated specimens. This is because the specimen geometry is easy to trim. The state of stress prior to shear replicates the K_0 condition, which properly captures many field situations. The devices are also designed to measure density changes during consolidation, as well as the time rate of deformation. The time rates are only applicable to consolidation in the apparatus, and should not be used to compute the coefficient of consolidation because drainage to the box separation may be occurring.

There are several shortcomings of translatory devices. The design of the device causes stress concentrations at the gap, which promote progressive failure. For these reasons, the peak friction angle is not reliable. As deformations become large, the area of contact between the two halves of the specimen reduces. Working in terms of force rather than stress adequately eliminates uncertainty in the frictional calculation. The area reduction does cause concern that the normal stress is increasing during shear, leading to unwanted densification in the shear zone. The top plate often tilts at larger deformation, accentuating the fact that the stress state is less than ideal and causes additional non-uniformities in the stress state.

When residual values are of primary interest, the large deformation data must be looked at more carefully. The shear deformation is generally limited to about half the height of the specimen, and this is often too little deformation to develop a stable, residual condition. Performing multiple shear cycles has been used in an attempt to achieve better particle alignment in clays. The reversals tend to allow the particles to retract toward their initial position, and thus this method is not very successful at reaching residual conditions.

Torsional Direct Shear Devices

Torsional direct shear devices have been developed specifically to measure the large deformation residual condition. This goal is achieved with a design that allows unlimited deformation between the two halves of the specimen, without causing a reduction in the contact area.

The solid torsional direct shear device consists of a disc of soil. In section view, the device looks similar to a circular translatory device. The specimen geometry is regular, so the device can be used on both intact and fabricated specimens. The specimen can be consolidated and rates of deformation collected to select the appropriate shearing rate. The specimen is sheared by rotating the top half of the box relative to the bottom. Refer to Figure 14.11 for a schematic drawing of this test set up and plan view showing the shearing action.

During shear, the stress state in the solid torsional direct shear device rotates and is unknown. This is the same situation as for the translatory device. The non-uniformities are significantly worsened by the rotational mode of loading. While Figure 14.11 gives the

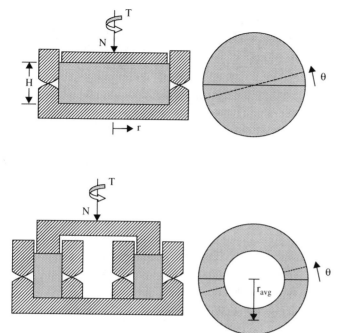

Figure 14.11 Solid torsional device.

Figure 14.12 Hollow cylinder torsional device.

impression that the shear deformation, θ, is uniform, the shear strain of interest (ignoring the shear surface for the moment) is actually the radius times the rotation divided by the specimen height ($\gamma = r\theta/H$).

Shearing is initiated on the outside perimeter and progresses toward the center of the device. For this reason, the solid rotational direct shear device should not be used to measure peak conditions. Once the shear surface is fully developed, the displacement rate will vary from zero at the centerline to a maximum at the perimeter. Unlike the translatory device, the ratio of the boundary measures of torque to normal force does not provide sufficient information to compute the friction angle. While the stress non-uniformity is less serious than at peak conditions, it is necessary to assume a radial distribution of stress to compute an average shear stress. This average can then be used with the average normal stress to calculate the residual friction angle.

The hollow cylinder torsional direct shear device provides unlimited shear deformation, while at the same time reducing the stress non-uniformity across the shear surface. The device is shown schematically in Figure 14.12. The hollow cylinder devices are normally used with laboratory-fabricated specimens since an intact specimen would need to be rather large, and the geometry is difficult to trim. In addition, the device is intended to measure the residual conditions, so there is no need to start with the intact structure. The specimen is consolidated as with the other direct shear devices, but the side shear is much more important due to having both an inside and an outside box wall.

During shear, the stress state increases from the K_0 condition and rotates as with the other direct shear devices. The relatively thin annular geometry greatly reduces the difference between the minimum and maximum shear deformation, but the gradient across the surface is still substantial. Once again, the shear failure initiates from the outside boundary and progresses inward. The device should not be used to evaluate the peak condition. Once the shear surface is fully developed, the displacement rate variation across the surface varies by the ratio of the inner radius to the outer radius of the specimen. The design greatly reduces the variation and makes the calculation of the average shear stress more reliable.

Both the solid and the hollow torsional devices experience problems with extrusion of material through the gap. The issue is more severe in the hollow torsional devices because both the inner and outer soil specimen surfaces have access to a gap.

Overview of Test Procedures

Depending on the device configuration, the test specimen can be trimmed from an intact sample or reconstituted from bulk material. The specimen is contained laterally in a two-piece, rigid container that is split at the mid-level of the specimen. The area (A) is constant during the consolidation process. When the two halves of the container are clamped together, the stress state is geostatic. The top and bottom of the specimen are contained by frictional, rigid platens. When testing coarse-grained materials, plates with grooves several grain diameters deep are used to promote transfer of the shear stress. Porous stones are generally used when testing fine-grained materials. The initial density of the specimen is computed from the soil mass and the dimensions of the box. The grooves in the plates are assumed to be completely filled with soil.

The shear box is placed in a container so the specimen can be submerged in water during the test, if desired. Submersion is not necessary for coarse-grained soils unless the test will be performed at very low stress levels. Under such conditions, surface charges and hygroscopic tension forces can become important, and submersion will eliminate these forces.

A normal force (N) is applied to the rigid top cap with weights on a hanger in order to consolidate the specimen. The two halves are still connected during consolidation. The consolidation stress is the applied force, including the top cap and top half of the shear box if not suspended, divided by the container area.

The normal force can be applied in increments or in one step. Coarse-grained materials will compress immediately. Fine-grained materials will require time to consolidate and consideration of the appropriate stress increments. If the material will remain overconsolidated throughout the consolidation process, then only one increment will be necessary. On the other hand, the load increment ratio must be limited to one whenever the material will be consolidated into the normally consolidated range. This limit is required to prevent excessive soil extrusion from around the top cap.

Consolidation rate data are required when testing fine-grained materials. These data are used to select the appropriate deformation rate to achieve drained conditions during the shearing phase of the test. The rate of deformation must be measured during consolidation of the last loading increment. These data should not be used to compute the coefficient of consolidation, rather only to select the appropriate shear rate. The rate of deformation determined is applicable to the boundary conditions within the test, and the impact of the drainage boundary at the gap is uncertain. The time rate of deformation can be analyzed using any of the interpretation methods presented in Chapter 17 in order to determine the time to 50 percent consolidation, t_{50}.

After the completion of consolidation, the two halves of the container must be separated for the shear phase of the test. When testing at low stress levels (whenever the applied normal force is less the 100 times the weight of the top half of the container), the top half of the container must be counterbalanced. Unclamp the two halves of the container and create a gap (0.65 mm or about one grain diameter) using the separation screws. Be sure to retract the screws after creating the gap.

The shearing rate must be slow enough to allow complete dissipation of the shear induced pore pressure. Allowing time for drainage is not difficult for sands, but can be for clays since they have a much smaller hydraulic conductivity. The measurements of normal displacement during consolidation are used to estimate the minimum time to failure. This minimum time to failure, t_f, is taken as 50 times t_{50}. It is also necessary to estimate the amount of shear deformation that will be required to cause failure. The failure deformation will depend on the type of material, stress level, and overconsolidation ratio. A reasonable starting estimate would be between 5 and 10 mm. Experience with a particular material will provide a better estimate. The estimates of failure time and displacement are used to set the displacement rate on the loading frame.

The shear displacement is applied and the reaction force and change in normal displacement are measured versus time while holding the normal force constant. The specimen is sheared until the shear force peaks then decreases, or until the limit of the equipment displacement has been reached.

The process of shearing the soil will rearrange particle contacts and cause volume change. This may be contraction or dilation, depending on the density and stress level. The rate of recording data will depend on the shear rate. It is reasonable to collect on the order of 500 data points throughout the test. The rate should be fast at the beginning of shear, and then can be slowed as the test progresses.

Important considerations for the direct shear test include:

- The density of the specimen is uniform at the end of consolidation, but the normal deformation measured during shear is only an indicator of behavior. The volume change occurs on a relatively thin shear layer within the specimen, making the boundary displacement measurements approximate.
- For coarse-grained materials, the drained behavior depends upon the initial density. Small details of the material change the resulting density. At the loosest state, the stress strain curve will monotonically approach the residual value. For dense coarse-grained materials, the stress will peak then lessen, and will also dilate during the test. Generally, specimens are tested at different normal loads, but at the same density.
- The peak condition of a load-deformation curve exhibiting a well-defined peak condition followed by significant softening should be considered approximate. The combined effects of progressive failure and rotation of the principle stress axes will make the peak force measurement imprecise. The peak condition should not be used when measured in a rotational device.
- Due to the large deformations imposed on the specimen, the residual friction angle should generally be reliable. The data from translational devices may not reach the residual condition, however, and must be checked.
- The direct shear test works well for interpretation of soil properties on predefined slide surfaces and artificial interfaces.
- Performing the test at a sufficiently slow shear rate is essential. Undissipated shear induced pore pressures in the shear zone will change the measured shear force. This error is especially severe when testing overconsolidated, fine-grained materials when the shear induced pore pressure will be negative, resulting in a strengthening of the shear zone.
- The mass of the floating portions of the specimen container (this is the top half of the shear box for most devices) must be less than 1 percent of the applied normal force used to consolidate the specimen. Since the force required to maintain the gap of the specimen box is transferred to the soil through side shear, an unknown normal force is applied and leads to uncertainty in the measured friction angle. Reducing the potential normal force associated with the upper half of the shear box effectively addresses this concern. When using older equipment, this requirement will limit the consolidation stress range unless the upper half of the shear box is counterbalanced. One easy, relatively low-cost solution is to drill two holes in the top half of the box and attach a threaded rod and a crosspiece. A hook in the center of the crosspiece is used to connect a spring to an adjustable support. The support is connected to a ring stand clamped to the loading frame. The support position is adjusted (without soil) until the top half of the box is suspended above the bottom half. See Figure 14.13 for a photograph of this particular setup.
- Despite many shortcomings, the direct shear test is relatively simple and inexpensive, and provides useful results for engineering practice. The test is intended to measure the friction angle of particulate systems or the shear resistance of the interface at a particular normal stress. It is also very effective for illustrating the basic drained behavior of coarse-grained materials.

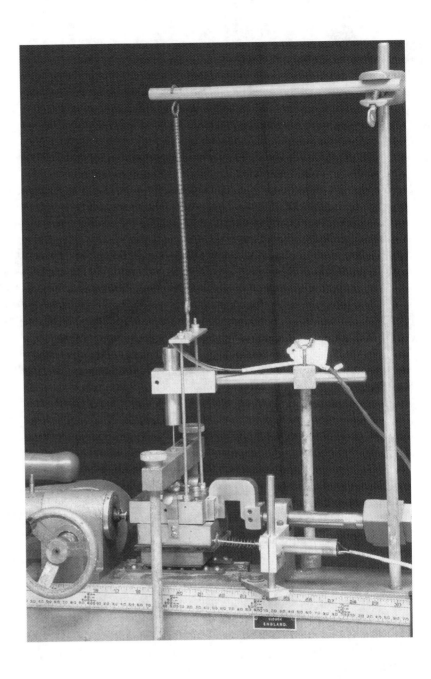

Figure 14.13 Example method of counterbalancing the direct shear equipment.

Typical results from the direct shear test on cohesionless soils are listed in Table 14.1.

CALIBRATION

Apparatus Compressibility

The device must be calibrated to account for apparatus compressibility due to the normal loads applied during the test if precise preshear densities are required.

1. Assemble the device with a "dummy specimen" made of a relatively incompressible material (such as steel) of similar dimensions to the specimen to be tested.

2. Place the "dummy specimen" in the device and set up the top cap and normal displacement transducer as will be done for the test.

3. Position the displacement transducer such that both compression and uplift can be recorded, then record the zero displacement reading.

4. Apply the normal load in increments up to the capacity of the equipment, then back down to zero while recording load and deformation for each increment.

5. Create a plot of the normal load versus deformation, called the compressibility curve.

6. Use the resulting apparatus compressibility curve to correct the calculation of normal displacements during testing.

Table 14.1 Typical results from the direct shear test on cohesionless soils.

Soil Type	Peak Friction Angle (°)	Soil Type	Peak Friction Angle (°)
Dense, well-graded, coarse sand*	37 to 60	Dense, rounded, well-graded sand**	40
Dense, uniform, fine sand*	33 to 45	Loose, angular, uniform sand**	35
Loose, rounded, uniform sand**	30	Dense, angular, uniform sand**	43
Dense, rounded, uniform sand**	37	Loose, angular, well-graded sand**	39
Loose, rounded, well-graded sand**	34	Dense, angular, well-graded sand**	45

*Lambe, 1951.
**Sowers and Sowers, 1951.

Equipment Requirements

1. Shear device (with counterbalance for low stress testing)
2. Split shear specimen container with grooved insert plates and top cap
3. Normal loading device. The normal force can be applied with calibrated dead weights or with a pneumatic loading device. If using a pneumatic loading device, the system must be capable of applying the load to within 1 percent of the desired value, but cannot exceed it.
4. Shearing device, with a motor, capable of imposing a uniform rate of displacement with less than +/− 5 percent deviation. The rate of displacement is dependent upon the material. A coarse-grained material is being used in this experiment, so a rate of 1 mm per minute is adequate.
5. Load cell to measure shear force, readable to 0.5 percent of the normal force. For this experiment, a load cell with a capacity of about 2,200 N (500 lbf) readable to 0.5 N (0.I lbf) should be adequate.
6. Normal and horizontal deformation transducers such as LVDTs, readable to 0.001 mm for normal deformation and 0.01 mm for horizontal deformation
7. Scale readable to 0.01 g with a capacity of at least 200 g
8. Calipers readable to 0.01 mm
9. Data acquisition system compatible with the three transducers.
10. Equipment for preparing specimens such as a tamping foot with dimension of approximately one-fourth of the shear box dimensions, container for pouring sand into the shear box, funnel with attached tube, and the like

Volume of Shear Apparatus

1. Assemble the box with the bottom spacers, two grooved insert plates, and top cap. Place the top grooved plate such that the ribs are touching but perpendicular to the bottom grooved plate.

2. Determine the initial depth of the specimen container to the four indicator points on the top cap, d_i.

3. Determine the area of the specimen container, A_b.

4. Measure the height (H_r), width (W_g), length (L_r), and number (N_g) of gaps between the ribs in the grooved insert plates.

5. Estimate the volume for the sand between the ribs in each of the grooved insert plates, V_g, using Equation 14.2.

$$V_g = \left(H_r \times W_g \times L_r \right) \times N_g \qquad (14.2)$$

Where

V_g = volume for the sand between the ribs of each grooved insert plate (cm^3)
H_r = height of the ribs in the grooved insert plate (cm)
W_g = width of the gaps in the grooved insert plate (cm)
L_r = length of the gaps in the grooved insert plate (cm)
N_g = number of the gaps

Apparatus Preparation

1. Locate all the transducers and record the calibration factors and data acquisition channel numbers.

2. Check the electronics including the input voltage, stability of signals, and direction of the output signals.

3. Obtain the mass of the top plate, top cap, and steel ball (M_c) to 0.01 g.

4. If using a counterbalance, adjust the support spring so the upper half of the box is suspended just above the lower half of the box, or obtain the mass of the top half of the box.

SPECIMEN PREPARATION

The laboratory assignment in this chapter will help develop experience with the device, the data acquisition system, and the mechanical characteristics of a coarse-grained material. For instructional purposes, the direct shear test is best performed at two different densities at three different stress levels each. Collectively the experiment will consist of performing six tests on dry sand.

As stated before, the stress strain curve and peak resistance of sand are a strong function of density and the applied normal (consolidation) stress. In this laboratory, specimens will be prepared at a dense and loose state. The dense sand will be rodded into place to achieve a very dense state without sophisticated equipment. The loose sand will be deposited with a funnel to allow the particles to settle at the angle of repose. Tests will be performed at normal loads of about 65, 160, and 250 N. Remember to account for the mass of the apparatus (top cap, ball, and sometimes the top half of the shear box) when computing the normal force applied to the specimen.

1. Assemble the box with the bottom spacers and one ribbed plate.

2. Place the dry sand in a pouring container and determine the initial mass of the sand and container, $M_{c,i}$.

3. Fill the box with about 40 g of sand.

   a. Use the funnel method to prepare a loose specimen. Remember the loose specimens are very delicate and slight vibrations will cause the sand to densify.

   b. For the dense specimen, pour the sand into the box and densify by tamping the surface of the sand to create a relatively dense specimen.

4. Determine the final mass of the sand and pouring container, $M_{c,f}$. Subtract the final mass from the initial mass to determine the specimen mass, M_d.

5. Level the sand surface and cover with the top ribbed plate and top cap.

6. Gently press the plate into the sand.

7. Measure the depth to the top of the top plate, d_s.

PROCEDURE

The direct shear test will be performed in general accordance with ASTM D3080. These procedures are specifically written for a Wykeham Farrance direct shear machine. The concepts will be the same for other devices, but specific details may need modification.

Apparatus Setup

1. Assemble the shear box in the frame.

2. Turn the hand crank until the box is in contact with the shear piston.

3. Place the ball and hanger on the top cap.

4. Align the equipment and set the normal displacement transducer in place.

5. Obtain zeros for the horizontal force and two displacement transducers (horizontal and normal deformation of specimen).

Apply Normal Stress to the Specimen

1. Compute the required normal force. Remember to account for the mass of the top cap, top plate, steel ball, hanger, and top half of shear box.

2. Apply the required masses (M_w) to the hanger.

3. Connect the counterbalance mechanism.

4. Measure the average depth of the four flats, d_f, on the top cap to compute the preshear height of the specimen.

5. Set up a voltmeter or other data recording device and record the normal deformation "zero" after the masses have been applied to the specimen.

6. Remove the screws that hold the two halves of the box together.

7. Simultaneously turn the two spacer screws to separate the box along the shear plane. Use one half turn and remember to return the screws to the original position before shearing.

Shear the Specimen

1. Since the test specimen is a dry sand, set the shear rate using the gears and lever position to the fastest rate.

2. Start the data acquisition system and take readings every second.

3. Turn on the motor and engage the clutch.

4. Shear the specimen for the full stroke of the load frame (about 1.25 cm).

5. At the end of the test, observe and record the condition of the top cap. Look specifically for rotation.

6. Pour the sand into a bowl and determine the final dry soil mass for confirmation of the setup calculations.

Calculations

1. Calculate the volume of the soil (V_t) using Equation 14.3:

$$V_t = \left(d_i - d_s\right) \times A_b + 2 \times V_g \qquad (14.3)$$

Where

V_t = total volume of the specimen (cm^3)
d_i = initial depth to the top of the top cap with the specimen container empty (cm)
d_s = depth to the top of the top cap with the soil specimen in place (cm)
A_b = area of the specimen container (cm^2)

2. Calculate the initial dry density of the specimen (r_d) using Equation 14.4:

$$\rho_d = \frac{M_d}{V_t}$$

(14.4)

Where

ρ_d = dry density of the specimen (g/cm^3)
M_d = dry mass of the specimen (g)

3. Calculate the preshear height of the specimen (H_{ps}) using Equation 14.5:

$$H_{ps} = \left(d_i - d_f\right) + H_r$$

(14.5)

Where

H_{ps} = preshear height of the specimen (cm)
H_r = height of the ribs (cm)

4. Verify that the maximum particle size in the specimen does not violate the criteria relative to specimen height and diameter.

5. Calculate the normal force (N) using Equation 14.6:

$$N = (M_c + M_w)g$$

(14.6)

Where

N = normal force applied to the specimen (N)
M_c = mass of top plate, top cap, steel ball, and hanger (kg) Note that this assumes a counterbalance is used. If not, add in the mass of the top half of the shear box.
g = acceleration due to gravity (m/s^2)

M_w = mass of the weights added to the hanger (kg)

6. Calculate the applied shear force at reading m (S_m) using Equation 14.7:

$$S_m = \left[\frac{V_{l,m}}{V_{in,l,m}} - \frac{V_{l,0}}{V_{in,l,0}}\right] \times CF_l$$

(14.7)

Where

S_m = shear force applied to the specimen at reading m (N)
$V_{l,m}$ = output voltage at reading m on the load cell (V)
$V_{l,0}$ = output voltage at zero reading on the load cell (V)
$V_{in,l,m}$ = input voltage to the load cell at reading m (V)
$V_{in,l,0}$ = input voltage to the load cell at zero reading (V)
CF_l = calibration factor for load cell (N/(V/Vin))

7. Calculate the horizontal (shear) displacement at reading m ($\delta_{H,m}$) using Equation 14.8:

$$\delta_{H,m} = \left[\frac{V_{hd,n}}{V_{in,hd,n}} - \frac{V_{hd,0}}{V_{in,hd,0}}\right] \times CF_{hd}$$

(14.8)

Where

$\delta_{H,m}$ = horizontal displacement at reading m (mm)

$V_{hd,m}$ = output voltage at reading m on the horizontal displacement transducer (V)
$V_{hd,0}$ = output voltage at zero reading on the horizontal displacement transducer (V)
$V_{in,hd,m}$ = input voltage to the horizontal displacement transducer at reading m (V)
$V_{in,hd,0}$ = input voltage to the horizontal displacement transducer at zero reading (V)
CF_{hd} = calibration factor for horizontal displacement transducer (mm/V/Vin)

8. Calculate the vertical (normal) displacement at reading m ($\delta_{N,m}$) using Equation 14.9:

$$\delta_{N,m} = \left[\frac{V_{nd,m}}{V_{in,nd,m}} - \frac{V_{nd,0}}{V_{in,nd,0}} \right] \times CF_{nd} \qquad (14.9)$$

Where

$\delta_{N,m}$ = normal displacement at reading m (mm)
$V_{nd,m}$ = output voltage at reading m on the normal displacement transducer (V)
$V_{nd,0}$ = output voltage at zero reading on the normal displacement transducer (V)
$V_{in,nd,m}$ = input voltage to the normal displacement transducer at reading m (V)
$V_{in,nd,0}$ = input voltage to the normal displacement transducer at zero reading (V)
CF_{nd} = calibration factor for normal displacement transducer (mm/V/Vin)

9. Calculate the dilation rate (the rate at which the sand expands or contracts) (μ_m) using Equation 14.10. Note that if readings are too close together, the scatter in the plot can be reduced by increasing the index spacing.

$$\mu_m = \frac{\Delta \delta_N}{\Delta \delta_H} = \frac{(\delta_{N,m+1} - \delta_{N,m-1})}{(\delta_{H,m+1} - \delta_{H,m-1})} \qquad (14.10)$$

Where

μ_m = dilation rate at reading m (decimal)
$\Delta \delta_N$ = change in normal displacement per increment (mm)
$\Delta \delta_H$ = change in horizontal displacement per increment (mm)
$\delta_{N,m+1}$ = normal displacement at reading m+1 (mm)
$\delta_{N,m-1}$ = normal displacement at reading m−1 (mm)
$\delta_{H,m+1}$ = horizontal displacement at reading m+1 (mm)
$\delta_{H,m-1}$ = horizontal displacement at reading m−1 (mm)

10. Create a plot of shear force versus horizontal displacement and normal displacement versus horizontal displacement. Determine the peak shear force, S_p, and the residual shear force, S_r.

11. Calculate the peak friction angle (ϕ_p) using Equation 14.11:

$$\phi_p = \arctan\left(\frac{S_p}{N}\right) \qquad (14.11)$$

Where

ϕ_p = peak friction angle (degrees)
S_p = peak shear force applied to the specimen (N)

12. Calculate the residual friction angle (ϕ_r) using Equation 14.12:

$$\phi_r = \arctan\left(\frac{S_r}{N}\right) \qquad (14.12)$$

Where

ϕ_r = residual friction angle (degrees)

S_r = residual shear force applied to the specimen (N)

13. Calculate the peak shear stress on the specimen, τ_p, using Equation 14.13:

$$\tau_p = \frac{S_p}{A_b} \qquad (14.13)$$

Note: Use the residual shear force to determine the residual shear stress on the specimen at failure.

14. Calculate the normal stress on the specimen, σ'_N, using Equation 14.14:

$$\sigma'_N = \frac{N}{A_b} \qquad (14.14)$$

15. Calculate the rate of deformation at reading m, $\dot{\delta}_{h,m}$, using Equation 14.15:

$$\dot{\delta}_{h,m} = \frac{(\delta_{H,m+1} - \delta_{H,m-1})}{(t_{m+1} - t_{m-1})} \qquad (14.15)$$

Where

$\dot{\delta}_{h,m}$, = rate of deformation at reading m (mm/min)
t_{m+1} = elapsed time at reading $m+1$ (min)
t_{m-1} = elapsed time at reading $m-1$ (min)

Report

1. Calculations for the phase relationships for each specimen.
2. Calculations for the peak force and corresponding displacement for each test.
3. Plot of shear force vs. horizontal displacement for each test.
4. Plot of normal displacement vs. horizontal displacement for each test.
5. Plot of dilation rate vs. horizontal displacement for each test.
6. Plot of shear stress vs. normal stress with friction angle for peak and residual conditions for each test.
7. A summary table including the initial dry density, the preshear void ratio, the peak and residual friction angle, and the maximum dilation rate for each test.

PRECISION

Criteria for judging the acceptability of test results obtained by this test method have not been determined through ASTM.

DETECTING PROBLEMS WITH RESULTS

When the Mohr's circles do not fall on a line for a series of tests, the initial density of specimens may be different. This could be due to sloppy handling of the specimen or poor specimen preparation procedures.

When testing at low stresses, it is crucial to counterbalance the top of the shear box. Not doing so results in large frictional forces compared to the applied normal load, and a higher interpreted friction angle.

If the force versus displacement curve is erratic, there may be large grains in the specimen impacting results, and the maximum particle size criteria have likely not been met. Low values of peak shear stress may be due to rotation of the top cap. Check the testing remarks for notation of a rotated cap. A negative cohesion intercept is impossible and therefore is a clear indicator of a calibration or calculation problem.

When testing clays, shearing so fast that pore pressure is not allowed to dissipate leads to a multitude of testing errors. In normally consolidated clays, the interpreted friction angle will be too low. For overconsolidated clays, the interpreted friction angle will be too high. The resulting cohesion intercepts could also be severely impacted.

ASTM D3080 Direct Shear Test of Soils Under Consolidated Drained Conditions

REFERENCE PROCEDURES

Refer to this textbook's ancillary web site, www.wiley.com/college/germaine, for data sheets, spreadsheets, and example data sets.

Lambe, T. W. 1951. *Soil Testing for Engineers*, John Wiley and Sons, New York.

Sowers, G. B. and B. F. Sowers. 1951. *Introductory Soil Mechanics and Foundations,* Macmillan, New York.

REFERENCES

Chapter 15

Strength Index of Cohesive Materials

SCOPE AND SUMMARY

This chapter provides background and detailed procedures to perform a hand-held shear vane, fall cone, miniature laboratory vane, pocket penetrometer, and unconfined compression test on a cohesive material. Performing each of the strength index tests on material that is judged to be similar allows comparison between various methods of testing. The pocket penetrometer, laboratory vane, fall cone, and handheld shear vane are all performed on the soil before it is extruded from the tube section. The specimen for the unconfined compression test is performed on soil after it has been extruded from the tube, and possibly trimmed.

TYPICAL MATERIALS

These strength index tests are performed on intact cohesive, fine-grained materials ranging in consistency from very soft to hard.

BACKGROUND

Strength index tests are performed on soil as received from the field, whereas Atterberg Limits are performed on remolded soil after altering the soil to specific states, such as between fluid and semisolid (LL) or solid and semisolid (PL). Strength index tests provide little control over boundary conditions, but provide an estimate of strength and are quick and fairly easy to perform. As such, these strength index tests should be used to assist in identifying and classifying soil, as well as an aid in determining the appropriate testing locations for engineering tests. The field tests, consisting of the handheld shear vane and pocket penetrometer tests, are extremely useful during sampling activities as a tool to identify potential problems leading to sampling disturbance.

Fall Cone

The fall cone was introduced in Chapter 9 as one of the methods to determine the liquid limit and plastic limit. For this specific application the geometry of the cone was selected to provide a particular penetration when the material was at the liquid limit. This was not a arbitrary process but rather guided by a theoretical analysis of cone penetration. Hansbo (1957) derived a relationship between the geometry of a cone and the undrained strength of an isotropic material. Based on this relationship, the fall cone can be configured to estimate undrained strength (Hansbo, 1957) in the laboratory for a range of consistencies. Refer to Figure 15.1 for a photograph of a typical fall cone device.

Figure 15.1 A fall cone device.

Table 15.1 Theoretical cone factors for smooth-faced and rough-faced fall cones.

Cone Angle, β (degrees)	Cone Factor, K_c	
	Smooth	Rough
30	2.00	1.03
45	0.84	0.50
60	0.40	0.25
75	0.22	0.15
90	0.12	0.09

Source: Adapted from Koumoto and Houlsby, 2001.

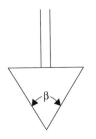

Figure 15.2 The cone angle (β) of a fall cone device.

To estimate strength, the fall cone is placed on the surface of intact soil, the cone is released for 5 seconds, and the penetration depth recorded. The presence of coarse particles adversely affects the results. The penetration depth is correlated to the strength based on cone geometry and cone mass, and the roughness of the cone surface. The cone geometry is characterized by the cone angle (β), as shown in Figure 15.2.

Equation 15.1 is used to determine the undrained strength (s_u) from the fall cone test.

$$ s_u = K_c \frac{M_c g}{d^2} \tag{15.1} $$

Where:

s_u = undrained strength (Pa)
K_c = cone factor (dimensionless)
M_c = mass of the fall cone (kg)
g = acceleration due to gravity (m/s^2)
d = penetration depth (m)

The cone factor is primarily dependent on the cone angle and surface texture. A table of cone factors for smooth and rough cones is provided as Table 15.1. This represents the extreme range in laboratory possibilities. It is most likely that the laboratory cone will be closer to the smooth condition but not perfectly smooth.

The undrained shear strength determined with the fall cone has been compared to the shear strength obtained using the miniature laboratory vane and remolded samples (Koumoto and Houlsby, 2001). The comparison is most favorable for the higher cone angles. In principle, the remolded material should have isotropic strength properties and one would expect comparable results for the two devices. However, anisotropy is an important strength characteristic of intact soils that can create appreciable differences between results obtained with the two devices when testing intact material.

Pocket Penetrometer

The pocket penetrometer is a useful and basic field test to determine the unconfined compressive strength of soil. The device is mentioned in Head (1980), but the authors could not locate a standardized procedure for the test. The device consists of a spring-loaded post. There are two different sizes of feet. Refer to Figure 15.3 for a photograph of a typical pocket penetrometer.

The method works by pressing the foot into the prepared, flat surface at the end of a thin-walled tube sample. The foot is advanced in a smooth motion until the indicator line meets the soil surface. The maximum reading is retained by a slip ring. The penetrometer creates an undrained bearing capacity failure in the soil and the scale is calibrated to provide the compressive strength of the material (i.e. the diameter of a Mohr's circle).

The scale of the pocket penetrometer reads in pounds/ft^2 (psf), tons/ft^2 (tsf), or kg/cm^2 (ksc) and has a scale of 1,000 to 9,000 psf when using the end of the rod, which has a diameter of 6.35 mm (0.25 in.). When the larger adapter foot is used which has a diameter of 25.4 mm (1 in.), the results on the scale must be divided by 16. The

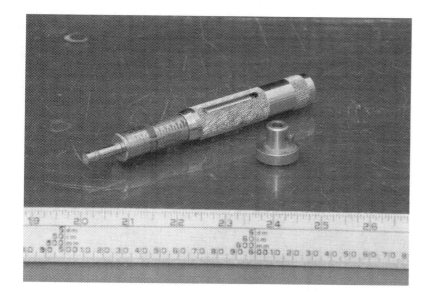

Figure 15.3 A pocket penetrometer.

pocket penetrometer is commonly used in the field to index intact soil samples as they are obtained, although the method can also be used in the laboratory. The presence of coarse particles adversely affects the results.

Handheld Shear Vane

The handheld shear vane, such as made by Torvane®, provides a simple and portable means of measuring undrained strength of saturated, fine-grained soils. The procedures for performing the handheld shear vane test are described in the Earth Manual (U.S. Bureau of Reclamation, 1998).

The device is comprised of a torque spring attached to a post. A shoe with blades is attached to the end of the post. Similar to the pocket penetrometer, the handheld shear vane has adapter shoes to measure strengths of varying ranges. Refer to Figure 15.4 for a photograph of a typical handheld shear vane.

To perform the handheld shear vane test, the device is pressed into a prepared, flat surface of soil, and rotated with a smooth rotation until the soil fails in shear. The failure surface is a disk defined by the diameter of the shoe and the depth of the blades. The maximum reading is retained by a pointer. Refer to Chapter 11 for a sequence of pictures depicting this type of measurement.

Figure 15.4 A handheld shear vane.

The handheld shear vane has three shoes, with each shoe measuring different ranges of strength. The smallest shoe has the least surface area and is used for the stiffer soils, while the standard and largest shoes have increasingly more surface area and are appropriate for softer soils. On a 76.2 mm (3 in.) diameter tube sample, about four handheld shear vane measurements can be made with the smallest shoe, three with the standard shoe, and only one with the largest shoe.

The handheld shear vane has a scale of 0 to 1 tsf for the standard sized shoe. The multiplier for the standard shoe is 1.0, while the multiplier for the small shoe and the large shoe is 2.5 and 0.2, respectively. While the larger shoe has a reduced capacity, it has increased resolution for use with very soft soils. Likewise, the smaller shoe has an increased capacity but lower resolution for use with stiff soils. The result is shear strength, which is the radius of a Mohr's circle.

This test is frequently used on the ends of thin-walled tubes as obtained in the field or as received in the laboratory. The presence of coarse particles adversely affects the results. The soil can also fail prematurely on silt or fine sand lenses. The handheld shear vane test must be performed at a location away from tube walls and other features that cause boundary effects.

Miniature Laboratory Vane

The miniature laboratory vane is a test that operates on the same principal as the handheld shear device, but allows for better control of the testing conditions. The procedures for performing a miniature laboratory vane test are standardized in ASTM D4648 Laboratory Miniature Vane Shear Test for Saturated Fine-Grained Clayey Soil. A similar test exists called the field vane test, which is performed within a borehole. The field vane test is described in ASTM D2573 Field Vane Shear Test in Cohesive soil.

The miniature laboratory vane is a simple apparatus comprised of a metal cross at 90°. This cross is connected to a rod that is attached at the other end to a torsional spring. The vane and the spring are connected to indicators that separately measure the angle of rotation of the vane and the spring rotation. The spring is easily replaced. Several springs are available to allow testing a range of soil strengths. Refer to Figure 15.5 for a photograph of a typical miniature laboratory vane.

To perform the miniature laboratory vane test, the vane is pressed into soil confined in a tube, ensuring that no torque is imparted on the soil. The vane spring is rotated at a rate of 1 to 1.5 degrees per second while recording the rotation position and the resistance of the spring until failure. The failure surface is a cylinder defined by the length and width of the blades. Like the handheld shear vane, the laboratory miniature vane result is the radius of a Mohr's circle.

The remolded strength can be measured by rotating the vane in the soil for a number of rotations, then performing a second strength test. Refer to Figure 15.6 for a sketch of a typical miniature laboratory vane and the face of the scale.

The miniature laboratory vane is only used in the laboratory and is typically performed on the ends of thin-walled tubes. The presence of coarse particles adversely affects the results. The soil can also fail prematurely on silt or fine sand lenses.

Unconfined Compressive Strength

The unconfined compressive strength can be determined on a cylinder of intact soil, as described in ASTM D2166 Unconfined Compressive Strength of Cohesive Soil. After removal from the sample tube, an intact soil specimen is placed in a compression loading frame, and the loading platen is advanced at a rate of 0.5 to 2 percent axial strain per minute. Refer to Figure 15.7 for a photograph of a typical setup for the unconfined compression test.

Readings of load, displacement, and elapsed time are recorded by hand or with a data acquisition system. The results are presented as a stress-strain curve. The maximum stress is calculated and recorded as the maximum unconfined compressive stress, which is equivalent to the diameter of Mohr's circle. The shear strength of the specimen is the radius of Mohr's circle and is calculated as one half of the unconfined compressive stress.

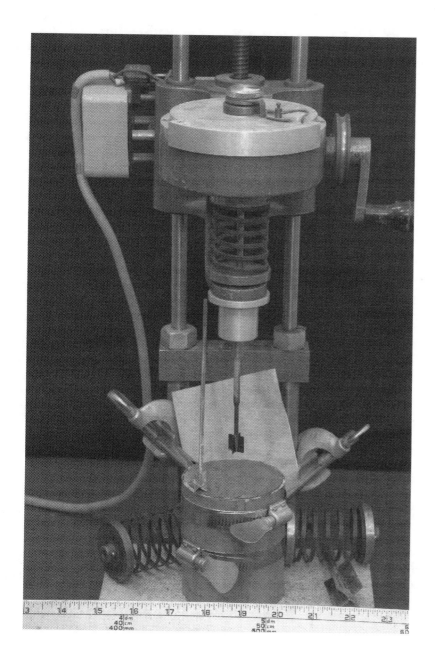

Figure 15.5 A miniature laboratory vane.

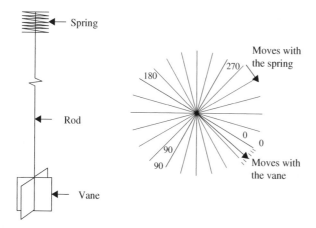

Spring

Rod

Vane

Moves with
the spring

Moves with
the vane

180

270

0

0

90

90

Figure 15.6 Schematic of a typical miniature laboratory vane.

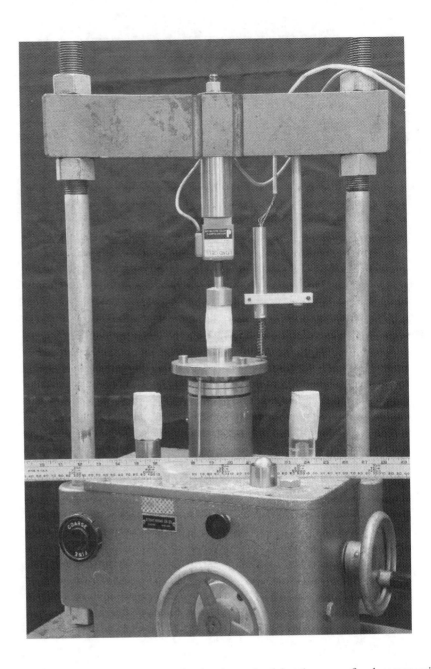

Figure 15.7 Typical setup for the unconfined compression test.

Remolded shear strength can also be determined by the unconfined compressive strength test. After measuring the intact strength, the structure of the specimen is completely destroyed while the same water content is maintained. This can be done by placing the intact material in a plastic bag and thoroughly reworking the soil without entrapping air. The material is fabricated into a specimen by some means that creates the same density as the intact specimen, usually by compaction into a mold. Ideally, the material used to determine the remolded strength is the specimen used to determine the intact strength by unconfined compression, provided the water content has been maintained.

Sensitivity

Intact materials often exhibit a reduction in shear stress after a peak value and with continued shear strain. This is called strain softening and is an important behavior when considering the implications for field performance. Once the peak strength has been surpassed, the material becomes softer with continued deformation, leading to progressive failure.

One quantitative measure of strain softening is sensitivity (S_t). Sensitivity is defined as the ratio of the unconfined compressive strength of an intact specimen (q_u) to the

Sensitivity (S_t)	Description*	Description
1	Insensitive	
1–2	Slightly sensitive	Low sensitivity
2–4	Medium sensitive	
4–8	Very sensitive	Sensitive
8–16	Slightly quick	
16–32	Medium quick	
32–64	Very quick	Quick
>64	Extra quick	

Table 15.2 Scales for sensitivity of soil.

Source: *Rosenqvist, 1953.

remolded unconfined compressive strength (q_r) at the same void ratio and water content, as expressed in Equation 15.2 (Terzaghi and Peck, 1967):

$$S_t = \frac{q_u}{q_r} \qquad (15.2)$$

Where:

S_t = sensitivity (dimensionless)
q_u = unconfined compressive strength of an intact specimen (kPa)
q_r = remolded unconfined compressive strength (kPa)

Sensitivity is commonly determined by a vane shear test, either in the field or in the lab. When using the vane shear test to determine this property, q_u and q_r are replaced by s_u and s_r. This revised relationship is expressed as Equation 15.3:

$$S_t = \frac{s_u}{s_r} \qquad (15.3)$$

Where:

s_u = undrained shear strength of an intact specimen (kPa)
s_r = remolded undrained shear strength (kPa)

The vane test has the advantage that the specimen does not need to be extruded from the tube or prepared, avoiding disturbance, especially in the cases of sensitive or quick soils. In addition, multiple miniature laboratory determinations can be made at the same depth in a sample tube. However, there will be numerical differences between the two measurements because the intact soil will have anisotropic strength and the remolded material should be isotropic. Numerous quantitative scales of sensitivity exist. Table 15.2 provides a scale that rates the sensitivity from insensitive to extra quick (Rosenqvist, 1953) along with a coarser scale that would more commonly be used in the United States where soils typically are not as sensitive.

Important considerations for determining strength index properties include:

- Cohesive soils are strain-rate sensitive. In general, strength index tests are performed very quickly and with very little control relative to rate consistency. Therefore, the rate of shearing must be considered when comparing various index measurements, comparing to measurements from more sophisticated strength tests, or using measurements in engineering calculations.
- Fall cone: The equation used to determine the undrained strength by the fall cone test has the depth of penetration term squared. Therefore, small errors

in the reading of penetration depth can cause a significant difference in the interpreted strength. The geometry and mass of the cone should be altered to produce a depth of penetration that is neither too large nor too small. A reasonable range is between 5 and 10 mm of penetration.

- Miniature Laboratory Vane: During insertion of the miniature laboratory vane into soil, a zone of disturbance is created around the vane blades. Excess pore pressure is also created by insertion and is not allowed to dissipate prior to the beginning of the test. These effects are built into the bias of the test results.

- Unconfined Compression Test: The data collected during an unconfined compression test should never be used to compute the Young's modulus of the material. The test does not have adequate control over the test conditions and the slope of the stress-strain curve will be unreliable.

- Strength anisotropy is a very important characteristic of intact cohesive soils. This directional dependence of the strength will cause differences between the various index tests. A perfectly remolded material should be isotropic and yield equal strength values for all the measurements, provided the rates are similar and the values properly converted to shear strength.

- Partial drainage will affect each index test to a different degree. In particular, comparing the hand shear vane and the pocket penetrometer can provide a useful measure of partial drainage. The strength of the penetrometer will be much higher as compared to the vane if drainage is important.

- The consistency descriptors of soil can be assigned based on the estimation of strength according to Table 15.3.

TYPICAL VALUES

Typical values from strength index tests are listed in Table 15.4.

CALIBRATION

The handheld shear vane will be calibrated initially by the manufacturer; however, checks should be performed periodically against another of the same device by interlocking blades and rotating, ensuring that the same reading results on both devices. Alternatively, a torque wrench could be used to initially measure the torque resulting

---

Equipment Requirements

1. Equipment necessary for determination of water contents: forced draft oven, desiccator, scale, water content tares, and so on
2. Scale with a capacity of at least 1 kg and readable to 0.01 g
3. Straight edge
4. Fall cone apparatus, preferably with a 60-degree polished stainless steel cone
5. Pocket penetrometer; include the larger, adapter foot when testing soft soils
6. Handheld shear vane with adapter shoes
7. Miniature laboratory vane with springs
8. Extruding equipment: hollow tube, bench vice, vice grips, piano wire, and so on (These procedures are covered in depth in Chapter 11.)
9. Deburring tool and metal file
10. Trimming supplies: split mold to trim the ends of the extruded specimen. If trimming the sides of the unconfined compression specimen, a miter box will also be needed.
11. Compression machine, with load cell transducer and displacement transducer
12. Tube cutting equipment: band saw and tube cutter with clamps

---

Table 15.3 Consistency of soil relative to strength.

Consistency of Clay	Undrained Shear Strength, kPa (TSF)	Unconfined Compressive Strength, kPa (TSF)
Very Soft	<12 (<0.125)	<24 (<0.25)
Soft	12–24 (0.125–0.25)	24–48 (0.25–0.5)
Medium	24–48 (0.25–0.5)	48–96 (0.5–1.0)
Stiff	48–96 (0.5–1.0)	96–192 (1.0–2.0)
Very Stiff	96–192 (1.0–2.0)	192–383 (2.0–4.0)
Hard	>192 (>2.0)	>383 (>4.0)

Source: Adapted from Terzaghi and Peck, 1948.

Table 15.4 Typical values from strength index testing.

Soil Type	Hand-held Shear Vane (kPa)	Laboratory Miniature Vane (kPa)			Unconfined Compressive Strength (kPa)		
		Intact	Remolded	S_t	Intact	Remolded	S_t
Maine Clay*	33 +/– 18						
Boston Blue Clay*	55 +/– 21						
San Francisco Bay Mud*	68 +/– 29						
Boston Blue Clay (OCR 1.3 to 3.6)**		36 to 108	13 to 31	3 to 4.9			
Boston Blue Clay (OCR 1 to 1.2)**		21 to 49	7 to 14	2.2 to 3.9			
Boston Blue Clay (OCR 1.3 to 3.6)***					26 to 93	4 to 19	2.8 to 10.8
Boston Blue Clay (OCR 1 to 1.2)***					34 to 52	7 to 10	5 to 6.8
Organic Silt***					29	7	4.3
Organic Silt****					17 to 20		

Note: A cohesive set of fall cone data and pocket penetrometer data could not be obtained.
*Unpublished laboratory data.
**After Ladd and Luscher, 1965.
***After Enkeboll, 1946, as appearing in Ladd and Luscher, 1965.
****After Lambe, 1962, as appearing in Ladd and Luscher, 1965.

at certain readings when the device is initially received, then periodically checking the device using the initial measurements as a basis. Check that the reading is zero when the indicator is returned to the stop position.

The pocket penetrometer has a direct readout of estimated compressive strength on the scale, which is the result of an empirical relationship with the spring constant in the device. Periodic checks of the calibration of the spring should be made using the information supplied by the manufacturer, which will include the spring constant and the corresponding load required to attain certain readings on the scale.

The fall cone device should be checked for wear at the point using the designated gage, and replaced when necessary. The mass of the fall cone and the sliding mechanism should also be checked for consistency with the value used in the calculations and to verify there is no friction impeding the sliding movement, respectively.

Formal calibration is required for the miniature laboratory vane and the transducers used in the unconfined compression test. Transducer calibration is covered in Chapter 11, "Background Information for Part II."

The torque springs for the miniature laboratory vane should be calibrated using a certified torque wrench or using a specialized torque applicator. The calibration factor (CF_T) is supplied by the manufacture but should be verified on a regular schedule. To do this, the dial and spring are attached such that the axis of rotation is parallel to the ground. The standard vane is replaced with a shaft fitted with a gravity loaded torque wheel. A series of masses are placed on the wheel hanger and the resulting rotation $(R_{c,n})$ of the spring is recorded. The torque applied by the mass is calculated using Equation 15.4:

$$T_{c,n} = m_n g l \tag{15.4}$$

Where:

$T_{c,n}$ = torque applied for reading n (N-m)
m_n = mass applied for reading n (kg)
g = acceleration due to gravity (m/s^2)
l = length of the lever arm (m), taken as the radius of the wheel

These values of applied torque are plotted against the spring rotation values, as shown in Figure 15.8. The slope of the line is the calibration factor for that spring.

The torque capacities of the springs that are typically provided with a miniature laboratory vane are listed in Table 15.4. The table also includes the resulting strength limits for each spring using a 1.27 cm (0.5 in) square vane.

SPECIMEN PREPARATION

For laboratory demonstration exercises, all of the tests should be performed in the same zone of the tube to allow comparison between the values. Since larger shoes must be used when performing pocket penetrometer and handheld shear vane testing on very soft soils, fewer of these tests can be performed at one depth within a tube. Figure 15.9 provides a suggested arrangement of the various index tests for a very soft and a stiff soil in order to make maximum use of the material and allow comparison of the various measurements.

The pocket penetrometer, laboratory vane, handheld shear vane, and fall cone are all performed on the soil before it is extruded from the tube section. The unconfined compression test requires a specimen with length-to-diameter ratio of 2 to 2.5. Extrude the soil from the tube using procedures described in Chapter 11. The section of tube must

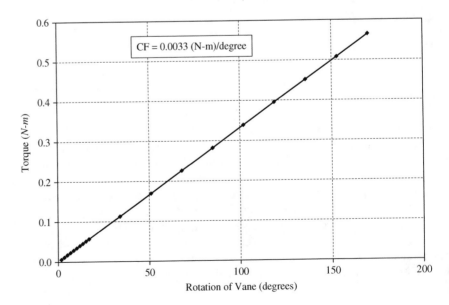

Figure 15.8 Example calibration curve for a miniature laboratory vane spring.

Spring No.	Capacity, (N-m)	Capacity, (in-lb)	Maximum Strength, kPa	Maximum Strength, TSF
1	0.194	1.72	48	0.5
2	0.390	3.45	96	1.0
3	0.599	5.30	144	1.5
4	0.813	7.20	192	2.0

Table 15.5 Typical spring capacities and maximum strengths for a 1.27 cm square vane provided with a miniature laboratory vane apparatus.

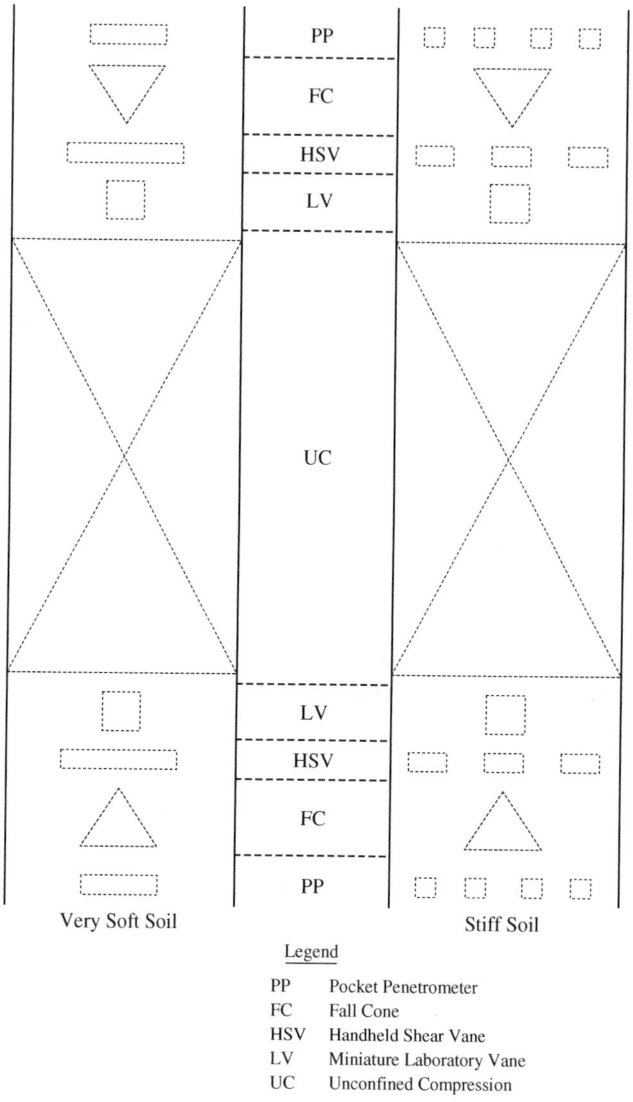

Note: The capacity is based on a rotation of 180 degrees.

Very Soft Soil

Stiff Soil

Legend

PP	Pocket Penetrometer
FC	Fall Cone
HSV	Handheld Shear Vane
LV	Miniature Laboratory Vane
UC	Unconfined Compression

Figure 15.9 Tube logs with example locations for strength index testing in order to use material efficiently; very soft soil (left); stiff soil (right).

have sufficient length to provide a specimen that meets the dimensional requirements untrimmed, or the specimen must be trimmed to the appropriate dimensions for testing.

PROCEDURE

The following procedure is provided for instructional purposes and to allow comparison between the various strength index measurements. The sequence should be performed on a section of the tube that is judged to contain similar material. The individual procedures are applicable to routine testing, but the tests would not normally be clustered as in this exercise.

Pocket Penetrometer

1. Use a straight edge to create a flat surface on each end of the tube.
2. Return the slip ring to zero.
3. Using one smooth motion, penetrate the shaft into the clay to the indication mark.
4. Record the strength value retained by the slip ring and the foot size (standard, large).
5. Repeat the measurement if possible at various locations on the surface.
6. Obtain the local water content by removing the material influenced by the pocket penetrometer test (e.g., to a depth of approximately one diameter below the pocket penetrometer depression).

Fall Cone

1. Use a straight edge to create a flat surface on each end of the tube.
2. Use the pocket penetrometers strength to estimate the required mass and cone angle to give approximately 5 to 10 mm penetration for the cone.
3. Record the cone angle and mass. Firmly constrain the tube section.
4. Set the cone on the surface of soil.
5. Set dial indicator and record zero depth.
6. Release cone for 5 seconds and record final penetration depth. If the penetration depth is less than 5 mm or greater than 10 mm, consider changing the cone geometry or mass and repeating the trial in order to obtain measurements with better resolution.
7. Obtain water content from the layer of material influenced by the cone.

Handheld Shear Vane

1. Use a straight edge to create a flat surface on each end of the tube.
2. Return the pointer to zero.
3. Penetrate the vane into the clay such that the base of the vane touches the soil.
4. Keep the vane straight and apply a small normal force to keep the handheld shear vane seated against the soil surface.
5. Using one smooth motion, rotate the vane until the soil fails. Use a rate of rotation that causes failure in about 5 to 10 seconds.
6. Record the shoe size and the strength value retained by the pointer.
7. Repeat the measurement if possible at various locations on the surface.
8. Obtain a water content by removing material to the base of the shear surface.

Miniature Laboratory Vane

1. Record the spring number and the calibration factor (CF_T) for the spring used on the miniature laboratory vane.
2. Use a straight edge to create a flat surface on each end of the tube.
3. Clamp the tube section to the apparatus with two ring clamps attached to the frame of the vane apparatus.
4. Adjust the needle indicator to the zero position for both the vane and the spring.
5. Push the vane into the soil until the top surface of the soil is at the indicator line on the vane shaft, which is 13 mm above the top of the vane blades.
6. Start rotating the vane and record the angle of rotation of the spring ($R_{T,n}$) and of the vane ($R_{R,n}$) in degrees versus time. The rate of spring rotation should be about 1 degree/second.
7. Watch the moment at which the torque applied to the soil through the spring is suddenly relieved.
8. Annotate the rotation of the spring and the vane, and the time from the beginning of the test. The spring needle indicator will remain at the maximum angle after the soil fails.
9. Manually rotate the vane 5 to 10 full rotations.

10. Adjust the vane position and indicator to read zero vane and spring.

11. Continue with the standard rotation rate until the spring reaches a constant value. This value will indicate the remolded strength of the soil. Record the constant value for the spring and the rotation in degrees.

12. Take a sample for water content determination by removing material within the sheared zone of the miniature laboratory test.

Unconfined Compression Test

1. Extrude the remaining portion of the sample from the tube using the procedure described in Chapter 11.

2. Trim a specimen from the sample using the procedure in Chapter 11. Ideally, it will only be necessary to trim the ends of the specimen.

3. Determine four water contents from trimmings, or if using an untrimmed specimen, from material above and below the specimen.

4. Obtain initial total mass (M_T) to 0.01 g.

5. Measure the initial height (H_0) and diameter (D_0) to 0.01 mm.

6. Place the cylindrical specimen in a compression frame. Although not required, if the frame has enlarged end caps, wax paper can be placed between the specimen and the end caps to reduce end effects.

7. Record the channel numbers, calibration factors, and input voltages for the force and displacement transducers.

8. Obtain the zero for the load cell to 0.01 mV.

9. Bring top cap into contact with specimen.

10. Record the "zero" reading for the displacement transducer.

11. Start the data acquisition system, recording vertical load and displacement versus time. Use a reading interval of approximately 5 seconds.

12. Load the specimen at 0.5 to 2 percent axial strain per minute.

13. Load the specimen until a failure plane develops, or 15 percent strain, whichever occurs first.

14. Remove the load. Measure the specimen height and diameter at several locations.

15. Measure angle of failure plane if one exists.

16. Sketch the final geometry.

17. Cut the specimen into three pieces (top, bottom, and middle) and obtain separate water contents.

18. Be sure to collect all the soil to obtain the total dry weight.

Calculations

Pocket Penetrometer

1. Convert the pocket penetrometer readings to kPa using the appropriate conversion factor and shoe factor.

2. Calculate the average and standard deviation of the pocket penetrometer readings.

Fall Cone

1. Calculate the undrained strength using Equation 15.1.

2. Convert the fall cone strength to kPa.

3. Calculate the average and standard deviation of the fall cone penetration readings.

Handheld Shear Vane

1. Convert the handheld shear vane readings to kPa using the appropriate conversion factor and shoe factor.

2. Calculate the average and standard deviation of the handheld shear vane readings.

Miniature Laboratory Vane 1. The area over which the shear strength is developed is the area of the cylinder (A_v) that circumscribes the vane. Calculate the area using Equation 15.5:

$$A_v = 2 \cdot \frac{\pi D^2}{4} + \pi D H \tag{15.5}$$

Where:
  A_v = area of the cylinder that circumscribes the vane (m²)
  D = diameter of the vane (m)
  H = height of the vane (m)

2. The vane constant (K) is calculated using Equation 15.6:

$$K_v = \frac{\pi D^2 H}{2} + \frac{\pi D^3}{6} \tag{15.6}$$

Where:
  K_v = vane constant (m³)

3. The torque (T) applied to the specimen is calculated using Equation 15.7:

$$T_n = \tau_n \cdot K_v \tag{15.7}$$

Where:
  T_n = torque applied to the soil at reading n (N-m)
  τ_n = shear stress applied to the soil at reading n (Pa)

4. To solve for the shear stress, the equation can be rearranged as Equation 15.8:

$$\tau_n = \frac{T_n}{K_v} \tag{15.8}$$

5. For each reading n, the torque (T_n) applied to the soil is expressed as Equation 15.9:

$$T_n = CF_T \cdot R_{T,n} \tag{15.9}$$

Where:
  $R_{T,n}$ = rotation on the torque meter at reading n (degrees)
  CF_T = calibration factor for the torque meter ((N-m)/degree)

6. Combining the previous two equations results in Equation 15.10:

$$\tau_n = CF_T \cdot R_{T,n} \cdot \left[\frac{1}{\dfrac{\pi D^2 H}{2} + \dfrac{\pi D^3}{6}} \right] \tag{15.10}$$

7. Simplifying the equation leads to Equation 15.11:

$$\tau_n = CF_T \cdot R_{T,n} \cdot \frac{6}{\left(3\pi D^2 H + \pi D^3 \right)} \tag{15.11}$$

8. Convert the shear stress to kPa.
9. Plot shear stress versus rotation angle for the intact measurement up to the peak stress value and add an indication line for the remolded stress.

1. Calculate the initial area of the unconfined compression specimen (A_0) in m^2.
2. Calculate the deformation (ΔH_n) at each reading n using Equation 15.12:

$$\Delta H_n = \left| \frac{V_{vd,n}}{V_{in,vd,n}} - \frac{V_{vd,0}}{V_{in,vd,0}} \right| \times CF_{vd} \qquad (15.12)$$

Where:

ΔH_n = vertical displacement at reading n (m)

$V_{vd,n}$ = output voltage at reading n on the vertical displacement transducer (V)

$V_{vd,0}$ = output voltage at zero reading on the vertical displacement transducer (V)

$V_{in,vd,n}$ = input voltage to the vertical displacement transducer at reading n (V)

$V_{in,vd,0}$ = input voltage to the vertical displacement transducer at zero reading (V)

CF_{vd} = calibration factor for vertical displacement transducer (m/V/Vin)

3. Calculate the axial strain ($\varepsilon_{a,n}$) at each reading using Equation 15.13:

$$\varepsilon_{a,n} = \frac{\Delta H_n}{H_0} \times 100 \qquad (15.13)$$

Where:

$\varepsilon_{a,n}$ = axial strain at reading n (%)

H_0 = initial height of the unconfined compression specimen (m)

4. Calculate the cross sectional area at reading n (A_n) using Equation 15.14:

$$A_n = \frac{A_0}{\left(1 - \dfrac{\varepsilon_{a,n}}{100} \right)} \qquad (15.14)$$

Where:

A_n = cross-sectional area at reading n (m^2)

5. Calculate the axial force applied to the specimen at each reading (P_n) using Equation 15.15:

$$P_n = \left| \frac{V_{l,n}}{V_{in,l,n}} - \frac{V_{l,o}}{V_{in,l,o}} \right| \times CF_l \qquad (15.15)$$

Where:

P_n = axial force applied to the specimen (N)

$V_{l,n}$ = output voltage at reading n on the load cell (mV)

$V_{l,0}$ = output voltage at zero reading on the load cell (mV)

$V_{in,l,n}$ = input voltage to the load cell at reading n (V)

$V_{in,l,0}$ = input voltage to the load cell at zero reading (V)

CF_l = calibration factor for load cell (N/mV/Vin)

6. Calculate the axial stress ($\sigma_{a,n}$) applied to the specimen at each reading n using Equation 15.16:

$$\sigma_{a,n} = \frac{P}{A_n} \tag{15.16}$$

Where:

    $\sigma_{a,n}$ = axial stress at reading n (Pa)

7. Convert the stress to kPa.
8. Plot the axial stress versus strain.
9. Determine the peak stress and report this value as the unconfined compressive strength ($2q_u$).
10. Divide $2q_u$ by two and report this value as the undrained shear strength (s_u).

Report

1. Plots and summary sheets from the fall cone, miniature laboratory vane, and unconfined compression tests.
2. If remolded determinations were performed for the miniature laboratory vane and/or unconfined compression tests, indicate the sensitivity results.
3. Summary of phase relations for unconfined test(s).
4. Sketch of the unconfined compression specimen(s) after failure.
5. Data sheets from the pocket penetrometer and handheld shear vane tests.
6. Table comparing the strengths and companion water contents from the various tests.

PRECISION

Criteria for judging the acceptability of test results obtained by test method D2166 (Unconfined Compressive Strength of Cohesive Soil) are given as follows as based on the interlaboratory study (ILS) conducted by the ASTM Reference Soils and Testing Program. Note that the ILS program was conducted using rigid polyurethane foam specimens in place of soil. The density of the foam used was approximately 0.09 g/cm^3 with an average determined peak compressive stress of 989 kPa and axial strain at peak compressive stress of 4.16 percent.

- *Within Laboratory Repeatability:* Expect the standard deviation of your results on the same soil to be on the order of 42 kPa for the peak compressive stress and 0.32 percent for the axial strain at peak compressive stress.
- *Between Laboratory Reproducibility:* Expect the standard deviation of your results on the same soil compared to others to be on the order of 53 kPa for the peak compressive stress and 0.35 percent for the axial strain at peak compressive stress.

Criteria for judging the acceptability of test results obtained by test method D4648 (Laboratory Miniature Vane Shear Test), the Fall Cone test method, the Handheld Vane Shear Test, or the Pocket Penetrometer test have not been determined within ASTM.

DETECTING PROBLEMS WITH RESULTS

For this set of tests, comparisons can be made directly among the compression tests (unconfined compression test and pocket penetrometer test) or among the shear tests (handheld shear vane, miniature laboratory vane, or fall cone). The results can also be compared between the two groups, but remember that different values of strength are expected for different modes of failure.

When the fall cone tests yield results that are not consistent with the other strength index tests or are generally unreasonable, the first item to check is the cone factor. Verify that the cone factor is correct for the cone geometry and roughness used in the test. Verify that the mass of the fall cone used in the equation to determine strength is correct. If the interpreted strength value seems too high, evaluate whether the specimen has been allowed to lose moisture. If so, prepare another soil surface and perform the test immediately. If the interpreted strength is too low, verify that the specimen has not been disturbed during handling. When the values within a set of fall cone measurements are variable, check that the penetration time is held constant over the various readings. Inconsistent results can also be caused by the presence of shells, stones, air voids, or other features.

When results of the handheld shear vane or pocket penetrometer are offset from the other index tests with the same mode of failure, evaluate the equipment for deficiencies or systematic technique problems. Verify that the devices are in good working order by compressing the pocket penetrometer against a hard surface and checking for any impedances. A similar check can be performed for the handheld shear vane by rotating the spring in the hand. When values of these two tests are dramatically different within a set, evaluate the techniques involved in performing the test. The motion involved in performing these tests must be smooth and consistent. Hold the handheld shear vane against the surface of the soil with a small, consistent normal formal force. Verify that the position of the hand while performing these two tests do not impact the position of the retaining ring on the pocket penetrometer or the pointer on the handheld shear vane. Another possible source of error is the spring constants of these devices. Check using the procedures described previously in the calibration section. The presence of inclusions can adversely impact the results of these tests as well.

When values for the miniature laboratory vane test do not seem reasonable, verify that the calibration factor used for the spring is correct. Another source of error could be a rate of rotation that is too fast or too slow, therefore the rate imposed must be compared to the recommended rates. A bent shaft can cause disturbance to the soil, and must be replaced. Inclusions can adversely affect the interpreted values and any isolated features observed in the soil specimen when obtaining the water content specimen should be noted on the data sheet and considered when interpreting the results. If multiple measurements are made and the results are variable, evaluate the technique used to insert the vane into the specimen. Make sure the motion is smooth and that rotation does not occur, causing disturbance around the blades of the vane. Also verify that the test is initiated immediately after insertion of the vane.

For the unconfined compression test, when values of stress or strain seem unreasonable, check the calibration factors for the load cell and the displacement transducer. Lower than expected values of sensitivity may be due to preexisting failure planes or excessive specimen disturbance. Check that the loading rate is within the recommended range.

REFERENCE PROCEDURES

ASTM D2166 Unconfined Compressive Strength of Cohesive Soil.

ASTM D4648 Laboratory Miniature Vane Shear Test for Saturated Fine-Grained Clayey Soil.

Earth Manual, Pocket Vane Shear Test.

Hansbo (1957), "Fall-cone Test."

MIT, Unpublished Procedure for Performing the Pocket Penetrometer Test.

REFERENCES

Refer to this textbook's ancillary web site, www.wiley.com/college/germaine, for data sheets, spreadsheets, and example data sets.

Enkeball, W. 1946. "Report on Soil Investigations for the Proposed New Library Building and Nuclear Physics Laboratory at MIT," unpublished.

Hansbo, S. 1957. "A New Approach to the Determination of the Shear Strength of Clay by the Fall-cone Test," *Proceedings of the Royal Swedish Geotechnical Institute*, no. 14.

Head, K. H. 1980. *Manual of Soil Laboratory Testing, Volume 1: Soil Classification and Compaction Tests,* Pentech Press, London.

Koumoto, T. and T. Houlsby. 2001. "Theory and Practice of the Fall Cone Test," *Géotechnique,* 51(8), 701–712.

Ladd, C. C., and U. Luscher. 1965. "Preliminary Report on the Engineering Properties of the Soils Underlying the MIT Campus," *MIT Research Report R65-58,* Cambridge.

Lambe, T. W. 1962. "Report of the Soil Engineering Investigation for the Center for Materials Science and Engineering, MIT," Unpublished report prepared for Skidmore, Owings and Merrill.

Rosenqvist, I. Th. 1953. "Considerations on the Sensitivity of Norwegian Quick Clays," *Géotechnique,* 8(1), 195–200.

Terzaghi, K., and R. B. Peck. 1948. *Soil Mechanics in Engineering Practice,* John Wiley and Sons, New York.

Terzaghi, K., and R. B. Peck. 1967. *Soil Mechanics in Engineering Practice,* John Wiley and Sons, New York.

U.S. Bureau of Reclamation, 1998. *Earth Manual: Part 1*, Denver, CO.

Chapter 16

Unconsolidated-Undrained Triaxial Compression

This chapter provides background information on the Unconsolidated-Undrained (UU) triaxial compression test and detailed procedures to perform a UU test on an intact, fine-grained soil. The test provides a means of obtaining undrained stress-strain behavior and the undrained strength of an unconsolidated cylindrical specimen in triaxial compression. A series of tests can be performed on companion specimens over a range of confining stress levels and the results interpreted to obtain the total stress parameters. Measurement of pore pressure during the test is discussed in the background section of this chapter. However, the laboratory instruction will be presented for a UU test as presented in ASTM D2850 Unconsolidated-Undrained Triaxial Compression Test on Cohesive Soils, which does not include pore pressure measurements.

In addition to UU testing, the laboratory experiment discusses preparing intact clay specimens for testing, introduces the details of a conventional triaxial cell, and provides the framework for the MIT stress path presentation of results.

TYPICAL MATERIALS

Unconsolidated-undrained triaxial testing is typically performed on soil materials ranging from clay to rock. Specimens can either be trimmed, intact samples or reconstituted in the laboratory. The test is applicable to specimens that are strong enough to maintain a cylindrical shape without lateral support on a bench. Special procedures and equipment must be used to test very soft, fine-grained materials.

BACKGROUND

The principles of soil strength are presented in many soil mechanics texts, such as Terzaghi, Peck, and Mesri (1996), Lambe and Whitman (1991), or Holtz and Kovacs (1981). The goal of this chapter is to focus on the testing details and the choices relative to the Unconsolidated-Undrained Triaxial Compression (UU[1]) test. Proper interpretation of the test results and appropriate selection of the test variables will be project specific and require an understanding of soil behavior.

The symbols "UU" are part of a designation system that separately recognizes each phase of the test. UU is actually a short-hand version of the full designation UUTC(L). The first "U" stands for the unconsolidated condition and denotes the fact that the specimen is not allowed to change in mass from the time it is assembled in the triaxial cell to the time of shearing. The second "U" stands for the undrained condition and denotes the fact that the specimen is not allowed to change in mass during the shearing stage of the test. The "T" stands for a triaxial test and is important because this defines the stress conditions being axisymmetric. The "C" stands for compression and defines the mode of loading in the triaxial test. The compression condition is one in which the specimen shortens during the shearing process. The "(L)" is for loading and indicates that the compressive shearing was performed by increasing the axial stress. While this explanation may seem overly complicated for the UU test, it introduces an extremely useful system applicable for general soil testing.

The UU test is the first strength test presented in this text where an attempt is made to impart uniform conditions of stress and strain on a specimen, and where the stress state in known. These are essential conditions for a test to be considered a simple element test, and potentially provide rigorous results for the characterization of mechanical properties. The initial state of stress is hydrostatic, meaning that all three principal stresses are equal. During shearing, the axial stress is the major principal total stress (σ_1) and the confining pressure applies the intermediate and minor principal stresses ($\sigma_2 = \sigma_3$). The confining pressure is maintained constant throughout the shearing process. The principal stress criteria are only satisfied if there is no shear stress on any of the specimen boundaries. In reality, the ends of the specimen are in contact with a stiffer material. As the specimen is compressed axially, it expands in the radial dimension, creating a surface traction on the ends. For more brittle materials, this can be an important factor, but for soft, plastic materials this will be a minor effect.

The test is performed on a cylindrical specimen having a height-to-diameter ratio between 2 and 2.5. This is a typical aspect ratio for a triaxial test specimen. The specimen is placed in a triaxial[2] testing apparatus, a confining pressure is applied without allowing drainage, and the axial deformation is increased at a relatively rapid rate until failure. Drainage is not allowed during the shearing phase of the test. Axial deformation and reaction force are measured during the shearing phase of the test. Pore pressure is not measured in the standard UU test.

---

[1] Other terms have been used for triaxial testing, such as the Q test (for Quick) for the Unconsolidated-Undrained Triaxial Test, Q_c test (Quick, Consolidated) for the Consolidated-Undrained Triaxial Test, and S test (for Slow) for the Consolidated-Drained Triaxial test. The term "R test" has also been used in place of Q_c, presumably because "R" falls between "Q" and "S" in the alphabet. This terminology is outdated and therefore is not used in this text.

[2] In soils testing, the term "triaxial test" is used in recognition of the fact that pore pressure is an important variable. The mechanics community would consider this a biaxial test because it is limited to two independently applied boundary stresses (axial stress and cell pressure).

The UU results must be analyzed in total stress space because there is no measurement of pore pressure. The deviator stress is defined as the major principal stress in the axial direction minus the minor principal stress. The deviator stress is not given a symbol but is simply designated as $(\sigma_1 - \sigma_3)$. The results for an individual test are presented as a stress-strain curve, as shown in Figure 16.1.

For an unconfined compression (UC) test, the compressive stress is calculated as the load divided by the corrected area of the specimen[3]. The compressive strength is the peak compressive stress $(2q_u)$, which is equivalent to the diameter of the Mohr's circle. In the UU test, the confining pressure is not zero, which shifts the stress state along the hydrostatic axis. The compressive strength is still the peak deviator stress. The undrained strength, s_u, is the radius of the Mohr's circle, which is the peak deviator stress divided by two. The stress state is completely defined in the UU test (ignoring the slight deviations caused by end friction), allowing the Mohr's circle to be constructed throughout the test.

The shearing is continued until 15 percent strain, the deviator stress has decreased by 20 percent of the peak, or the axial strain is 5 percent greater than the strain at the peak deviator stress. A well-defined failure surface usually becomes evident when the deviator stress decreases. The failure condition is taken as either the peak in the deviator stress versus axial strain plot or the stress condition at 15 percent strain. The Mohr's circle at the failure condition is presented in Figure 16.2 using the same test data as presented in the stress-strain curve of Figure 16.1. As a reminder, stress state plots must have equal

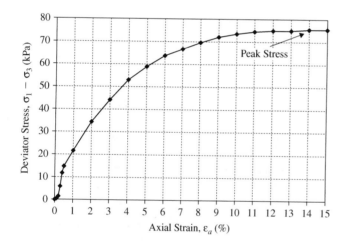

Figure 16.1 Stress-strain curve generated during a UU test.

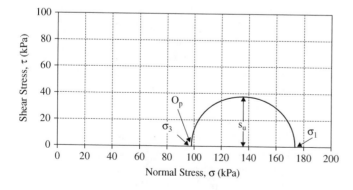

Figure 16.2 Mohr's circle representing the results generated during a UU test.

---

[3] The Unconfined Compression (UC) test is a special case of the UU test. In the UC test, a confining pressure is not applied to a specimen and no membrane is required. The UC test is presented in Chapter 15.

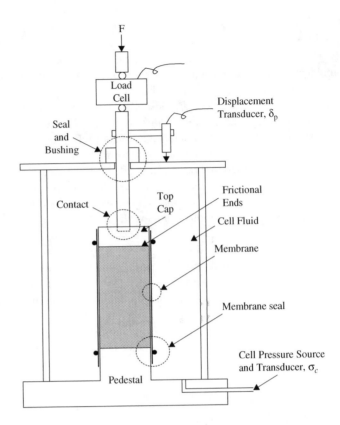

F

Load
Cell

Displacement
Transducer, δ_p

Seal
and
Bushing

Contact

Top
Cap

Frictional
Ends

Cell Fluid

Membrane

Membrane seal

Cell Pressure Source
and Transducer, σ_c

Pedestal

Figure 16.3 Basic elements of a triaxial apparatus necessary for performing a UU test.

spacing of increments on the x and y axes. Adding the origin of planes to the Mohr's circle provides physical alignment of the stress state. In the case of the triaxial compression test, the maximum shear stress is on a 45-degree plane through the specimen.

The UU test is performed using a very simple version of the triaxial apparatus. Figure 16.3 presents a schematic diagram of the essential details, but there are many variations commercially available. Figure 16.4 presents a picture of a typical apparatus along with the load frame.

The following items discuss the most important elements of the equipment.

- The triaxial chamber is essentially a pressure vessel consisting of a base plate, a cylinder, and a top plate. These are usually held together with rods on the outside of the cylinder. The plates are made of noncorrosive metal. A pedestal is connected to the base, providing an elevated support for the specimen. The top of the pedestal must be flat and the same diameter as the specimen or larger. The base is normally fitted with a connection to fill the chamber and apply pressure. The cylinder can be acrylic or metallic, depending on the pressure levels. The top plate contains a vent hole and a center opening for the piston rod.

- The seal and bushing connected to the top plate provide alignment of the piston shaft and prevent the pressurized fluid from leaking out of the chamber. The bushing and shaft combination must be stiff enough to prevent lateral movement of the top cap during the test. The specimen must be concentrically loaded. Friction between the moving shaft and the pressure seal can be important. This is a particular concern for weak soils.

- The top cap must be rigid, the same diameter as the specimen or larger, have a flat surface, and apply a small force to the specimen. The top connection with the piston rod must not allow rotation of the top cap during the test. In most devices, the shaft is not attached to the top cap but rather rests in an alignment depression.

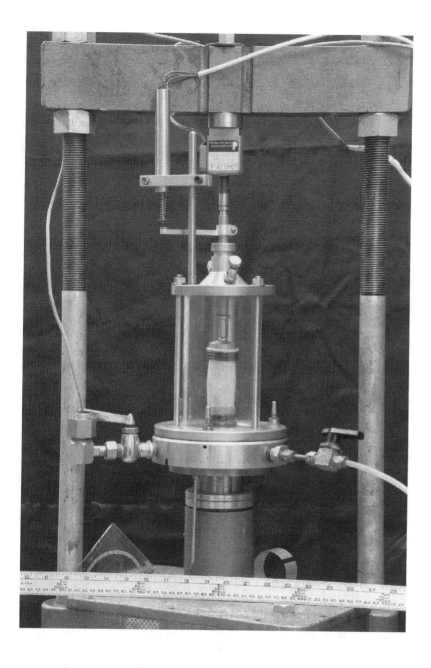

Figure 16.4 Typical equipment setup for performing a UU test.

- The membrane seals the specimen from the chamber fluid. This seal is an essential design detail. A leak in the membrane will severely impact the results, yet there is no accommodation to detect a leak during a test. Membranes are made of everything from thin rubber to Viton® to copper, depending on the magnitude of the applied cell pressure and the consistency of the soil. The membrane is sealed to the pedestal and top cap with o-rings, clamps, or heat-shrink tubing. Some of the force applied to the top cap is taken by the membrane. This is accounted for with the membrane correction, which is a function of the thickness and modulus of the membrane material.
- The cell fluid is used to apply the hydrostatic pressure to the specimen. The fluid can be water, glycerin, silicone oil, or hydraulic oil. Air is not advisable for confinement because rubber is permeable to air and air leakage into the specimen is impossible to detect. Water is acceptable for lower pressures, but oils are better options for high-pressure testing. Glycerin has been used successfully with very weak clays without the need for a membrane.

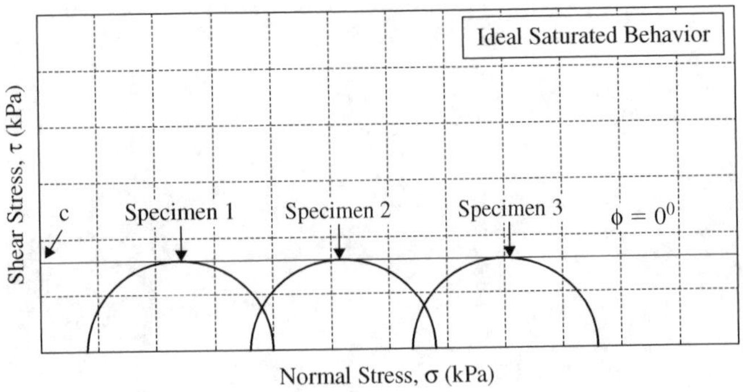

Figure 16.5 Mohr's circles for Unconsolidated-Undrained Triaxial Compression results for completely saturated specimens in total stress space.

- The cell pressure is controlled by a manual regulator and measured with a pressure transducer. When an air compressor is used as the energy source, an air/water interface external to the triaxial apparatus is required to prevent significant air diffusion into the specimen.
- The axial force is generally measured outside the chamber. The measured force is affected by the seal friction and by the uplift force of the cell pressure acting on the piston. The uplift force can be obtained by calibration, but the piston friction becomes an uncertain error in the test. Standard practice is to start the test with a gap between the piston and the top cap, start advancing the piston, and measure the combined affect of the piston friction and uplift force. This is treated as a shift in the load cell zero.
- The piston displacement is measured relative to the top plate of the apparatus using an LVDT. The start of specimen deformation requires interpretation of the force-displacement relationship. Due to seating errors of the piston with the top cap and the specimen with the platen and top cap, the start of loading is often difficult to define.
- The axial deformation is applied with a hydraulic or screw driven load frame. The rate is set at 1 percent per minute for most materials, and the specimen is strained to at least 15 percent axial strain. The strain rate should be reduced to 0.3 percent per minute for brittle materials.

ASTM D2850 Unconsolidated-Undrained Triaxial Compression Test on Cohesive Soils and the above discussion are focused on a single UU test. Several tests must be performed on companion (similar) specimens over a range of confining pressures in order to measure the total stress strength envelope. This envelope is often characterized by a linear model, parameterized by the cohesion intercept (c) and the friction angle (ϕ). A line that is tangent to all the circles will provide the parameters of the total stress envelope. The slope of the line is the friction angle. The envelope does not a priori have to be a linear relationship. The actual shape is a matter of soil behavior. Interpretation of the data to select the most appropriate relationship requires engineering judgment, which is beyond the scope of this text. For discussion, it will be assumed that the envelope can be represented by a linear Mohr-Coulomb relationship.

There are two possible outcomes from a series of tests on identical specimens. If the material is fully saturated and relatively soft as compared to the stiffness of water, then all the tests will yield the same strength, independent of the confining pressure. The friction angle will be equal to zero and the cohesion intercept will be equal to the undrained strength. Only one test is actually required in this situation. This is considered the *ideal saturated behavior* and is illustrated in Figure 16.5.

Given any other material condition, the size of the Mohr's circle will increase as the confining pressure increases. Figure 16.6 illustrates this behavior. The intercept will, in all cases, be less than the strength of any individual specimen. In this case, multiple tests are required to completely define the relationship.

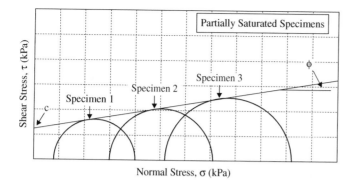

Figure 16.6 Mohr's circles for Unconsolidated-Undrained Triaxial Compression results for partially saturated specimens.

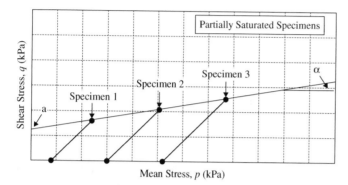

Figure 16.7 MIT Stress Path Space presentation for Unconsolidated-Undrained Triaxial Compression results for partially saturated specimens in total stress space.

Representation of the stress state in terms of shear stress versus normal stress provides a visually appealing portrayal of the condition at failure and makes it easy to construct the Mohr-Coulomb failure criteria. The Mohr's circle also provides additional information concerning the stress state on various planes, as well as the orientation of these planes relative to the physical geometry. This stress space is limited to the stresses in a plane and is very difficult to use when considering changes in stress over time.

The MIT stress space is one of several options that have been developed for use in more general situations. The stress state at any time (or condition) can be represented by a single stress point instead of a Mohr's circle. This greatly simplifies the visualization and allows changes in stress state to be represented by a trace of points called a stress path. The MIT stress system represents the stress state using a mean stress, p, and a shear stress, q. The stress path traces the maximum shear stress and the mean stress at any particular time during the test. The definitions of p and q are given by Equation 16.1 and Equation 16.2, respectively:

$$p = \frac{\sigma_1 + \sigma_3}{2} \tag{16.1}$$

$$q = \frac{\sigma_1 - \sigma_3}{2} \tag{16.2}$$

Where:

p = mean stress and center of Mohr's circle (kPa)
σ_1 = major principle stress (kPa)
σ_3 = minor principle stress (kPa)
q = shear stress and radius of Mohr's circle (kPa)

Like the Mohr's circle space, the MIT system is limited to the stress state in a plane. It sacrifices the stress orientation information in favor of being able to track the stress path. The results of Figure 16.6 appear as Figure 16.7 when represented by the MIT stress path space. The failure envelope is parameterized by the angle, α, and cohesion intercept, a. Like all stress spaces, equal increments in stress on the x and y axes must be dimensionally equal.

Conversion of the failure envelopes between the two stress systems is a simple matter of geometry. The Mohr's envelope parameters are related to the MIT envelope parameters as shown in Equation 16.3 and Equation 16.4:

$$\sin \phi = \tan \alpha \tag{16.3}$$

$$c = a/\cos \phi \tag{16.4}$$

Interpretation of UU test results is a difficult task because the lack of pore pressure information leaves the story incomplete. In fact, the UU test is both embraced and shunned by the profession. On the positive side, the test provides a numerical measure of the strength under controlled conditions. More importantly, it is fast, easy to perform, and applicable to all strength ranges. This reality has contributed to the persistence of the test. On the negative side, a combination of sampling disturbance, specimen variability, and the lack of pore pressure information make the test results highly variable and impossible to interpret rigorously. It can be reasonably argued that the strength from a UU test is no more reliable than a handheld shear vane measurement.

The ideal saturated behavior can be explained in terms of effective stress. Consider the changes in stress for an intact sample as it goes from the field to the triaxial apparatus. In the ground, the sample is acted on by a total stress and has some positive pore pressure. The difference in these stresses results in the effective stress. The sample is extracted from the ground and then removed from the sampling tube. While the sample sits on the laboratory bench, the applied total stress is reduced to zero. The pore pressure will be a negative, and unpredictable, value. This negative pore pressure is generated by expansion of the sample due to the total stress reduction, shear-induced pore pressure, and surface tension at the air-water interface over the boundary of the specimen. The sample is held together by the *sampling effective stress*, which is equal to the negative of the (negative) pore pressure. Submerging the sample in water will negate the surface tension, increase the pore pressure to zero, and the sample will completely disintegrate. This is proof that the material has no cementation and the strength is derived by the effective stress.

In theory, a completely saturated soil having a drained bulk compressibility considerably higher than the bulk compressibility of water will have the same failure stress, regardless of the confining pressure. Application of an increment in total stress will cause a change in the pore pressure of a completely sealed specimen. The Skempton B parameter is used to quantify the ratio of the increment in pore pressure to the increment in applied total stress, as shown in Equation 16.5:

$$B = \frac{\Delta u}{\Delta \sigma_c} = \frac{1}{1 + n \frac{C_w}{C_{sk}}} \tag{16.5}$$

Where:

B = pore pressure parameter (dimensionless)
Δu = increment in pore pressure (kPa)
$\Delta \rho$ = increment in applied total stress (kPa)
n = porosity (dimensionless)
C_w = compressibility of water (kPa^{-1})
C_{sk} = compressibility of soil skeleton (kPa^{-1})

The value of the compressibility of water is about $5.4 \times 10^{-3} kPa^{-1}$. The B value for soft to medium stiff soils is expected to be very near unity. For these materials, the applied increment in confining pressure will produce an equal increment in pore pressure and the effective stress will remain constant at the sampling effective stress value. This is shown as Equation 16.6:

$$\sigma' = \sigma + \Delta\sigma - (u + B\Delta\sigma) = \sigma - u + (1 - B)\Delta\sigma \qquad (16.6)$$

Where:

σ' = effective stress (kPa)

σ = applied total stress (kPa)

u = pore pressure (kPa)

Since the strength is controlled by the effective stress rather than total stress, the measured strength of identical saturated, soft clay specimens tested at different confining pressures will be constant, leading to a friction angle (ϕ) is equal zero interpretation, such as shown previously in Figure 16.5.

For a material that has ideal saturated behavior, it is only necessary to perform one UU test to obtain the necessary parameters. Performing a series (usually three) of tests at different confining pressures allows evaluation of testing errors or identification of deviations from ideal behavior. Some agencies require as a rule a series of three tests at different confining pressures, and further require a fourth test to be run if the results of any of the first three tests deviate by more than a specified factor from the average. If the pore pressure was measured accurately during the UU test, it would be possible to evaluate the results in terms of effective stress and to consider the consequences when the sampling effective stress differs from the in-situ stress state.

Deviations from ideal saturated behavior can be the result of: (1) drainage from the specimen into the triaxial measurement system, (2) lack of initial specimen saturation, (3) finite interparticle contact within the soil skeleton, or (4) cavitation of the pore fluid during shearing. Samples from below the water table are normally saturated but can become unsaturated with occluded air bubbles due to stress relief, temperature changes, bacteria activity, or by exceeding the air entry pressure, among other reasons. Samples from the vadose zone are normally unsaturated and typically contain continuous air voids. Cemented, highly compressed or lithified materials can have appreciable interparticle contact areas. Very stiff materials will dilate during shear, resulting in large negative shear-induced pore pressures. All the above conditions result in a nonzero friction angle and require careful consideration when interpreting the results from a series of UU tests.

One of the essential requirements of a UU test is that the water content remains constant throughout the various stages of the test. Therefore the specimen must be properly sealed in the triaxial apparatus. The membrane, in combination with well-designed connections to the top cap and pedestal, provides the appropriate seal against the chamber fluid. The pedestal and top cap must also prevent drainage. For a standard UU test without pore pressure measurement, it is important to use smooth, solid end caps. Drainage of water out of the specimen will result in consolidation and an increase in effective stress. The importance of unwanted flow out of the specimen depends on the stiffness of the material. The increase in effective stress with confining pressure will result in a small and erroneous friction angle.

Testing specimens that have developed occluded gas bubbles in the pore fluid are a particular problem. These materials are saturated in nature and the purpose of the test is to measure the properties of the saturated state. The presence of pore gas makes the specimen compressible and changes the pore pressure response to shearing. Increases in the confining pressure will not create equal changes in pore pressure (i.e., the B value will be less than unity), so the effective stress will increase. As the confining pressure increases, the bubbles will decrease and the response will approach the ideal saturated behavior. This will result in a curved envelope of decreasing slope.

UU testing is also performed on naturally unsaturated materials, such as compacted materials or intact samples from above the water table. The properties of unsaturated soils will depend on the confining pressure because the pore pressure will not necessarily respond to changes in the confining pressure. As such, the material will contract and the interparticle contact forces will increase. The strength will increase with higher confining pressures due to increased interparticle forces. For unsaturated materials, the explicit goal of the testing program is to obtain the relationship between measured properties and the confining pressure. The total stress envelope obtained from UU testing is appropriate for these materials, with the understanding that disturbances. still exists

One common application of testing partially saturated specimens is compacted clay liners for landfills. The compacted clay specimen is not saturated during placement and will typically not become saturated in situ. Further, the liner will remain at fairly low overburden stresses. The confining pressure for testing is specified based on the overburden pressure. Some agencies prefer to use a confining pressure of the in-situ earth pressure coefficient (K_0) value times the effective vertical stress. Others use an estimate of mean stress, meaning two times the horizontal earth pressure coefficient plus one times the vertical earth pressure coefficient all divided by three (for three axes), with the result multiplied by the effective vertical stress. Hence, the UU test provides a means for obtaining an undrained strength applicable to this particular condition.

Many materials have substantial contact area between individual grains. These materials will not strictly obey the effective stress principle because the pore pressure does not act on the contact area. UU testing will measure an increase in strength with increasing confinement. Interpretation of the UU results will be problematic because both the magnitude of the pore pressure and the confining stress properly influence the strength. Once again, not knowing the pore pressure will compromise the interpretation.

Finally, stiff or dense saturated materials will dilate during shearing. Under undrained conditions, dilation is resisted by the development of negative pore pressures. In the laboratory, water usually cavitates when the absolute pressure drops to about negative 80 kPa (negative 0.8 bar). Upon cavitation, gas will expand rapidly and the pore pressure will remain constant at this limiting value. Since negative pore pressure is the source of positive effective stress increments in the sealed specimen, cavitation prevents further strength gain. As the confining pressure is increased, the decrement in pore pressure to cause cavitation also increases, leading to an increase in measured strength with confining pressure. For these materials, the UU test will measure a friction angle greater than zero. The applicability of these results will depend on the field situation and stress condition.

Interpretation of UU test data is considerably enhanced through the measurement of the pore pressure during shearing. When pore pressure measurements are made in a UU test, it is called a UU bar test. This test requires a special base pedestal that provides connectivity to the pore fluid. The rate of strain must be decreased to allow equalization of the excess pore pressure developed during shear. The base must measure pore pressure, yet not allow flow out of the specimen. Since the specimen cannot be back-pressure saturated (else violating the conditions of the test), the measurement system must be very rigid and fully saturated before the specimen is put on the base. Generally, this is accomplished using a high air entry stone epoxied permanently into the base. The proper strain rate should be computed based on an estimate of the coefficient of consolidation.

Other important considerations on unconsolidated-undrained triaxial testing include:

- Soil strength is anisotropic. The triaxial compressive loading condition is generally the strongest direction. The UU strength will be higher than the average strength due to anisotropy effects.
- Soil strength is sensitive to the shearing strain rate. The strain rate in the UU test is much faster than would occur in the field. The resulting interpreted undrained strength will be higher than actual due to strain rate effects.
- Alteration of the sample due to sampling and handling is generalized as disturbance. Disturbance will lower the measured undrained shear strength as compared to the field condition. The amount of disturbance is highly variable, resulting in the large scatter of results measured by the UU test.
- Anisotropy and rate effects provide compensation for disturbance. Depending on the balance of these factors, the UU test will underpredict the undrained

strength in some cases, and overpredict the undrained strength in other cases. The UU test relies on compensation of uncontrolled errors, making it a risky measurement.

- Some testing agencies measure pore pressure during this test in order to develop effective stress parameters. For a UU test with pore pressure, effective stresses are plotted, as well as total stresses in some cases. Specialty equipment is required to measure pore pressures in the UU test. It is not possible to use a conventional triaxial apparatus and apply a back pressure to saturate the specimen and pore pressure measurement system. This changes the test from a UU test to a consolidated-undrained, CU, test.

- Failure surfaces are a common occurrence in the UU test. As soon as a failure surface is identified in the specimen, the test should be terminated. Stress results computed after the shear surface has formed are unreliable.

- The strain to failure in a UU test is often greater than 10 percent. This is much larger than expected for field conditions, and is a result of sample disturbance, starting from a low effective stress compared to the field condition, and shearing from a hydrostatic stress state.

- The axial force is generally measured externally in the UU test. The friction loss in the piston pressure seal can create a substantial error in the strength measurement. This is a particular concern for soft materials. The cell pressure will also cause uplift of the piston. A correction due to both factors is normally measured on a test-by-test basis by taking the zero of the load cell as the piston is advanced at the standard rate, but before contacting the top cap. Refer to Figure 16.8 for a depiction of this process. For a particular triaxial setup and cell pressure, the value of the correction should be approximately the same from test to test. Other, more explicit methods are available for determining the reasonableness of the correction. While the correction described herein adjusts for the pressure uplift force and the dynamic seal friction, it ignores changes in friction with lateral loading.

- The membrane has a stiffness and ASTM D2850 requires a correction to be made to account for the force carried by the membrane if it contributes greater than 5 percent of the peak deviator stress. One method for determining the modulus of the membrane is described in the calibration section of this chapter, while incorporation of the membrane correction into the results is explained in the calculation section.

- End caps are frictional and are a significant contributor to the initiation of failure surfaces. The interpretation of the results must consider these effects.

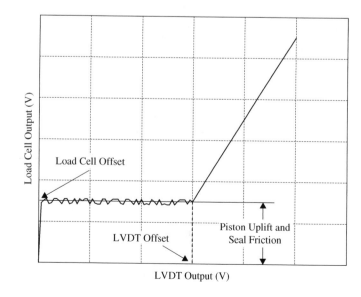

Figure 16.8 Offset in load cell reading due to the piston uplift and seal friction.

- Most stress-strain curves from the UU test are curved slightly upward at the start of loading. This is due to seating and misalignment of contact surfaces, resulting in stiffening, and is not a reflection of soil behavior. For this reason, the modulus from the UU test is not reliable.
- Even though the UU is relatively inexpensive and easy to run, a handheld vane shear device or similar index test would provide comparable information and is much cheaper, quicker, and easier to run. The handheld vane shear device is incorporated in tube sample processing procedures for many laboratories. For compacted clays, the match between UU results and the typical application is more appropriate.
- The aspect ratio of specimens must be kept within a range of 2 to 2.5. A specimen must be tall enough to avoid the influence of the surface traction on the ends and short enough to eliminate instability effects due to buckling.
- In the same way maximum particle size requirements have been discussed previously in this text to avoid size effects, there is also a limit for strength testing. For the UU test, ASTM D2850 limits the maximum particle size to less than one-sixth of the specimen diameter. In addition, particles on the surface of the specimen may puncture the membrane after application of the cell pressure. Such particles must be removed, the hole filled with material from the trimmings during specimen preparation, and a pertinent note added to the specimen preparation remarks.
- In many cases, the interpretation of a friction angle not equal to zero is due to a partially saturated specimen that is intended to be fully saturated. In these situations, the results are difficult to interpret. Even small amounts of drainage can increase the interpreted strength of a specimen.

TYPICAL VALUES

Typical results from unconsolidated-undrained triaxial testing are listed in Table 16.1.

CALIBRATION

The transducers must be calibrated. Refer to Chapter 11, "Background Information for Part II," for detailed procedures on calibration. For the UU test, the correction for piston uplift and seal friction will be determined during testing.

The modulus of the membrane must be determined. One approximate method of determining the modulus of the membrane is provided below. For comparison, a typical elastic modulus for a latex membrane is 1.4×10^6 Pa.

1. Cut a piece of membrane across the diameter to a width of at least 10 mm. Measure and record the initial width, w_0.
2. Measure and record the unstretched thickness, t_m, of the membrane.
3. Calculate the cross-sectional area of the membrane, A_m, using Equation 16.7:

$$A_m = 2t_m w_0 \qquad (16.7)$$

Where:

A_m = cross-sectional area of the membrane (m^2)
t_m = unstretched thickness of the membrane (m)
w_0 = initial width of cut section of membrane (m)

Table 16.1 Typical results from unconsolidated-undrained triaxial testing.

Soil Type	S_u (kPa)
Boston Blue Clay (OCR 1.3 to 3.6)	39 to 68
Boston Blue Clay (OCR 1 to 1.2)	23 to 38
Organic Silt	24 to 60

Source: After Ladd and Luscher, 1965.

4. Obtain two rigid (but light) rods with lengths at least as long as the length of membrane.

5. Position one rod inside the membrane, along the length, and support the rod in a horizontal position.

6. Position the second rod inside the membrane, letting it hang on the bottom.

7. Measure the initial length of the membrane, L_0, in the stretch direction (i.e., the distance between rods).

8. Select a mass or masses to apply to the bottom rod. Apply the load such that it is applied evenly over the bottom rod. Hanging two equal masses on the rod, one on each end evenly spaced from the edge of the membrane, is appropriate.

9. Measure and record the change in length, ΔL, of the membrane.

10. Calculate the elastic modulus of the membrane, E_m, using Equation 16.8:

$$E_m = \frac{mgL_0}{A_m \Delta L} \qquad\qquad (16.8)$$

Where:

E_m = elastic modulus for the membrane (Pa)
m = total mass applied to the specimen (kg)
g = acceleration due to gravity (m/s^2)
L_0 = initial length of the membrane (m)
ΔL = change in length of the membrane (m)

Equipment Requirements

The equipment described below is only one possible combination and has been selected based on the equipment commercially available, along with the type of soil being tested in the experiment (soft, marine clay). Refer to ASTM Standard D2850 for more options. Although not typically done in practice, for instructional purposes soil specimens are trimmed to allow for more specimens to be obtained from the same tube.

1. Axial loading device: A screw jack driven by a compression device. The loading device must be capable of providing the required loading rate. The loading frame capacity and loading rate are selected according to the type of material being tested (1 percent per minute for plastic materials and 0.3 percent per minute for brittle materials).

2. Axial load measuring device: An external load cell capable of measuring the load accurately within 1 percent of the maximum load. Typically, a load cell with a capacity of 2200 N (500 lbf) readable to 0.4 N (0.1 lbf) is sufficient.

3. Triaxial compression chamber with the ability to fill the chamber from the bottom, a vent valve at the top, and a capacity greater than the applied cell pressure. A transparent cylinder, such as made of acrylic, is usually desired in order to be able to observe the specimen setup and shearing process. Check the cylinder for defects, such as cracks or crazing, that will adversely impact the integrity of the structure. Do not use a cylinder that has been damaged.

4. Axial load piston: A piston passing through the top of the chamber, and outfitted with a seal. The seal must be designed to contain

the applied cell pressure, without excessive friction such that the variation in axial load is greater than 0.1 percent of the axial load at failure. The piston must be rigid enough such that lateral bending does not occur. Generally, a piston with a diameter of at least one-sixth of the specimen diameter is sufficient.

5. Pressure control device: The pressure control device must be capable of applying and controlling the chamber pressure to within 2 kPa (0.25 psi) for pressures less than 200 kPa (28 psi), and to within 1 percent for pressures greater than 200 kPa.

6. Specimen cap and pedestal: The top cap and bottom pedestal must be solid, rigid, and planar, and of a diameter equal to or greater than the initial specimen diameter. The pedestal must be attached to the base to prevent lateral movement. The top cap must apply a pressure of 1 kPa (0.15 psi) or less to the specimen. The contact between the top cap and the loading piston must prevent tilting, and the top cap must be aligned within 1.3 mm (0.05 in) of the vertical axis of the specimen. Machined acrylic caps work well for UU testing purposes.

7. Deformation indicator: LVDT with a range of at least 20 percent of the initial specimen height, and an accuracy of 0.03 percent of the initial specimen height. An LVDT with a range of 50 mm (2 in) and readable to 0.025 mm (0.001 in) usually meets these requirements.

8. Rubber membrane: A rubber membrane is required to seal the specimen from the applied pressure, while not creating excessive resistance. The thickness of the membrane must be less than or equal to the initial diameter of the specimen. The diameter of the membrane must be in the range of 90 to 95 percent of the initial diameter of the specimen. Before each use, inspect the membrane for damage. A damaged membrane will likely allow a leak during the test, invalidating the results.

9. Membrane stretcher: A stretcher capable of supporting the membrane away from the specimen while sliding the membrane over the specimen. A cylinder with holes and attached tubes capable of applying and releasing a vacuum on command works well for this purpose.

10. O-rings: Flexible o-rings with an unstretched inside diameter of 75 to 85 percent of the diameter of the specimen caps. Inspect each o-ring for scratches or vitrification, and discard any o-rings with signs of damage. Clean and apply a lubricant, such as vacuum grease, to the o-rings before each use.

11. Sample extruder: Use a manual method of a cylindrical block of approximately the same diameter as the inner diameter of the tube to push the sample out of the section of tube after breaking the tension between the soil and the tube. This procedure has been discussed in detail in Chapter 11, "Background Information for Part II." Other pieces of equipment are required, such as a small diameter hollow tube, vice grips, two washers, a vice, and so on.

12. Specimen size measurement devices: Digital calipers with a range of at least 300 mm (12 in) and a resolution of 0.001 mm (0.0001 in) or smaller.

13. Digital timer: readable to the nearest 1 second or smaller increment.

Use an intact specimen of clay. If necessary, trim to the dimensions of the specimen pedestal as indicated below. In all cases, maintain the required height to diameter ratio of between 2 and 2.5 and the specimen ends must be squared off perpendicular to the specimen axis. Refer to Chapter 11, "Background Information for Part II," for additional information regarding trimming triaxial specimens.

SPECIMEN PREPARATION

1. Rough-trim the specimen with a wire saw and obtain several water contents from the trimmings.
2. Place the specimen in a miter box.
3. Rough-cut the cylinder using the miter box.
4. Final-cut the specimen using the miter box.
5. Wrap the specimen in one layer of wax paper that is about 3 mm shorter than the final specimen height. Use a split mold to size the wax paper.
6. Place specimen in split mold and gently tighten.
7. Trim off the specimen ends to height of mold with wire saw and finish with straight edge.

The Unconsolidated-Undrained test will be performed in general accordance with ASTM Standard Test Method D2850.

PROCEDURE

1. Determine the mass of the specimen. If trimmed, measure the mass of the specimen, mold, and wax paper, remove the specimen and wax paper from the mold, then obtain the mass of the mold and wax paper. Subtract the two masses to calculate the mass of the specimen. If untrimmed, measure the mass of the specimen directly, taking care to not touch the specimen with fingertips, but rather use supporting plates or other means of handling the soil.
2. Measure the diameter of the specimen at several locations using the calipers.
3. Measure the specimen height at several locations using the calipers or a ruler.

Specimen Measurements

1. Clean and apply silicone grease to the sides of the top cap and bottom pedestal. Note that applying grease provides a better seal between these surfaces and the membrane.
2. Locate the specimen on base pedestal.
3. Place the top cap on the specimen.
4. Place the membrane on a membrane stretcher and apply a vacuum.
5. Position the membrane stretcher with membrane around the specimen and release the vacuum. Remove the stretcher.
6. Fix the membrane to the top and bottom caps with o-rings.
7. Assemble the cell, cylinder, and top plate.

Prepare Cell and Setup Specimen

8. Fill with cell fluid.

9. Record the zero of the cell pressure transducer.

10. Place the displacement gauge bar on the piston and attach the LVDT.

11. Lower the piston until it is seated on the top cap.

12. Record the zero output on the external load cell and LVDT.

13. Raise the triaxial cell into contact with the piston, then lower the cell by several millimeters to create a small gap between the piston and top cap.

14. Apply the confining pressure.

Shear Undrained

1. Use a strain rate of 1 percent per minute for plastic soils and 0.3 percent per minute for brittle soils.

2. Start the data acquisition system measuring time, displacement, force, cell pressure, and input voltage at a reading interval of 2 seconds. The time interval can be increased several times during shear by factors of 2 to 3 up to twenty seconds.

3. Shear the specimen by engaging the axial motor drive system.

4. Continue to shear until a failure plane has developed or 15 percent strain.

Dissemble

1. Turn off the axial motor.

2. Drain the cell fluid back into the storage container.

3. Disconnect the gage bar.

4. Remove the clear cylinder.

5. Dry off the specimen.

6. Roll the membrane down and remove the specimen.

7. Measure total mass, dimensions, and failure plane information. Take a photo of the failed specimen, if desired.

8. Cut the specimen into three pieces and measure water contents.

Calculations

1. Calculate the phase relationships of the test specimen. Refer to Chapter 2 for further guidance.

2. Calculate the specimen compression due to the cell pressure confinement, ΔH_{cp}, using Equation 16.9:

$$\Delta H_{cp} = \left[\frac{V_{dt,os}}{V_{in,dt,os}} - \frac{V_{dt,0}}{V_{in,dt,0}} \right] \times CF_{dt} \qquad (16.9)$$

Where:

ΔH_{cp} = specimen compression due to the cell pressure confinement (mm)
$V_{dt,os}$ = output voltage at the offset reading on the LVDT (V)
$V_{in,dt,os}$ = input voltage to the load cell at the offset reading (V)
$V_{dt,0}$ = output voltage at the LVDT zero reading (V)
$V_{in,dt,0}$ = input voltage to the LVDT at the zero reading (V)
CF_{dt} = calibration factor for the LVDT (mm/V/Vin)

3. Calculate the axial strain at each reading m, $\varepsilon_{a,m}$, using Equation 16.10:

$$\varepsilon_{a,m} = \frac{\Delta H_m}{H_0} \times 100 \qquad (16.10)$$

Where:

$\varepsilon_{a,m}$ = axial strain at reading m (%)

ΔH_m = change in height of the specimen determined at each reading as compared to the offset using the LVDT (mm)

H_0 = initial height of the specimen (mm)

4. Calculate the area at each reading m, A_m, using Equation 16.11.

$$A_{c,m} = \frac{A_0}{(1 - \varepsilon_{a,m}/100)} \qquad (16.11)$$

Where:

$A_{c,m}$ = area of the specimen at reading m (m^2)

A_0 = initial area of the specimen (m^2)

5. Calculate the correction to the axial load due to piston uplift and seal friction, $P_{LC,c}$, using Equation 16.12:

$$P_{LC,c} = \left[\frac{V_{l,os}}{V_{in,l,os}} - \frac{V_{l,0}}{V_{in,l,0}} \right] \times CF_l \qquad (16.12)$$

Where:

$P_{LC,c}$ = axial load correction due to piston uplift and seal friction (N)

$V_{l,os}$ = output voltage at the offset reading on the load cell (mV)

$V_{in,l,os}$ = input voltage to the load cell at the offset reading (V)

$V_{l,0}$ = output voltage at the load cell zero reading (mV)

$V_{in,l,0}$ = input voltage to the load cell at the zero reading (V)

CF_l = calibration factor for the load cell (N/mV/Vin)

6. Calculate the corrected axial load at each reading m, P_m, using Equation 16.13:

$$P_m = P_{LC,m} - P_{LC,c} \qquad (16.13)$$

Where:

P_m = corrected axial load at reading m (N)

$P_{LC,m}$ = axial load measured by the load cell at reading m, using Equation 16.12, but replacing the offset reading values with those at reading m (N)

7. Compute the correction for the membrane at each reading m, $\Delta(\sigma_1 - \sigma_3)_m$, using Equation 16.14:

$$\Delta(\sigma_1 - \sigma_3)_m = \frac{4 E_m t_m \varepsilon_{a,m}}{D_m} \qquad (16.14)$$

Where:

$\Delta(\sigma_1 - \sigma_3)_m$ = membrane correction at reading m (Pa)

E_m = elastic modulus of the membrane (Pa)

t_m = thickness of the membrane (m)

D_m = diameter of the specimen at reading m (m)

8. Calculate the corrected deviator stress at each reading m, $(\sigma_1 - \sigma_3)_m$, using Equation 16.15:

$$(\sigma_1 - \sigma_3)_m = \frac{P_m}{A_{c,m}} - \Delta(\sigma_1 - \sigma_3) \qquad (16.15)$$

Where

$(\sigma_1 - \sigma_3)_m$ = corrected deviator stress at reading m (Pa)

9. Plot the stress-strain curve for the test and the Mohr's circle for the failure condition.

Report

1. A summary of the phase relationships for the specimen(s) at each phase of testing.
2. A summary of shear results including:
   - Tabular summary data
   - Compressive strength, σ_1, and σ_3 at failure and a plot of the Mohr's circles
   - Rate of axial strain
   - Axial strain at failure
   - Plot of stress versus strain

PRECISION

Criteria for judging the acceptability of test results obtained by this test method have not been determined. However, it is reasonable to expect that two tests performed properly on the same material will produce an undrained strength value within about 20 percent of the average value.

DETECTING PROBLEMS WITH RESULTS

There are many possible sources of error in the UU test. Problem solving is made all the more difficult because pore pressure is not measured. Most notable deviations from ideal behavior include different values of deviator stress for different values of confining pressure for a series of specimens that are believed to be saturated. If deviations from ideal behavior are detected, the following suggestions are made to isolate the effects.

Results that fall outside of typical ranges may be due to improper determination of the axial load correction. Review previous testing records for the value of the axial load correction. This value should remain constant from test to test at similar confining pressures. Dramatic changes to the value can be caused by a bent piston or improper maintenance of the equipment, in which case repair or maintenance is necessary. Make sure that the seal on the piston is properly lubricated, and that the piston surface is smooth.

The interpreted undrained shear strength can be too low if a failure plane has developed prematurely. This can be caused by tilting of the top cap, the diameter of the top cap or bottom pedestal being too small, or a preexisting failure plane. Verify that the top cap is not allowed to tilt during testing. Review the testing remarks for failure planes initiating at the cap or pedestal location, a sign that the specimen is too large for the setup. Also verify that the specimen failure surface does contain a silt seam or a preexisting failure surface, which would lead to a low interpreted undrained shear strength.

Abnormally high friction angles can be the result of testing a series of specimens at confining pressures spaced too closely together. Confining pressures should normally bracket the total stress of interest and be spaced sufficiently apart. A reasonable spacing of confining pressures might be at least the expected undrained shear strength for the specimen.

Verify that the specimen is saturated by checking the dimensions, the water contents, and the value of specific gravity used to determine the level of saturation for accuracy. Compare the handheld shear vane and/or pocket penetrometer testing results performed adjacent to the specimen to determine whether the material is cemented. Verify that the specimen is not compressing due to the application of cell pressure. This can be the case for unsaturated specimens.

Make sure that drainage into the specimen is not occurring by checking the data sheets for observations of the specimen being wet after removal of the membrane. Also check the equipment for signs of leakage, including changes in cell pressure or abnormal changes in the volume of confining fluid used during the test. Note that changes in the volume of confining fluid may not be observable, depending on the type of equipment used to introduce the fluid and to apply confining pressure.

Check for possibility of dilatancy by reviewing the material description, if the Visual-Manual procedures were followed. Otherwise, specialty equipment would be required to measure pore pressures during testing and verify that negative pore pressures are not being generated. If they are, higher confining pressures will be required to prevent cavitation during shearing.

REFERENCE PROCEDURES

ASTM D2850 Unconsolidated-Undrained Triaxial Compression Test on Cohesive Soils.

REFERENCES

Refer to this textbook's ancillary web site, www.wiley.com/college/germaine, for data sheets, spreadsheets, and example data sets.

Holtz, R. D. and W. D. Kovacs. 1981. *An Introduction to Geotechnical Engineering*, Prentice-Hall, Englewood Cliffs, NJ.

Ladd, C. C. and U. Luscher. 1965. "Preliminary Report on the Engineering Properties of the Soils Underlying the MIT Campus," *MIT Research Report R65–58*, Cambridge.

Lambe, T. W., and R. V. Whitman. 1991. *Soil Mechanics*, John Wiley and Sons, New York.

Terzaghi, K., R. B. Peck, and G. Mesri. 1996. *Soil Mechanics in Engineering Practice*, John Wiley and Sons, New York.

Chapter 17

Incremental Consolidation By Oedometer

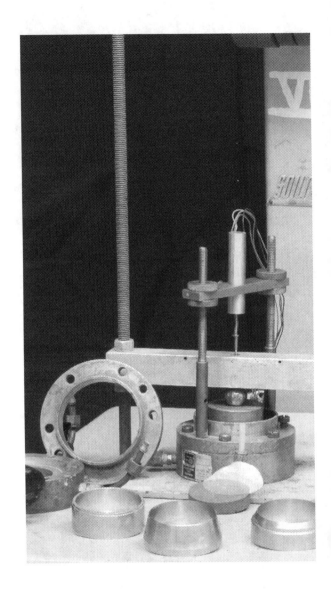

SCOPE AND SUMMARY

This chapter provides background material on consolidation and detailed procedures to perform the incremental consolidation test using an oedometer. Important factors that impact interpretation of measurements are discussed. Methods to interpret the consolidation properties, such as preconsolidation stress, compression ratios, rates of deformation, and so on, are described.

TYPICAL MATERIALS

Consolidation tests are typically performed on intact, cohesive soil materials or soils that consolidate slowly enough such that the time rates of deformation are of engineering concern. Equipment is commonly available to perform consolidation tests on soils with a value of coefficient of consolidation less than 10^{-7} m^2/s. The oedometer test is also performed on noncohesive materials to obtain the one dimensional compression characteristics.

BACKGROUND

Volume change is an important behavior of particulate materials. Volume change can occur by compaction, consolidation, compression, and secondary compression. Unfortunately, the terminology used in the profession is somewhat vague on this topic.

Compaction has already been presented in Chapter 12. The following discussion provides a brief overview of the other volumetric related processes.

Primary consolidation is the process of progressive volume change due to dissipation of excess pore pressure. Consolidation generally occurs at a constant degree of saturation. The material does not need to be fully saturated, but the air voids cannot be continuous. The fluid (plus occluded air) is forced out of the pore space under a pressure gradient. Consolidation is a coupled process between flow and volume change. The rate of consolidation depends on the amount of volume change, or compressibility, as well as the ability of water to flow through the soil matrix, called hydraulic conductivity. When the process causes a decrease in volume it is called consolidation, and when the process causes an increase in volume it is called swell. In either case, the excess pore pressure can be caused by a change in total stress, a change in the boundary pore pressure, or changes in the soil structure due to shearing.

Compression is a rate-independent process, and is volume change associated with a change in effective stress. It is a component of consolidation and characterizes the movement of the particles due to the reaction in interparticle forces. When the volume change causes contraction, it is called compression. The opposite of compression is rebound.

Secondary compression is time-dependent volume change at constant effective stress. Secondary compression is one specific type of creep. As used in the context of consolidation, the term highlights the volume change aspect, and distinguishes this process from undrained and drained shear creep, which are time-dependent shear deformations. Secondary is the logical adjective to separate secondary compression changes in volume from primary consolidation changes. Secondary compression will be used in this text for a volume decrease and secondary rebound for volume increase.

The four processes are applicable to general, three-dimensional stress conditions. The standard incremental consolidation test limits the boundary conditions to one dimensional flow and strain. This greatly simplifies the experimental measurement and also simulates a very common situation encountered in engineering practice. ASTM D2435 One-Dimensional Consolidation Properties of Soils Using Incremental Loading covers incremental consolidation testing using an oedometer.

Results of a consolidation test can be used in many different ways. The interpreted parameters can be used to predict the magnitude and rate of settlement, as well as to estimate the percent consolidation at various times under a single increment in the field. The stress history of the deposit can be defined, which in turn can be used to develop strength parameters through the Stress History and Normalized Soil Engineering Properties (SHANSEP) approach (Ladd and Foott, 1974; Ladd, 1991).

The incremental consolidation test (often referred to as an oedometer test) is performed by applying a series of constant stress levels to the specimen. The deformation is measured versus time during each constant stress increment. Consolidation theory is used to interpret each stress increment. A photograph of typical incremental loading oedometer equipment is provided as Figure 17.1.

Axial deformation is the single measure of the soil's response to the stress increment. The deformation results can be presented in terms of settlement, strain, or void ratio. All three values are linear variations of the same deformation measurement. The ensuing discussions in the chapter will be presented only in terms of strain; however, the theory and applications apply equally well in terms of settlement or void ratio. Compression and volume decreases are taken as positive. The axes will be reversed when working in void ratio space because as strain increases, void ratio decreases.

A typical test might extend over several weeks and require more than a dozen stress increments. Each stress increment yields a "time curve," which is a plot of strain versus time over the duration of the increment. The curve must be interpreted to determine the rate of consolidation and the strain associated with the level of stress.

Figure 17.2 presents a typical time curve for a positive stress increment (loading). In this case, the time is plotted on a log scale to illustrate the important features of the process. While the actual strain rate is always decreasing with time, this transformation

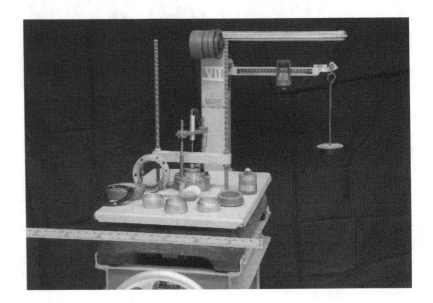

Figure 17.1 Photograph of typical incremental oedometer and load frame.

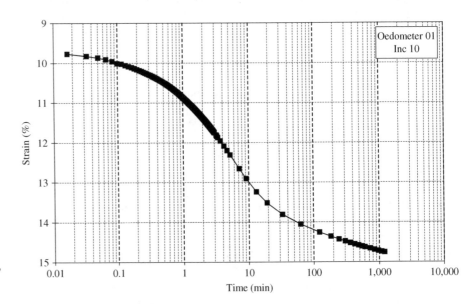

Figure 17.2 Typical "time curve" for a load increment.

of the time axis gives the impression that the rate first increases and then decreases. At early times, the strain is controlled by primary consolidation, which is described by consolidation theory. This phase continues up to the point when the curve becomes log-linear. This log-linear process is secondary compression. The transition between primary consolidation and secondary compression is generally gradual. In theory, this would be a well-defined point in the test when the pore pressure is zero.

The time curve is used to determine the strain at the end of primary consolidation, which is a value that depends on soil behavior. The strain at the end of the increment depends on how long the increment remains on the specimen. The stress increment is usually maintained for at least one log time cycle past the time corresponding to the end of primary consolidation. This provides sufficient information to properly interpret the measurements. The rate parameter is the coefficient of consolidation. The magnitude of deformation and the coefficient of consolidation are characteristics of the material. The details of the interpretation will be discussed later in this chapter.

The strain at the end of primary consolidation is plotted versus the axial effective stress, σ'_a, for each increment. Figure 17.3 presents results, including an unload-reload

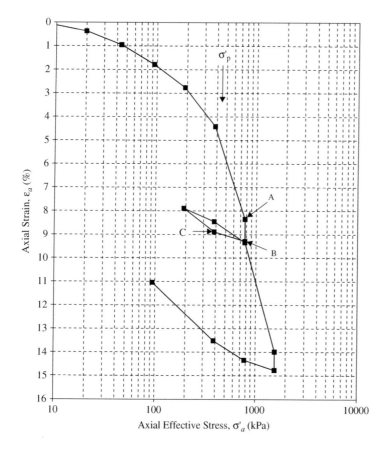

Figure 17.3 Typical compression curve measured on a Boston Blue Clay specimen.

cycle, for a test on Boston Blue Clay. The curve combines the results from all the time curves, providing a complete representation of the compressibility of the material.

Two data points must be plotted for every stress increment prior to a stress reversal. (See point A and B on Figure 17.3.) This accounts for the secondary compression (or secondary rebound) that occurred in the increment prior to reversal. The secondary compression strain is permanent and a curve connecting end of primary points will erroneously yield a negative slope at each stress reversal, such as if points A-C were connected instead of A-B-C.

The compression curve is used to interpret the compressibility characteristics of the specimen. Loading behavior is separated into two domains: small strain "elastic" compressibility and large strain "plastic" compressibility. These domains are separated by the yield stress, which in one-dimensional loading is called the preconsolidation stress, σ'_p. The material is normally consolidated when stressed beyond the preconsolidation stress. The compressibility is nearly constant when plotting strain versus log stress in the normally consolidated domain. This linear portion of the plot is sometimes referred to as the virgin compression line (VCL). Anytime the material is inside (to the left of) the normally consolidated line, it is referred to as overconsolidated. The compressibility in the overconsolidated domain will depend on the distance away from the normally consolidated condition. It will also be different in loading and unloading conditions.

The volumetric response of soils is the result of several mechanisms. Elastic deformation occurs due to the particle deformation, particle bending, and changes in particle spacing. Particle deformation can occur at stress concentration points as well as within the grain structure. Bending is likely limited to platy-shaped clay particles. Particle spacing changes will occur at contact points and between aligned particles of fine-grained materials as double layers are pressed together. The elastic deformations are recoverable and cause the rebound strain upon stress removal.

Plastic deformation occurs due to particle reorientation, particle crushing, and destruction of cementation bonds. Particle reorientation can be relative slip at particle contacts or rotation of individual particles. Crushing can be of individual grains or chips off the edges of grains. These plastic mechanisms result in irrecoverable strain. The plastic deformation is the difference in strain at a given stress between two unload curves. The nonlinear behavior of the soil causes a difference in plastic strain with the amount of unloading.

The incremental oedometer test provides a wealth of information about the specimen. It is rather surprising just how much is obtained from a test that only deforms the material in one dimension. A test performed with time-deformation measurements for every increment will provide the following information:

- Plots of strain, ε_a, versus time for each stress increment
- Interpretation of the time curves to get the strain associated with each stress level
- Interpretation of the time curve to get the coefficient of consolidation, c_v
- Interpretation of the time curves to get rate of secondary compression, $C_{\alpha\varepsilon}$
- Plot of strain versus applied axial effective stress, σ'_a
- Preconsolidation stress, σ'_P
- Compression ratio (CR), recompression ratio (RR), and swelling ratio (SR)
- Hydraulic conductivity, k_v, for each stress increment

Consolidation Theory

The process of consolidation is often described in terms of a piston and spring analogy, as illustrated in Figure 17.4. The analogy provides a descriptive visual tool to develop an understanding of the process and the important factors.

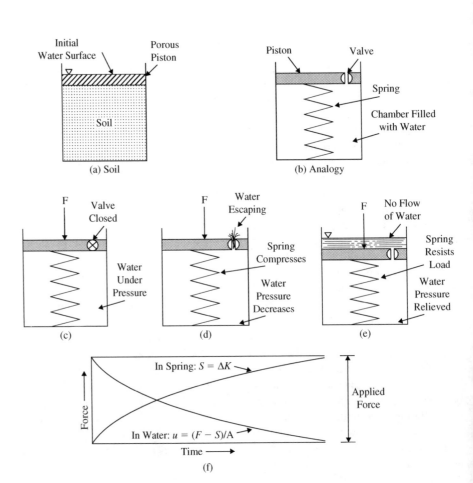

Figure 17.4 Spring analogy for consolidation.(Adapted from Lambe and Whitman, 1969. Copyright John Wiley and Sons. Printed with permission.)

In this analogy, the valve represents the hydraulic conductivity of the soil, the spring represents the soil matrix, and the water represents the pore fluid, as in (a) and (b) in Figure 17.4. A larger valve mimics higher hydraulic conductivity and a stiffer spring would be analogous to a stiffer soil skeleton. Starting with the piston full of water, the valve open, and no force applied to the piston would result in no water pressure and no spring deflection, as in (b) of Figure 17.4. If the valve is closed and a load is placed on the piston, the water would develop a pressure equal to the force divided by piston area to balance the force increment, as in (c) of Figure 17.4. Provided the water is stiffer than the spring, the spring force would be zero and no deflection would result. Opening the valve will allow water to flow from the chamber and the piston to move down, compressing the spring, as in (d) of Figure 17.4. The flow rate will be driven by the water pressure and, as the water leaves the chamber, the chamber pressure will decrease as force is redistributed to the spring, as shown graphically in (f) of Figure 17.4. Clearly, the flow rate will decrease with time until all the pressure is completely relieved and the force is fully transferred to the spring. The final state is shown in frame (e) of the figure. The time duration required for the water pressure to return to zero depends on two elements of the system: the size of the valve and the stiffness of the spring. The duration of flow would be similar if the valve were small and the spring stiff as when the valve was large and the spring soft. This illustrates the reason the consolidation is a coupled behavior between two fundamental material properties.

The time curves generated from consolidation are most commonly interpreted using Terzaghi's theory of consolidation (Taylor, 1948). There are other, more complex, theories that account for factors such as large strain deformation and secondary compression. Terzaghi's theory is relatively simple to apply and captures the essentials of the consolidation process. The following discussion is put in terms of the test specimen, but applies equally to a layer of fine-grained material in the field.

Consolidation theory is necessary to make sense of the fact that strain varies continuously with time for every stress increment. Without a method of interpretation, it would not be possible to construct a stress-strain curve or to quantify the rate of strain going from one stress point to the next. Terzaghi's theory of consolidation provides a powerful tool to guide this interpretation, but the theory makes a number of assumptions. The assumptions are as follows:

- The material within the test specimen is uniform.
- Soil grains and water molecules are incompressible.
- Flow and deformation are one-dimensional.
- The pore space is completely saturated with fluid.
- The specimen does not change in dimension.
- Darcy's Law is valid.
- Soil hydraulic conductivity is constant.
- Soil compressibility is linear.

The first three assumptions are relatively straightforward and easily satisfied in the laboratory. Saturation of the specimen is often a problem and is certainly cause for concern. The classic interpretation does not consider the change in thickness of the specimen, and is referred to as the small strain assumption. For compressible soils, this will create a small variation in the rate curve. The last three assumptions are extremely important. Flow gradients in the specimen are very large, which may alter the rate of consolidation as energy is dissipated in turbulent flow. It is clear that both the hydraulic conductivity and the compressibility of the soil change with effective stress. As the stress increment becomes larger, these assumptions become more problematic. While the assumptions are not strictly true, the theory remains an invaluable contribution to the profession. It is necessary to evaluate these assumptions when applying the results of a consolidation test to the field situation.

Based on these simplifying assumptions, Terzaghi developed the very elegant governing equation shown as Equation 17.1 to describe the process of consolidation:

$$c \frac{\partial^2 u_e}{\partial z^2} = \frac{\partial u_e}{\partial t} - \frac{\partial \sigma}{\partial t} \tag{17.1}$$

Where:

  c = coefficient of consolidation (m^2/sec)
  u_e = excess pore pressure (kPa)
  z = position within specimen (m)
  t = time (sec)
  σ = total stress (kPa)

When applied to the consolidation test, the variation in total stress with time (the last term on the right) is zero because the applied stress is constant throughout the increment. This reduces the equation to a second order differential equation describing the variation in excess pore pressure, u_e, with time and position within the specimen. The material is characterized by a single parameter: the coefficient of consolidation. This is not a material property, but rather a lumped parameter derived from two independent material properties: the hydraulic conductivity and the compressibility. This critical simplification is the source of the last two assumptions in Terzaghi's theory of consolidation.

The solution to Equation 17.1 for the boundary conditions of a specimen with perfect drainage at the top and bottom boundaries provides the excess pore pressure as a function of time and position. This result is provided in Equation 17.2:

$$u_e = \sum_{m=0}^{m=\infty} \frac{2u_i}{M} \left(\sin MZ \right) e^{-M^2 T} \tag{17.2}$$

Where:

  m = series counter (integer)
  u_i = initial excess pore pressure (kPa)
  M = dummy variable (dimenionless)
  Z = position factor (dimensionless)
  T = time factor (dimensionless)

M is given by Equation 17.3:

$$M = \frac{\pi}{2}(2m + 1) \tag{17.3}$$

The number of terms required to obtain a stable value of the excess pore pressure varies greatly, and depends on the time and position. Convergence must be checked when using Equation 17.2. Time is represented in the solution by a dimensionless time factor, T, given by Equation 17.4. The time factor couples the coefficient of consolidation with the drainage conditions and time.

$$T = \frac{ct}{H_d^2} \tag{17.4}$$

Where:

  c = coefficient of consolidation (m^2/s)
  H_d = drainage height of the specimen (m)

The dimensionless position factor, Z, given by Equation 17.5, scales a given location in the specimen by the drainage height.

$$Z = \frac{z}{H_d} \tag{17.5}$$

Where:

z = distance from the top of the specimen (m)

The consolidation ratio at any location within the specimen, U_z, is calculated as Equation 17.6. This is the amount of pore pressure that has dissipated and is normalized by the initial excess pore pressure.

$$U_z = 1 - \frac{u_e}{u_i} \qquad (17.6)$$

Where:

U_z = consolidation ratio at a particular location and time (dimensionless)

The solution to Equation 17.6 is typically given graphically, as shown in Figure 17.5. The figure on the left shows the soil specimen with free-draining top and bottom boundaries. In this case, the drainage height would be half the specimen thickness because gravity forces are ignored in the solution and water can flow to either boundary. The consolidation ratio varies from zero to unity as consolidation progresses. Each contour line corresponds to a constant time factor and is called an isochrone. The isochrones are symmetric about the midheight of the specimen for double drainage. At time equal to zero, the pore pressure is uniform throughout the specimen. It is interesting to note that the rate of pore dissipation at the middle of the drainage height is independent of specimen thickness until after the time factor reaches about 0.05.

While Equation 17.6 and Figure 17.5 are informative, the basic solution is not useful for test interpretation. It would require measuring pore pressure within the specimen at a particular location and then matching the rate of change with the theory in order to compute the coefficient of consolidation and the strain corresponding to the end of consolidation. This is not useful in the laboratory for test interpretation, but it is extremely valuable for field assessment of the degree of consolidation.

The average degree of consolidation is a similar concept to the consolidation ratio. In the calculation, excess pore pressure is integrated over the specimen height in order to obtain an average value at each value of the time factor. The result is then normalized to the initial average value. The average degree of consolidation for the specimen, U, is given as Equation 17.7:

$$U = 1 - \frac{\int_0^{2H_d} u_e \, dz}{\int_0^{2H_d} u_i \, dz} \qquad (17.7)$$

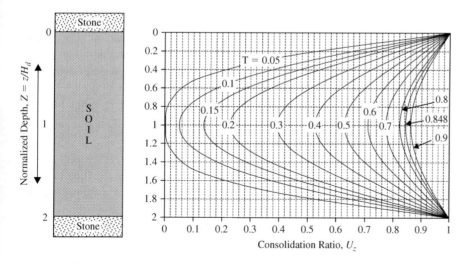

Figure 17.5 Percent consolidation as a function of position for double drainage.

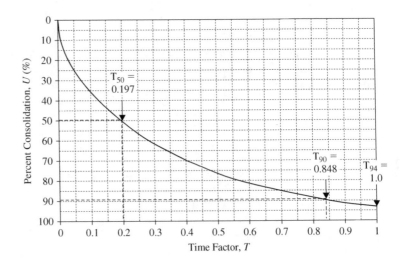

Figure 17.6 Percent consolidation, U, as a function of Time Factor, T.

The relationship between the average degree of consolidation and the time factor is presented in graphical format as Figure 17.6. Strictly speaking, this does not help with the interpretation of the consolidation test. In order to compute U, it would be necessary to measure pore pressure at a number of locations in the specimen, fit a distribution through the measurements, and then compute an average value of pore pressure. Clearly, the pore pressures are not going to be measured at multiple locations over time in a specimen. Interpretation of the test data is made possible by assuming that degree of consolidation is proportional to the specimen deformation. This assumption is only true if the soil has a linear stress-strain relationship.

The relationship between the average degree of consolidation and the time factor is an essential tool for interpretation of the consolidation process. It predicts that the rate of consolidation proceeds at an ever decreasing rate (i.e., the curve is concave upward). The curve asymptotically approaches unity. This means that, in theory, consolidation never truly finishes. Two characteristic points are commonly used in the interpretation for the coefficient of consolidation: T_{50} and T_{90}. These points are indicated in the figure.

Since the relationship is often used to compare laboratory and field settlement measurements, it is useful to have equations rather than a graph. When the value of U is less than 0.6, the time factor can be calculated using Equation 17.8:

$$T = \frac{\pi}{4}U^2 \tag{17.8}$$

Equation 17.8 highlights another very important characteristic of the consolidation process. The equation shows that the degree of consolidation (or strain) should be linear when plotted against the square root of time for the first 60 percent consolidation. When the value of U is greater than 0.6, the time factor can be calculated using Equation 17.9. This is an approximate fit to the solution.

$$T = -0.933 \log(1 - U) - 0.085 \tag{17.9}$$

Material Properties

The above theoretical solution greatly simplifies the representation of material properties. When working with soils, it is necessary to recognize and differentiate the reality that material behavior is directional specific. This is normally handled by adding a subscript (or multiple subscripts) to the parameter of interest. The coefficient of consolidation, c, in the Terzaghi solution has no subscript. In practice, the symbol should always have a subscript to indicate the direction of the flow and deformation. In the following definitions, the subscript "v" is used to denote a specimen tested in the vertical direction.

The coefficient of consolidation is the material parameter used to scale the theoretical solution, and has units of length squared per time. Through the consolidation test, the coefficient of consolidation can be determined by evaluating the change in height of a specimen versus time during an increment of loading. The definition of the coefficient of consolidation is shown as Equation 17.10:

$$c_v = \frac{T_v H_d^2}{t} \qquad (17.10)$$

Where:

c_v = coefficient of consolidation in the vertical direction (m²/s)
T_v = time factor (dimensionless)
H_d = drainage height in the test at 50% consolidation of the increment (m)

The coefficient of volume change, m_v, is the slope of the compression curve when stress is plotted on a natural scale, and is given by Equation 17.11[1]. In practice, judgment is required to select the most appropriate value. One could use the values from individual stress increments, or fit a compression curve through all the measured points and the take the slope of the interpreted curve.

$$m_v = \frac{-\Delta \varepsilon_a}{\Delta \sigma'_a} \qquad (17.11)$$

Where:

m_v = coefficient of volume change in the vertical direction (kPa⁻¹)
ε_a = axial strain in the test specimen (decimal)
σ'_a = axial effective stress applied to the test specimen (kPa)

The compression ratio, CR, is the slope of the virgin compression line on a plot of strain versus the log of the axial effective stress. The relationship is given by Equation 17.12:

$$CR = \frac{\Delta \varepsilon_a}{\Delta \log \sigma'_a} \qquad (17.12)$$

Where:

CR = compression ratio (decimal)

The recompression ratio, RR, is the slope of the recompression line on the same plot, whereas the swelling ratio, SR, is the slope of the swelling line on the plot of strain versus axial effective stress.[2] Determination of these parameters involves interpretation and the application of engineering judgment. As a matter of convention, there are no subscripts for the compressibility terms in the log stress space, and the vertical direction is implied unless specified otherwise. Compression behavior can not be represented precisely in both natural and log stress space. The suitability of the representation depends on the stress range of interest.

---

[1] Alternatively, this parameter can be described in void ratio space as the coefficient of compressibility, a_v.

[2] The compression index, C_c, is the slope of the virgin compression line on a plot of void ratio versus log of the vertical effective stress. Likewise, the recompression index, C_r, is the slope of the recompression line (i.e., loading curve from start of loading approaching the preconsolidation pressure) on the same plot, and the swell index, C_s, is the slope of the swelling line. To convert any of the indices into ratios, the index must be divided by one plus the initial void ratio.

The hydraulic conductivity, k_v, describes the rate at which a fluid is able to move through the soil matrix, and has units of length per time, as shown in Equation 17.13. The hydraulic conductivity is indirectly computed from the compressibility and the coefficient of consolidation for each consolidation increment. The calculation is considered indirect because the computed value depends on all the approximations embedded in the theoretical solution. Equation 17.13 establishes that the coefficient of consolidation is a coupled parameter related to the hydraulic conductivity of the soil, the compressibility of the soil, and the mass density of fluid passing through the material. Hydraulic conductivity and compressibility are the material properties.

$$k_v = c_v m_v \rho_w g \qquad (17.13)$$

Where:

k_v = hydraulic conductivity in the vertical direction (m/s)
ρ_w = mass density of water (g/m^3)
g = acceleration due to gravity (m/s^2)

The rate of secondary compression, $C_{\alpha\varepsilon}$, is the slope of the straight line portion of the curve after primary consolidation on a plot of strain versus log of time plot.[3] The rate of secondary compression is given as Equation 17.14. Once again, the directional subscript is not included for secondary compression.

$$C_{\alpha\varepsilon} = \frac{-\Delta\varepsilon_a}{\Delta \log t} \qquad (17.14)$$

The preconsolidation stress is the maximum consolidation stress previously experienced by the soil. The preconsolidation stress occurs at the rounded portion of the consolidation curve, and its value can be determined through graphical constructions performed on various representations of strain versus axial stress plots. In reality, the transition is gradual. Two constructions will be described in detail later in this chapter. Conceptually, the preconsolidation stress separates the elastic and plastic range of loading. During a laboratory test, once the consolidation stress exceeds the in-situ preconsolidation stress, the maximum stress applied to the specimen becomes the new preconsolidation stress. The overconsolidation ratio, OCR, is the preconsolidation stress divided by any lesser axial effective stress on an unload-reload curve, as shown in Equation 17.15:

$$OCR = \frac{\sigma'_p}{\sigma'_a} \qquad (17.15)$$

Where:

OCR = overconsolidation ratio (dimensionless)
σ'_p = preconsolidation stress (kPa)

When characterizing the in-situ stress history, the preconsolidation stress from the test specimen would be divided by the in situ vertical effective stress, σ'_{v0}, obtaining the overconsolidation ratio. The preconsolidation stress in the field may be greater than the maximum consolidation stress. The additional increment can be the result of secondary compression, cementation, chemical alteration, and so on.

Interpretation of Time Curves

The strain versus time measurements from each increment must be analyzed in an attempt to obtain the soil characteristics. The purpose for evaluating each increment is five-fold:

---

[3] The rate of secondary compression in terms of void ratio has the symbol $C_{\alpha e}$.

- To determine the end of primary consolidation
- To calculate the coefficient of consolidation
- To calculate the hydraulic conductivity
- To determine the coefficient of secondary compression
- To determine the primary compression ratio

It is often challenging, and sometimes impossible, to obtain all the values from a time curve. There are many reasons for the task to be difficult. There are procedural difficulties in applying the change in stress at the same time as measuring the start time of the load increment. In many situations, the material does not satisfy all the assumptions embedded in the theory resulting in a different curve shape. At times, the stress increment does not produce sufficient pore pressure to generate a characteristic curve, or secondary compression strain alters the curve shape. The procedural details of the test are designed to create interpretable curves, but complications are common.

Many procedures have been developed to aid in the interpretation of the consolidation curve. The reasons are simple. First, the theoretical curve of degree of consolidation versus the time factor does not have any particular identifying features. It is simply a smooth curve with a decreasing negative slope. Second, the theory does not include secondary compression, which is part of soil compression behavior. The various procedures transform the axes in ways that enhance particular features of the theoretical relationship.

Two common methods are used to interpret time curves for the coefficient of consolidation and the end of primary consolidation. They are the Square Root of Time method (Taylor, 1948) and the Log Time method (Casagrande, 1936). In addition, the 3-t method is presented as a third method to estimate the end of consolidation. The methods of construction will be presented, as well as the important considerations relative to testing.

Square Root of Time Method

The Square Root of Time construction transforms the time scale by plotting strain versus the square root of time. The method is based on the fact that the first 60 percent of the theoretical curve is a parabola and will be linear when plotted in this format. A 15 percent offset line from this initial slope will intersect the theoretical curve at 90 percent consolidation. The method attempts to exploit this feature of the theory by performing a similar construction on the strain versus square root of time measurements.

A few details must be addressed before jumping into the interpretation. The exact time of load application is critical to proper interpretation of the time curves. Changing the value corresponding to zero time will have a measurable effect on the slope and intercept of the curve. The first data point on the plot corresponds to ε_0, which is equal to the final strain point of the previous increment. This strain value is not the start of consolidation.

The line drawn through the linear portion of the data will be extrapolated back to the axis to determine the strain corresponding to the start of consolidation, ε_s. The difference between ε_s and ε_0 is initial compression of the specimen and is a deviation from consolidation theory. A large value would indicate an unsaturated specimen or an apparatus calibration error. The 15 percent offset line is drawn to pass through ε_s and with a slope that is 1.15 times the best fit line through the data. The point at which the offset line intersects a curve through the data represents the square root of time to 90 percent consolidation, t_{90}, and the strain at 90 percent consolidation, ε_{90}. Refer to Figure 17.7 for an example of the Square Root of Time construction.

Given the characteristic points from the graphical procedure, it is now possible to compute the values of interest. The strain at the end of primary consolidation, ε_{100}, is computed using Equation 17.16:

$$\varepsilon_{100} = \left(\frac{\varepsilon_{90} - \varepsilon_s}{9} \right) \times 10 + \varepsilon_s \qquad (17.16)$$

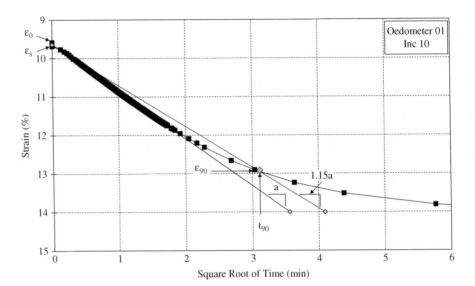

Figure 17.7 Square Root of Time method to determine the time to the end of primary consolidation.

The strain at 50 percent consolidation, ε_{50}, is computed using Equation 17.17:

$$\varepsilon_{50} = \left(\frac{\varepsilon_{100} + \varepsilon_s}{2} \right) \tag{17.17}$$

Consolidation theory does not consider changes in the height of the specimen, but this is an important factor in the test since the strain can be very large and the drainage height is squared in the calculation. The average drainage height during the increment is taken as the average height of the specimen (i.e., the height of the specimen at 50 percent consolidation) when the boundary conditions consist of one drainage boundary, called single drainage. When there are two drainage boundaries, called double drainage, the average height of the specimen is divided by two to obtain the average drainage height of the increment. The drainage height is calculated using Equation 17.18:

$$H_d = \left(1 - \frac{\varepsilon_{50}}{100} \right) \times \frac{H_0}{f} \tag{17.18}$$

Where:

H_0 = initial height of the specimen (m)
f = drainage condition factor (dimensionless)

The value of f is one for single drainage and two for top and bottom drainage.

The coefficient of consolidation based on the Square Root of Time fitting method is then calculated using Equation 17.19:

$$c_v = \frac{0.848 H_d^2}{t_{90}} \tag{17.19}$$

Where:

t_{90} = time to 90 percent consolidation for the increment (s)
0.848 = time factor corresponding to 90 percent consolidation (dimensionless)

The primary compression ratio is not a material property, but is used when evaluating the test results. The ratio provides a measure of the amount of strain in an increment that is due to primary consolidation. This is useful when looking for deviations from consolidation theory. The primary compression ratio is calculated using Equation 17.20:

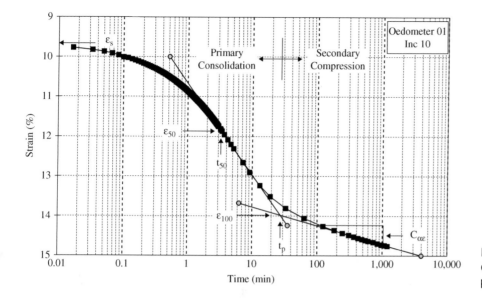

Figure 17.8 Log Time method to determine the time to the end of primary consolidation.

$$r = \frac{\varepsilon_{100} - \varepsilon_s}{\varepsilon_f - \varepsilon_0} \qquad (17.20)$$

Where:

ε_f = strain at the end of the increment (%)

The square root of time method has the advantage that it will always yield a result. In some situations, the start of the curve will not be very linear, but it is always possible to estimate a linear fit. This advantage is also a serious weakness of the method. Since the data will always produce a decreasing slope, there is no way to critically evaluate the reasonableness of the intersection point. For this reason, the square root of time method should be used with caution anytime the linear range is poorly defined. The method also has the advantage that it can be used to predict the end of primary consolidation. This is very useful in field applications, and when performing accelerated tests.

Log of Time Method

The Logarithm of Time construction transforms the time scale by plotting strain versus the time on a log scale. The method is based on the observation that the theoretical curve is S-shaped when plotted with the time factor on a log scale. Further, the central portion of the curve has a fairly extended log-linear section. The curve then asymptotically approaches 100 percent consolidation. Casagrande suggested using this feature of the theoretical curve and then rotating the coordinate to make the asymptote approach the rate of secondary compression. The method then provides a procedure to determine the end of primary strain. The following paragraphs present the details and calculations for the method.

The first task is to find the start of consolidation. Since the log plot does not have a zero time, many publications have presented a method to determine the start of consolidation using a construction in the log time representation. This part of the curve is often difficult to evaluate and it is much easier and more insightful to use the starting time and strain, ε_s, from the root time construction. The next step is to draw a tangent line to the steepest, downward slope of the time curve. This section should be well defined using several data points. The last line is drawn through the straight-line portion of the long-term section of the data. These data should also form a straight line. The intersection of the two lines represents the strain at the end of primary consolidation, ε_{100}, and the time corresponding to the "end of primary" consolidation, t_p. Remember, in theory the end of primary is at infinity. The final piece of information from the plot is the time corresponding to 50 percent consolidation, t_{50}. Refer to Figure 17.8 for an example of the Log of Time construction.

Given the characteristic information from the two curves, it is now possible to proceed with the calculations. The average drainage height, H_d, is computed using Equation 17.18, with the root time value of ε_s and the ε_{100} from the log construction. The coefficient of consolidation based on the Log Time fitting method is calculated using Equation 17.21:

$$c_v = \frac{0.197 H_d^2}{t_{50}}$$

(17.21)

Where:

t_{50} = time to 50 percent consolidation for the increment (s)
0.197 = time factor corresponding to 50 percent consolidation

The Log Time method is much more rigorous than the Root Time method because it places more requirements on the shape of the consolidation curve. If a curve results in the characteristic S-shape, then it is more likely that the material is conforming to the theory. The price for this additional level of confidence is time. Collecting enough information to establish the linear secondary slope adds significantly to the test duration.

Even on ideally shaped curves, the Root Time method yields a higher value of the coefficient of consolidation in laboratory scale tests. The coefficient of consolidation determined by the Square Root of Time method is usually about 2 +/- 0.5 times the value determined by the Log Time method. (Ladd, 1973, as appearing in Ladd et al., 1977) The reason for this discrepancy is still a topic for research, but is likely due to a combination of scale effects and the influence of secondary compression. In theory, there is no end to primary consolidation. The log time construction generally results in a slow transition into secondary compression, introducing concern as to what time to use as the end of primary consolidation. For consistency with the interpretation of e_{100}, the time for the end of primary should be taken at the intersection point, as shown in the figure.

3-t Method

In many situations, it is impossible to determine the end of primary consolidation for a stress increment. This will happen in the consolidation test when using small stress increments to better define the compression behavior, and is extremely common when consolidating test specimens in shear tests. The Root Time method will provide an estimate of the end of primary consolidation, but there is no way to tell if this overestimates or underestimates the time required for full pore pressure dissipation. The 3-t method provides an effective solution to this dilemma. The strain data are plotted on a log time scale and a line is drawn through the steepest portion of the data. A parallel line is drawn shifted by a factor of three larger in time. The intersection of this shifted line and the test data is taken as the strain at the end of primary consolidation. This method should not be used to compute the coefficient of consolidation, but does provide a predictive method to decide when to increment the load, to shear a test specimen, or obtain a strain value. The method is illustrated in Figure 17.9.

Rate of Secondary Compression

The rate of secondary compression is determined as the slope of the long-term portion of the time curve for an increment, as measured over a log cycle of time past the end of primary consolidation. An example of the determination of the coefficient of secondary compression is included in the Log Time interpretation of the time curve, as shown in Figure 17.8. The transition between primary consolidation and secondary compression is often gradual. Data should be available for at least one log cycle of time beyond the end of primary before accepting the slope as the rate of secondary compression. The rate of secondary compression will be overestimated if interpreted too early on the time curve.

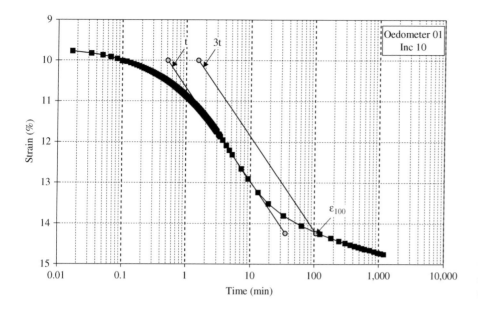

Figure 17.9 3-t method to determine the end of primary consolidation.

The compression relationship is the stress-strain response of the soil to one-dimensional loading. For each stress increment, there will be a strain value corresponding to the end of the increment and, if the time interpretation is successful, a second strain value corresponding to the end of primary consolidation. The end of primary (EOP) strain values will result in a unique compression curve, whereas the end of increment (EOI) strain values will depend on the amount of secondary compression in each individual increment. When constructing the EOP curve it is important to plot the EOI strain for each increment prior to a stress reversal. Interpretation of the compression measurements will provide:

Interpretation of Compression Behavior

- The preconsolidation stress
- The compression ratio, recompression ratio, and swelling ratio
- The coefficient of volume change to determine the coefficient of hydraulic conductivity

The ASTM standard D2435 does not require time readings to be taken for every increment. This testing method will produce a compression curve that includes some amount of secondary compression. The shift will result in more strain at any level of stress and a decrease in the interpreted preconsolidation pressure.

Many methods have been proposed to estimate the preconsolidation pressure from the compression measurements. Once again, this is a problem of transforming the plotting dimensions to create a better defined transition between the normally consolidated and over consolidated regions. Two methods will be presented in the following discussion. The Casagrande construction is simple and the time-honored procedure. The second method is the Strain Energy procedure, which is more objective and can be generalized to other test methods.

The Casagrande construction is performed on a plot of strain versus axial effective stress on a log scale. While it is very common to plot one-dimensional compression data in the log effective stress space, one should always review the compression results when stress is plotted on a natural scale. This often provides a very different impression of the measurements, and it is the natural stress scale that maps directly to depth.

Casagrande Construction

The first step in the interpretation is to draw a smooth curve through the data. Normally, the measurements will be spaced evenly across the log stress scale and it is unlikely that the preconsolidation stress will correspond to an individual point. Select a point on the curve that appears to have the most curvature (the minimum radius of

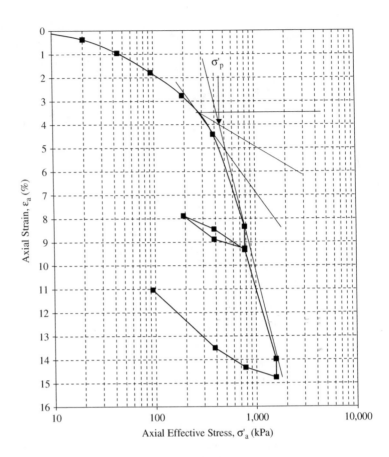

Figure 17.10 Casagrande Construction to determine the preconsolidation stress.

curvature). Draw a line tangent to the interpreted curve that passes through the selected point. Next, draw a horizontal line through the same point on the curve. Draw a third line bisecting the tangent and the horizontal lines. Finally, extend a line that represents the steepest portion of the interpreted normally consolidated compression curve. The intersection of the extension of the normally consolidated line and bisecting line gives the preconsolidation stress. Refer to Figure 17.10 for an example of the Casagrande construction.

Strain Energy

Strain energy is the work done per unit volume on the specimen. Becker et al. (1987) proposed a work-based procedure to determine the preconsolidation stress. The method plots work against the consolidation stress, with both plotted on a natural scale. The quantity of work is calculated for each increment as the sum of the average force for each increment multiplied by the increment in deformation and divided by the current volume. The work calculation can be performed under general stress and strain conditions as shown in Equation 17.22:

$$W = \int (\sigma_1 d\bar{\varepsilon}_1 + \sigma_2 d\bar{\varepsilon}_2 + \sigma_3 d\bar{\varepsilon}_3)$$

(17.22)

Where:

W = work done per unit volume on the specimen (kN-m/m^3)

$\bar{\varepsilon}_i$ = natural strain ($= \Delta l/l$) in each principle direction

σ_i = average stress in each principle direction (kPa)

In the case of the one-dimensional consolidation test, ε_2 and ε_3 are by definition equal to zero and Equation 17.22 reduces to a single term. The natural strain can be

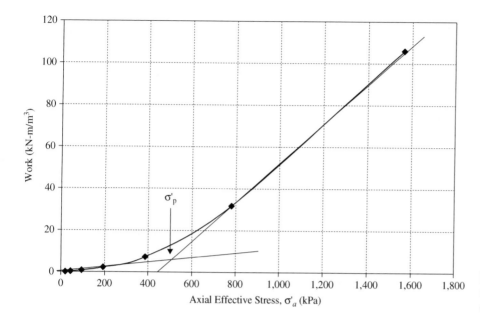

Figure 17.11 Strain Energy method to determine the pre-consolidation stress.

replaced by engineering strain ($\Delta l/l_0$) and the integral can be replaced by a summation over the increments of the test resulting in Equation 17.23:

$$W_j = \sum_{m=1}^{j} \left(\frac{\sigma_{a,m} + \sigma_{a,m-1}}{2} \right) \left(\frac{\varepsilon_{a,m} - \varepsilon_{a,m-1}}{1 - \varepsilon_{a,ave}} \right)$$

$$= \sum_{m=1}^{j} \left(\frac{\sigma_{a,m} + \sigma_{a,m-1}}{2} \right) \left(\ln \frac{1 - \varepsilon_{a,m-1}}{1 - \varepsilon_{a,m}} \right) \qquad (17.23)$$

Where:

W_j	= work per unit volume of the specimen up to increment j (kN-m/m³)
j	= index value for stress increment (integer)
$\sigma_{a,m}$	= stress at current stress increment (kPa)
$\sigma_{a,m-1}$	= stress at the previous increment (kPa)
$\varepsilon_{a,m}$	= strain at current increment (decimal)
$\varepsilon_{a,m-1}$	= strain at the previous increment (decimal)
$\varepsilon_{a,ave}$	= average strain during the increment (decimal)
m	= index used in summation (integer)

The work calculation is not unique. The value depends on the size of the increment. This is because the stress-strain (or force-deformation) curve is not linear. This is an important consideration when applying the calculation to tests having an unload-reload cycle. The Strain Energy construction involves plotting the work versus axial effective stress on a natural scale. The curve is geometrically similar to the strain versus log stress plot. A line is drawn through the high stress linear portion of the data. The low stress range should also approximate a straight line. This is often a matter of some judgment. The stress at the intersection of these two lines represents the preconsolidation stress. Refer to Figure 17.11 for an example of the Strain Energy construction.

The strain energy method has the advantage of plotting the stress on natural scale. This reduces the variation in the preconsolidation pressure due to subjective decisions about the slope of the initial straight line. The method can also be used to approximate upper and lower estimates by varying the slope of the initial line. The high stress line is generally very stable.

Equipment

The incremental consolidation test is performed using commercially available equipment. Several device variations are available to measure one-dimensional compression and consolidation properties of soils. The equipment is a combination of two independent components: the consolidometer and the loading device. The simplest form of a consolidometer is the oedometer. The standard oedometer is usually combined with a single-axis, gravity-actuated loading machine. More sophisticated versions of the oedometer are called consolidometers and are fitted with lateral stress measurements, or can be back-pressure saturated.

The loading frame must apply a constant force, in discrete steps. During each increment, the deformation is measured versus time. The load can be applied using masses on a lever arm, masses on a piston, a pneumatic load system, a hydraulic load system, or a screw-driven piston. The pneumatic and hydraulic load systems are advantageous in that computer control can be used to apply the force and sequence the loading. The pneumatic systems have the ability to apply low pressures, but have a relatively small maximum pressure that can be achieved. The hydraulic systems can apply high pressures, but not low pressures. The addition of computer control provides test automation as well as flexibility to perform controlled gradient or constant rate of strain consolidation testing. The screw-drive systems have the advantage (as do the gravity systems) of maintaining the load if power is lost.

The standard oedometer is by far the most common device used for the incremental loading consolidation test. The device has been modified over the years to simplify the design, but is functionally identical to the original devices. A schematic of the basic elements of the two oedometer geometries are provided in Figure 17.12.

The following items discuss the most important elements of the equipment.

- The oedometer is composed of all the elements that contain the soil and provide the appropriate boundary conditions. It consists of a base, drainage stones, the specimen ring, top cap, and water bath. The specimen is normally drained from both the top and bottom surfaces to atmospheric pressure.
- The water bath keeps the specimen from drying during the several days to several week long test. The standpipe provides drainage to the bottom surface. This need not be connected to the water bath, but must be kept full of water when performing rebound stress increments.
- The specimen ring contains the soil. The ring must be made of a noncorrosive material and have a smooth inside surface. Brass and stainless steel are common. The ring must be thick enough to prevent lateral strain of the specimen. The inside surface is normally coated with a thin layer of grease to reduce friction between the ring and soil.
- The specimen ring controls the geometry of the specimen. The diameter-to-height ratio of the specimens typically ranges from 2.5 to 4. The lower limit is to prevent excessive side wall friction and the upper limit comes from practical

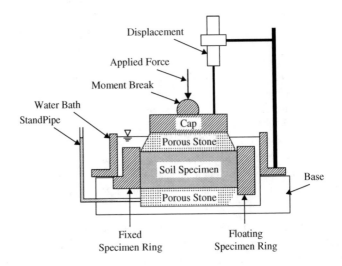

Figure 17.12 Sketch of typical oedometer with boundary conditions and applied forces.

concerns for trimming the specimen. The height of the specimen must be greater than about ten times the maximum particle size.

- The specimen ring can be either a floating ring design (right side of Figure 17.12) or a fixed ring design (left side of Figure 17.12). Different base and stone configurations are required of each ring design. The floating ring reduces the side wall friction by a factor of 2.

- The specimen is drained through the top and bottom surface. These drainage boundaries must be at least ten times more permeable than the specimen. The stones are normally made of carborundum, porous brass, or porous stainless steel. Carborundum is preferable because it does not bend. Metal stones can warp with use, which will cause errors in the deformation measurement. The top (and bottom when using a floating ring) stone diameter should be slightly less than the inside diameter of the ring. In addition, the stone should be tapered at about a 15 percent angle to a smaller diameter away from the soil. This prevents the stone from locking in the ring if tilted during the test. It should be possible to rotate the stone 30 percent about its axis while inside the ring. Stones should be cleaned on a regular interval and stored in water to extend useful life. Allowing the stone to dry will draw clay particles into the stone and eventually cause clogging.

- An interface filter is normally located between the stone and the soil. This thin layer prevents intrusion of particles into the stone. Traditionally, filters have been Whatman #54 paper. Monofilament nylon filters are available in a range of opening sizes and provide a better alternative. The paper filters are thicker and have complex compressibility behavior. The apparatus compressibility must be measured using the same loading sequence as in the test. Apparatus compressibility can be represented with a generic equation when using nylon filters.

- The top cap functions to distribute the concentrated load across the surface of the porous stone. It must be rigid and noncorrosive. The top cap should be about the same diameter as the matching surface of the top cap to facilitate alignment. The top surface of the cap should have a depression for the loading ball to provide a moment break. Traditionally, the top cap is allowed to rotate during the test. This is called a free top cap configuration. The design accommodates surface imperfections when trimming the specimen and allows non-uniform deformations during the test. More recent trends are to use a fixed top cap design, which prevents this rotation and forces the specimen to deform uniformly during the test.

- Deformation of the specimen is measured with a displacement transducer that uses the base of the oedometer as a reference surface. While Figure 17.12 shows this measurement being made off axis to the top cap, it is actually made concentrically to the top of a loading bar that rests on top of the moment break. Deformation measurements must be corrected for apparatus compressibility. This accounts for elastic compression of the apparatus, the stones, and the filter. Since the specimen is relatively thin, small errors in deformation will convert to large strain values. Consider the fact that the thickness of a piece of paper converts to about 0.5 percent strain! The system must have the ability to measure very small deformations (on the order of 0.0002 cm, or about 0.01 percent strain) while also having the capacity to measure large strains of up to 50 percent.

- The loading frame must be capable of applying constant forces over a range of three orders of magnitude, applying the force quickly, and still accommodate large deformations of the specimen. The required capacity will depend on the stiffness of the specimen and this controls the size of the frame. No matter what the capacity, the smallest controllable force should be 0.1 percent of the capacity providing measurements over three orders of magnitude. The challenge is then to provide good stability at the low stress levels, while at the same time having the high-stress capacity.

- The load must be applied rapidly. The actual time depends on the rate of consolidation. The log time plot (Figure 17.8), shows considerable deformation as early as 0.5 percent of the time to primary. Ideally, the load would be applied and stable by this time in the increment. For a specimen that reaches EOP in 15 minutes, this would translate to less than 5 seconds.

- Stability of the load throughout the increment is also an important consideration. Increment durations on the order of days are common, and weeks when trying to evaluate secondary compression. During primary consolidation, slight variations (<5 percent) will not cause serious variation in the time curve. However, variations in load will seriously impact the interpreted rate of secondary compression, and the adverse effects will become more severe the later they occur in the curve. This is an important limitation of load frames that require manual adjustment to maintain constant load.

- Stability of the room temperature can also be important. Thermal strain of both the soil and the apparatus will result in errors in the deformation measurement. Obvious sources, such as sunlight, heating units, air conditioner vents, and the like, should be avoided. Temperature stability is especially important during secondary compression. Systematic testing on a 24-hour load cycle can easily superimpose a daily temperature cycle on the secondary curve.

Overview of Procedures

Consolidation testing of intact specimens is a detailed process that requires many decisions prior to testing as well as during the test. It is best to have an estimate of the in-situ effective stress, the preconsolidation pressure, and sampling effective stress before starting the test. These values will impact the method of trimming, the loading schedule, and the capacity of the load frame. The following discussion provides an overview of the important steps in the test and the reasons for various decisions.

The test should be performed on the highest-quality material that has been sampled. Selecting a representative material type is important, but biasing the measurement toward the more plastic material will provide a better measure of the preconsolidation pressure. Radiography is a most valuable tool in selecting the test location. Chapter 11 provides a detailed discussion on extracting the soil from the tube and trimming the material into the specimen ring. The specimen must fit tightly into the ring and the exposed surfaces must be knife-edge flat. All holes created by removing inclusions on the specimen surface must be filled with similar material. Voids will dramatically increase compressibility. The specimen should be recessed below the top edge of the ring to allow the stone to sit inside the ring. This assures proper alignment. The recess should also be created at the bottom of the specimen when using a floating ring configuration. At the end of the trimming process, measurements should provide the moist mass and dimensions of the specimen.

At this point, the specimen is held together by the sampling effective stress, which is generated by negative pore pressure. Releasing this pore pressure will cause the specimen to swell and also decrease the preconsolidation stress. The test can either be setup using the dry or the wet method. The dry method would be used for materials with high swell potential or large sampling effective stresses. The wet method can be used with soft soils. Cover both surfaces of the specimen with the filter discs. Assemble the specimen into the oedometer using the device specific instructions. When following the dry setup method, the porous stones should be drained by placing them on a paper towel for several minutes. This will remove all the free water, leaving the stones damp but not air dry. When following the wet setup method, the base should be filled with water, and the top of the bottom stone blotted surface dry. In no situation should the specimen be allowed to swell.

Assemble the oedometer into the load frame and apply a small seating load. This load should be enough to place all the contacts into compression, but not so large as to cause consolidation. This value will depend on the stiffness of the specimen. Next obtain the displacement zero reading, which will correspond to the initial dimension of the specimen. At this point, the displacement should be stable. Immediately increase the load if the specimen begins to swell. A reasonable seating load would be about 5 N.

The pore pressure must be zero for the compression data to be valid. The negative pore pressure is eliminated by flooding the specimen with water. If the applied stress is below the sampling effective stress, then the specimen will begin to swell. This is the reason to have an estimation of the sampling effective stress or the stress history of the specimen. For low OCR materials, a reasonable estimate of the sampling effective stress would be 10 to 25 percent of the in-situ effective stress. This provides a good value for the first stress increment.

Apply the first stress increment and begin recording time deformation data. After about two minutes, inundate the specimen from both the top and bottom surfaces. Watch carefully to see if the specimen starts to swell, is stable, or continues to compress. If the specimen starts to swell, immediately proceed to the next stress increment. This increase in load will cause compression, which may be followed by further swelling if the stress is still below the sampling effective stress. If necessary, apply more increments until the specimen is stable and compressing. The smallest stress that prevents swelling is the first usable stress increment, and the test can proceed according to the intended time schedule.

The schedule of stress levels may be project specific or standard for the laboratory. The preferred schedule should be available prior to the start of testing. The stress step between each stress level is commonly called a load increment. The Load Increment Ratio (LIR) is the change in load (stress) for an increment divided by the current load (stress) applied to the specimen. The LIR should be limited to 0.5 to 1.0. The two concerns when choosing the LIR are extrusion and secondary compression effects. Large LIRs will cause soil to squeeze past the gap between the specimen ring and the porous stone. This extrusion of soil is due to the high hydraulic gradients caused by the stress increment relative to the current effective stress. Extrusion causes an error in the measured deformation due to loss of soil from the control volume. Very low LIRs will distort the shape of the time deformation curve. The small stress increment will reduce the strain associated with primary consolidation relative to secondary compression. This will make interpretation of the curve for c_v and ε_{100} difficult or even impossible. In general, a test will be performed with a constant LIR for all loading increments, resulting in equally spaced data points when plotted in log stress space. For unloading, the LIR can be reduced to between –0.5 and –0.75. This is possible because extrusion is not a factor, and fewer data are required to define the unloading curve where the material behavior is much stiffer.

In some circumstances, it may be desirable to reduce the LIR to provide more definition of the shape of the compression curve. A good example would be to reduce the LIR to 0.3 near the preconsolidation stress to provide more definition of the rounded portion of the consolidation curve. This will preclude interpretation of the time curves. These stress increments should all have the same time duration, and must be at least as long as the time to primary of one of the normally consolidated increments. The ε_{100} strain for each increment at a reduced LIR is taken as the strain at the end of that constant time interval. The 3-t method should be used to decide when to increase the load in this situation.

The time schedule for the load increments should also be determined prior to the test. The duration must be long enough to allow primary consolidation to finish and include at least one cycle of secondary compression prior to a loading reversal. Deformation versus time data should be recorded for every increment. ASTM D2435 allows the test to be performed using 24-hour increment durations with one reading taken at the end of each increment. One complete time curve must be measured in the normally consolidated range to verify that the 24-hour duration is sufficient to achieve full consolidation.

After the last stress increment is complete, the apparatus is disassembled and the checked for potential problems. Remove all the excess water from the water bath. Leave the stone on the specimen and remove the specimen ring. Remove the stone, leaving the filter in place. Scrape any material from the sides of the stone and around the inside perimeter of the ring above the filter. This material is the extruded soil. Place this material in a separate tare and determine the oven-dried mass, M_{extr}. Measure the final height of the specimen (H_f), the moist mass ($M_{t,f}$), and the oven-dried mass (M_d). Clean the stones and return them to the storage container.

Important considerations for the determination of consolidation properties include:

- The coefficient of consolidation characterizes the time variation in deformation during primary consolidation. Scaling the laboratory measurements to the field application is done by changing the drainage height in Equation 17.10. Since this term is squared, increasing the drainage height by a factor of two increases the time to a particular percent consolidation by a factor of four. As might be expected, the drainage height on a field scale can far exceed 500 times that in the laboratory, leading to tens of thousands of multiples in clock time when scaling from consolidation in the laboratory to the field. Great care should be used when interpreting the time curves to obtain the most consistent c_v values considering the various choices possible when performing the constructions. Experience has shown that field settlement rates can be much greater than predicted by laboratory-determined rates of consolidation. At least part of this effect can likely be attributed to lateral drainage effects in the field. (Ladd et al., 1977; Simons, 1974). The fact that the root time and the log time methods consistently yield different laboratory c_v values, combined with the field observations, suggest that more research is needed on this topic.
- Field settlements are computed from the laboratory strain measurements. Here again, the scaling factor is very large. Laboratory strain values are multiplied by very large numbers to compute surface deformations. A 0.5 percent strain error applied to a 10 m layer results in a change of 5 cm in predicted settlement. This demands careful consideration of the errors related to strain calculations. The flatness of the prepared specimen surface and the apparatus compressibility calibration are prime sources of error.
- Apparatus compressibility must be subtracted from each deformation measurement. Apparatus compressibility is the measured relationship between applied force and deformation of the equipment. Apparatus compressibility is normally measured by assembling the equipment in the same configuration as will be used during testing, while replacing the soil specimen with a solid disc of brass or stainless steel (called a dummy specimen). The filters must be wet during the calibration. When using nylon filters, the calibration can be performed quickly and the force-deformation relationship represented by an equation. This will provide sufficient precision. When using paper filters, the compression behavior of the filter depends on loading sequence and time. The papers are rather thick, so the correction becomes more significant. Measure the apparatus compressibility using the same load sequence in the test including the unload-reload cycles. Take a reading of deformation at 15 seconds and use this for the apparatus deformation corresponding to the particular load.
- Disturbance is a serious concern when interpreting consolidation test results. Disturbance has many causes; perhaps the most discussed is the sampling process. These issues have been discussed in Chapter 11, "Background Information for Part II." While there is no way to completely prevent sampling disturbance, methods are available to minimize the effects. Stress relief causes disturbance and occurs due to bringing the sample from a depth to the ground surface. Stress relief cannot be prevented. Sample and specimen handling cause disturbance, but can be largely avoided by using careful procedures. Trimming will cause a certain amount of disturbance and can be kept small using appropriate trimming methods. Voids around the perimeter of the specimen would be considered laboratory-caused disturbance. Evaporation causes disturbance to the specimen; however, the effects are opposite to those described above. The natural water content of the material should be maintained until the specimen is set up in the

oedometer. Swelling of the specimen during the initial stages of the test is also a form of disturbance. Disturbance has a distinct and measurable impact on the consolidation curve. Refer to Figure 17.13 for an example of four compression curves. One of the curves is for a consolidation test performed on an intact specimen of Boston Blue Clay. Above this curve is a conceptual curve representing ideal conditions if it were possible to measure the deformation on a specimen without changing any of the in-situ conditions. This would be the "field" relationship. The lowest-lying curve is also conceptual and would be expected on a completely remolded specimen starting at the in-situ water content. This would represent the worst-case situation of disturbance. Varying disturbance would result in a series of curves between the field and remolded limit conditions. The conceptual curves are joined with a projection of the laboratory curve at a strain corresponding to 0.42 times the initial void ratio. This convergence point has been proposed by Schmertmann (1955) as a method to reconstruct the field curve from laboratory measurements. The most obvious impact of disturbance is to increase the measured strain at every stress level. This results in an increase in the interpreted recompression slope (recompression ratio) but a decrease in the virgin compression slope (compression index). Depending on the application (stress increment in the field), disturbance can lead to either an overestimation or underestimation of field behavior. The second effect of stress history is to change the interpreted preconsolidation stress. The conceptual behavior presented in Figure 17.13 is a gradual decrease in the preconsolidation stress. Comparison of preconsolidation stress values from tests with varying degrees of disturbance does not support such a clear trend. Figure 17.14 presents a conceptual comparison of the

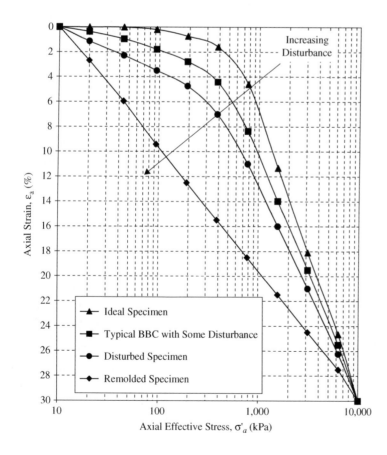

Figure 17.13 Effects of disturbance on the compression curve.

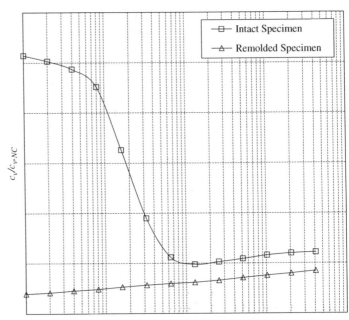

y-axis: $c_v/c_{v\text{-}NC}$

Legend:
- Intact Specimen
- Remolded Specimen

x-axis: Log of Axial Effective Stress, Log(σ'_a)

Figure 17.14 Effects of disturbance on the determination of the coefficient of consolidation.

coefficient of consolidation over stress level for an intact specimen and for a completely remolded specimen. The coefficient of consolation will be altered by disturbance because the property depends on a combination of the compressibility and hydraulic conductivity. The compressibility clearly changes due to disturbance, but the hydraulic conductivity will only change slightly due to the change in void ratio. In general, the c_v values will be lower at all stress levels. The impact will be highest during reloading when the material is overconsolidated. In fact, the c_v versus stress level trend can provide an indication of disturbance.

- Strain continues to increase (or decrease for an unloading increment) with time after primary consolidation due to secondary compression. The longer the load duration, the more secondary deformation that will be recorded, but recognize that the rate is decreasing with time. Mesri and Castro (1987) observed that the rate of secondary compression is proportional to the compression index. This constant is between 2 and 4 percent for most geo-materials. Based on this simple relationship, it is possible to construct a conceptual set of curves representing the unique EOP compression behavior along with the shift in the curve due to secondary compression. Refer to Figure 17.15 for a schematic of this concept. The effect of secondary compression is to increase the strain at every stress level, increase the recompression ratio, and decrease the preconsolidation pressure, but it will not change the virgin compression ratio, provided the time increments are the same. Plotting the deformation at one day (as allowed by the ASTM standard) will most likely decrease the preconsolidation pressure as compared to the unique EOP value. The magnitude of this shift will depend on the time required for primary consolidation. The shift can be as large as 10 percent per log cycle of secondary compression. For a material with a time of 10 minutes to the end of primary consolidation, the shift could be as large as 20 percent. Figure 17.16 presents Boston Blue Clay test results with the end of increment strain measurement added for each stress level. The load duration is much less than 24 hours in this test, except for the 800 and 1600 kPa increments. The graph clearly shows the final effect of secondary compression, which is the permanent strain

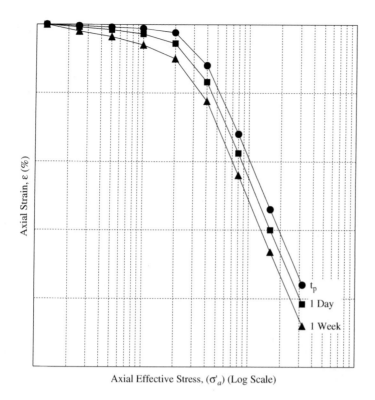

Figure 17.15 Effects of length of the time increment on the compression curve.

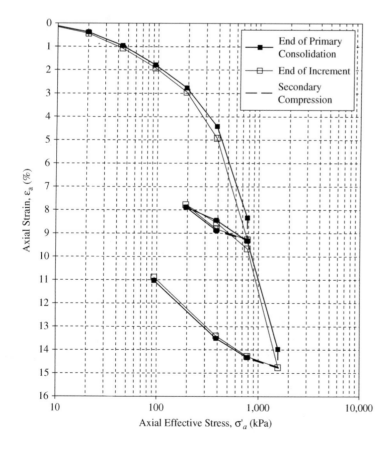

Figure 17.16 Compression curve results plotted at the end of primary consolidation and the end of an increment.

offset at the load reversal points. Secondary compression strain must continue far enough so that the correct swell ratio is measured. This is also true, but to a lesser extent, on a stress reversal toward reloading. In contrast, having a load increment ratio greater than 0.5 will generate enough primary strain to return to the EOP curve when continuing to load or unload.

- The temperature at the time of consolidation testing impacts the rate of consolidation and the compression curve. Temperature has a significant impact on secondary compression measurements. This is more related to variations in the thermal strain of the apparatus and the soil than due to fundamental changes in soil behavior. During secondary compression, the strain rate is very low and the strain increments are very small. Slow laboratory temperature variations (especially diurnal) can create serious errors in the interpreted rate of secondary compression. The interpreted coefficient of consolidation is for the viscosity of water at room temperature. Since the coefficient of consolidation is used to calculate the hydraulic conductivity, the hydraulic conductivity is also for room temperature conditions. Both values should be corrected to the field temperature by the change in viscosity. The laboratory values would be multiplied by about 0.75 to match a field condition of 10°C.

- As the temperature increases, the interpreted preconsolidation stress decreases due to changes in balance between interparticle forces and contact strengths (Mitchell, 2005). Thus particles rearrange in response to the temperature change, causing a shift in the compression curve. Figure 17.17 presents data from a specimen of Arctic Silt tested in an oedometer submerged in a temperature batch. The specimen was temperature-equilibrated in a brine solution at 0°C and then loaded through the first cycle. The temperature was then increased and the

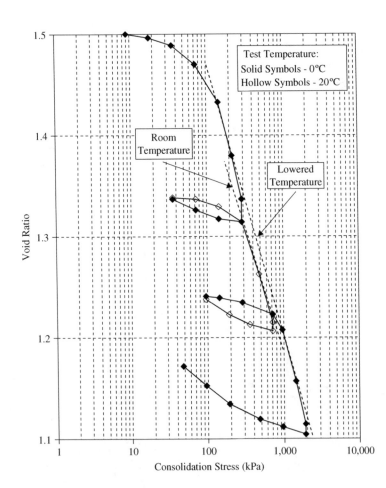

Figure 17.17 Effects of testing temperature on the consolidation curve.

(Based on the concepts presented in Ladd et al., 1985)

specimen equilibrated at 20°C. The compression behavior was then measured at room temperature. A third cycle was repeated at 0°C. The measurements clearly show the shift in the curve due to the temperature change.

- Fine-grained materials are sensitive to the salt concentration of the pore fluid. The sensitivity increases with plasticity. Normally the consolidation test is performed using distilled water. Salt water is a better option if in-situ salt concentrations are high, if the material is very stiff, or if the soil will be rebounded to high OCR values.
- Friction between the inside of the specimen ring and the soil causes a reduction in the average axial stress. Using a thinner specimen reduces the amount of friction; however, a smaller specimen is less representative of the sample and has larger disturbance effects due to trimming and inclusions. The friction can be evaluated using the boundary conditions applied to the specimen, as shown in Figure 17.18.

In the fixed specimen ring configuration, the average applied force is given by Equation 17.24:

$$F_{ave} = F - F'H_d \qquad (17.24)$$

For a floating specimen ring configuration, the average applied force is given in Equation 17.25:

$$F_{ave} = F - \frac{F'H_d}{2} \qquad (17.25)$$

Where:

F_{ave} = average applied force (N)
F = applied force (N)
F' = interface force per height (N/mm)
f = interface friction per unit area (N/mm^2)

The interface force per height is the inside perimeter of the ring multiplied by the interface friction per unit area.

The friction reduces the vertical effective stress applied to the soil and shifts the measured compression curve to higher stress levels. This leads to an overprediction of the preconsolidation stress (Taylor, 1942). For normally consolidated soils, the ratio of the friction to the vertical effective stress is about 17 percent when the specimen ring is made of steel and about 7 percent when the specimen ring is coated with a lubricant. For overconsolidated soils, the horizontal stress ratios are higher, and the resulting friction ratios are much larger.

A correction should be applied to account for the effects of friction, or at the very least a lubricant should be applied to the inside of the specimen ring to reduce the friction effects. Friction is commonly ignored when interpreting

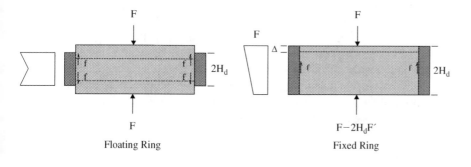

Floating Ring   Fixed Ring

Figure 17.18 Distribution of vertical force in a floating and fixed specimen ring configuration. (Modified from Lambe, 1951)

the results from a consolidation test. This may be justified in that the friction provides partial compensation for disturbance effects. However, the magnitude of the compensation is not measurable.

- The analysis of the rate of consolidation assumes that the top and bottom boundaries of the specimen have perfect drainage. In fact, the porous stones and the filter discs do provide some resistance to flow. The effective hydraulic conductivity of the stone/filter combination should be at least ten times more permeable than the soil specimen. The effective hydraulic conductivity can be measured using a simple version of a constant head hydraulic conductivity test as presented in Chapter 13.

- The Load Increment Ratio should never be larger than 1 when increasing the stress. Larger values will likely create excessive extrusion around the stones. This limit does not apply to decreasing stress increments. LIR values less than about 0.5 will become very difficult to interpret, resulting in loss of the coefficient of consolidation and rate of secondary compression data. It will still be possible to estimate the strain at the end of primary consolidation, so smaller LIR values can be used to better characterize the compression curve. This is especially useful when trying to define the preconsolidation stress of sensitive or cemented materials.

- The standard oedometer apparatus does not provide a means to saturate the test specimen. Lack of saturation has two impacts on the test results. First, air in the pore space will increase the compressibility of the pore fluid. This increases initial compression (ε_s-ε_0) when the stress is applied, and reduces the magnitude of the initial pore pressure. The second effect of air in the pore space is to reduce the hydraulic conductivity, and hence slow down the rate of consolidation. Testing with equipment that provides the capability to back pressure saturate should be considered when testing unsaturated samples.

- The loading duration is not important provided the time curves are measured and interpreted for every stress increment. This technique will yield the EOP compression curve. If the time curves are not measured, then it is important to keep the same load duration for each increment (except the reversal increments) to include a consistent amount of secondary compression. The duration must be sufficient to complete primary consolidation. ASTM D2435 allows 24-hour durations, but this could also be 12 hours for most soils. Constant durations are typically used in commercial laboratories, regardless of measuring the time curve due to the practicalities of operating a business. Computer automated equipment can also make use of the 3-t or the square root of time interpretation methods to load just after the end of primary.

- The coefficient of consolidation is computed from the strain versus time data. During the consolidation increment, the consolidation process is controlled by the hydraulic conductivity and the compressibility of the soil. The hydraulic conductivity changes with void ratio for the specific specimen. The compressibility is the change in stress divided by the strain increment, which during consolidation is ε_{100}-ε_0. Depending on the amount of secondary compression in the previous increment and the amount of initial compression in the current increment, this compressibility may be measurably different from the soil compressibility, which is the stress increment divided by $\Delta\varepsilon_{100}$.

- The method of placing the specimen into the oedometer can have significant impact on the measured behavior of stiff materials. Preventing swelling while getting control of the effective stress is essential to measuring the correct reloading stiffness. After applying the load, the soil must have access to water when using the dry-mounting method to be sure the negative pore pressure is dissipated at the start of the test. Excessive swelling will increase the reloading slope and decrease the preconsolidation stress.

- In theory, it is possible to measure the specimen height and mass at the end of the test and compare these measurements to the initial measurements as a check on the measured deformation during the test. The specimen must be unloaded to the seating value and allowed to swell to the end of primary consolidation under this small stress level. This can take a very long time. While this comparison is helpful in identifying gross errors, it simply takes too much time and is insufficiently precise to warrant use in routine practice.

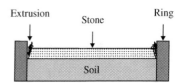

Figure 17.19 Soil extrusion.

- Extrusion of soil between the specimen ring and the stone may occur during the test when applied gradients are too high or the equipment is faulty. This can happen if loads are applied too roughly or there is a chip along the edge of the stone. Refer to Figure 17.19 for a schematic diagram of soil extrusion. Extrusion is most likely to occur in the normally consolidated stress increments and will cause an increase in the measured strain. Extrusion should not be significant in a properly run test. During the final disassembly of the specimen, any material squeezed around the stones should be collected for evaluation. The dry mass of the extruded soil, $M_{d,e}$, is obtained at the end of the test. The volume of the extruded material can be computed based on the average void ratio throughout the test using Equation 17.26:

$$V_e = (1 + e_{ave}) \frac{M_{extr}}{G_s \rho_w}$$
(17.26)

Where:

V_e = extruded volume of material at average void ratio (cm^3)
e_{ave} = average void ratio during test (dimensionless)
M_{extr} = dry mass of extruded material (g)
G_s = specific gravity of solids (dimensionless)
ρ_w = mass density of water (g/cm^3)

The average error, E_a, is then calculated using Equation 17.27:

$$E_a = \frac{\Delta e}{e_{ave}} \cdot 100 = \frac{V_e}{A \cdot (H_{ave} - H_s)} \cdot 100$$
(17.27)

Where:

E_a = average error in test (percent)
Δe = change in void ratio during test (dimensionless)
A = specimen area (cm^2)
H_{ave} = average specimen height during test (cm)
H_s = height of solids (cm)

An average error greater than 1 percent would be a serious concern and cause to evaluate the procedures and equipment. The test may have to be rerun.

- Creating a several millimeters recess to provide positive alignment of the porous stone may seem trivial, but is one of the key details toward achieving consistent results.
- Assessment of the compressibility is a matter for the engineer and must be done in the context of the engineering application. The test result provides a measure of strain for a specific stress. There is clearly some judgment involved in selecting these strain values, but the procedures are rather well defined. Connecting these stress-strain points with a series of straight lines is not a proper interpretation of the measurement. The engineer must decide on the most appropriate representation (strain versus log stress, strain versus stress, and so on) of the information and then choose a method to represent the results. The most appropriate method may be a hand curve fit, or a straight line, or a specific function.
- There are several empirical correlations for typical values of compression index. One of the most commonly used for intact soils is Terzaghi and Peck

(1948), shown as Equation 17.28. This relationship was developed empirically from data on low- to medium-sensitivity clays.

$$C_c = 0.009 \cdot (LL - 10) \qquad (17.28)$$

Where:
LL = liquid limit (integer)

TYPICAL VALUES

Typical values obtained during incremental one-dimensional consolidation testing are listed in Table 17.1.

CALIBRATION

The load frame should be calibrated to be sure it is functioning within specification. Depending on the type of load frame, this can be done by checking the lever arm ratio or by formally calibrating against a force transducer.

The deformation measurement must be corrected for apparatus compressibility if the correction is greater than five percent of the measured deformation during the test, or when paper filter discs are used, according to ASTM D2435. The method of calibration depends on the type of filter disc. In either case, the disc should be soaked in water (at about the same salinity as to be used in the test) and blotted dry prior to the calibration. Then assemble the apparatus as in the test, but with a stainless steel or brass dummy specimen in place of the test specimen.

Table 17.1 Typical values obtained during incremental one-dimensional consolidation testing.

Soil Type	CR (C_c)	RR (C_r)	SR (C_s) [unload 1 log cycle]	$C_{\alpha e}/C_c$ or $C_{\alpha e}/CR$	c_v (x10^{-8} m^2/s) [Virgin Compression]	k(m/s)
Boston Blue Clay*	(0.35)		(0.07)			
Boston Blue Clay** (OCR 1.2 to 3.75, Compression at σ'_{vo})	0.14–0.25	0.016–0.044	0.019–0.039		40 +/− 20	
Boston Blue Clay** (OCR 1 to 1.2, Compression at σ'_{vo})	0.14–0.22	0.016–0.039	0.014–0.032		20 +/− 10	
Maine Clay***	(0.5)				20 to 40	
Mexico City Clay***	(4.5)				0.2 to 2.5	
San Francisco Bay Mud****					2 to 4	
San Francisco Bay Mud*****				0.05		3×10^{-9}
Harrison Bay Arctic Silt******	0.168	0.027				
Inorganic Clays and Silts*******				0.04 +/− 0.01		
Organic Clays and Silts*******				0.05 +/− 0.01		
Peat and Muskeg*******				0.06 +/− 0.01		

*Mitchell, 1956, as appearing in Lambe and Whitman, 1969.
**After Ladd and Luscher, 1965.
***After Lambe, 1951.
****After Holtz and Kovacs, 1981.
*****After Mesri and Choi, 1985.
******After Yin, 1985.
*******Mesri et al., 1994.

Equipment Requirements

1. Scale readable to 0.01 g with a capacity of at least 1000 g
2. Equipment to measure water contents
3. Trimming supplies, including recess spacer
4. Calipers
5. Dial gage comparator
6. Filter fabric or filter discs, trimmed to the same diameter as the specimen or up to 1 mm smaller
7. Distilled water
8. Clock and timer
9. Oedometer with specimen ring, porous stones, top cap, and so on
10. Appropriate set of weights for the stress increments
11. LVDT or LST, readable to 0.0025 mm (0.0001 in), to measure deformation
12. Consolidation load frame
13. Data acquisition system

When using paper filter discs (Whatman no. 54), use the following procedure

1. Determine the sequence of loads to be applied to the specimen.
2. Apply the seating load and record the displacement transducer zero.
3. Apply the same loads to the apparatus as will be used when testing the specimen.
4. At each load increment, record the deformation reading at 15 seconds, 30 seconds, 1 minute (and 2 minutes and 5 minutes, if necessary).
5. Determine the stable voltage reading for each load and subtract the zero reading.
6. Convert the voltage increment into displacement in millimeters.
7. Tabulate the change in displacement ($\delta_{a,j}$) for each increment as the apparatus deflection.

When using nylon filter discs, use the following procedure to obtain a generic calibration relationship

1. Apply a reasonable seating load and record the load (F_j) and displacement reading ($V_{a,j}$).
2. Apply loads to the capacity of the loading frame using a LIR of 1.
3. At each load level, record the force and displacement reading.
4. Unload the apparatus to the seating load using the same increments while recording the displacement readings.
5. Repeat the process for a second complete load-unload cycle.
6. Convert each voltage reading to a displacement ($\delta_{a,j}$).
7. Use the F_j, $\delta_{a,j}$ pairs to fit an equation to be used as the apparatus calibration. A simple exponential function is generally adequate.

SPECIMEN PREPARATION

The consolidation test should be performed on an intact specimen of fine-grained soil. For instructional purposes, it is best to have a fairly compressible, medium-stiff material.

This will make trimming and set up relatively easy, yet provide enough deformation during the test to make interpretation of the time curves less challenging. It is also helpful to have a material that consolidates in the 10-minute range. Much faster rates of consolidation require some training to apply loads smoothly and quickly.

This is an ideal test to link the radiography of a sample to proper sample selection and careful extrusion techniques outlined in Chapter 11. A test can easily be performed on material from a tube section 5 cm long.

PROCEDURE

The consolidation test by incremental oedometer will be performed in general accordance with ASTM Standard Test Method D2435. Since many steps and stages are involved in this test, the experiment will serve to illustrate the reduction, summary, and presentation of many sets of data. The following procedure is for the wet-mounting method. The dry-mounting method would require slight modifications to the initial apparatus setup.

Apparatus Preparation

1. Grease specime.n ring and cutting shoe (if not part of the ring).
2. Determine the mass of the greased specimen ring with one filter disc (M_{rf}).
3. Measure the dimensions of the specimen ring (H_r and D_r).
4. Boil or ultrasound the stones and store underwater.
5. Measure the insert distance of recess tool (H_{rt}).
6. Measure the thickness of one filter disc (H_{fd}).
7. Record the combined mass of the top cap, top stone, and moment break (M_{tp}).

Specimen Preparation

1. Trim the specimen into the cutting ring, using techniques described in Chapter 11. Remember to maintain specimen orientation during trimming.
2. Recessing the soil into the ring is very important for the consolidation test. Cover the surface with the filter disc and use the recess tool to create a gap at the top of ring. Trim excess soil from the bottom with a wire saw and finish with the straight edge. If using a floating ring, there should be a recess on both the top and bottom.
3. Determine the mass of specimen, ring, and filter ($M_{srf,i}$).

Apparatus Assembly (Wet Method for Non-swelling Soils Only)

1. Fill the base with water.
2. Insert bottom stone and filter. Keep stone moist.
3. Remove excess water from the stone surface with a paper towel.
4. Place the specimen ring with soil on the stone and filter.
5. Cover the rim of the trimming ring with a gasket.
6. Tighten with the locking collar.
7. Drain the top stone on a paper towel and place on the specimen (the filter is already in place).
8. Place the top cap on the stone.
9. Locate the assembly in the loading frame along with the displacement transducer and the balance arm. Determine and record the tare force of the assembly, (F_{tr}).
10. Apply a seating load of about 5 N and set the displacement transducer to the limit of the linear range. Check to be sure the transducer core moves freely. Record the zero value ($V_{disp,0}$) and the input voltage at this reading ($V_{in,0}$).

Note: For the dry mounting method, allow stones to air dry and eliminate steps 1, 3, and 7.

1. Consolidate the specimen using a load increment ratio ($\Delta\sigma'/\sigma'$) between 0.5 and 1.0 for loading and –0.75 and –0.50 for unloading. One recommended loading schedule of stress (kPa) levels is: Seating, 12.5, 25, 50, 100, 200, 400, 800, 400, 200, 400, 800, 1600, 800, 400, 100, Seating.

2. Data can be recorded in one data file or into a separate data file for each stress increment. Separate files are used for the worksheets provided.

3. Confirm that the test clock is consistent with the data acquisition system clock.

4. Start recording the displacement transducer with the data acquisition system using a 1-second reading interval. Record the time, force (F_j), starting displacement reading ($V_{disp,j,0}$), and file name on the data sheet, and apply the first load increment. After about 1 minute, fill the water bath. If the specimen begins to swell, immediately apply the next increment. It will be necessary to increase the stress level until the specimen does not swell (about one-fourth of the estimated overburden stress). Slow down the reading rate to limit the data set to a reasonable size.

5. For each increment, record the displacement readings ($V_{disp,j,i}$) versus time using the data acquisition system. Remember that the initial portion of the curve is very important to define the start of consolidation. Start recording at a 1-second interval before touching the equipment to change the load. Record the time and apply the increment. Then slow down the reading rate.

6. For instructional purposes, plot both the square root of time and the log of time curves during at least one increment.

7. Apply increments after the end of primary consolidation has been reached or exceeded.

8. Allow one cycle of secondary compression to occur under the maximum load and before the unload-reload cycle.

9. For instructional testing, at the end of the test, unload the specimen to the seating load, but be sure the displacement transducer remains in place and allow time for swelling.

10. Record the final displacement reading ($V_{disp,f}$).

11. Remove the water from the bath and remove the oedometer from the load frame.

12. Remove the top cap and stone.

13. Scrape off any extruded material from the top of the specimen and sides of the stone, and oven-dry (M_{extr}). If using a floating ring, repeat the process with the bottom stone.

14. Dry the surface of the specimen, and determine the mass of the specimen, ring, and filter ($M_{srf,f}$).

15. For instructional testing, use a dial comparator to measure the final gage height of the specimen plus one filter ($H_{f,g}$).

16. Extrude the soil from the ring and remove filter.

17. Measure the moist mass of the specimen ($M_{t,f}$).

18. If material is needed for index tests, then use a pie slice of the specimen to obtain the final water content (ω_f).

19. Otherwise, oven-dry the entire specimen and obtain dry mass (M_{ds}).

20. Collect washings from the filter and the inside of ring and oven-dry (M_{dw}).

---

1. Determine the dry mass (M_d) of the specimen using either Equation 17.29 or Equation 17.30: **Calculations**

$$M_d = M_{ds} + M_{dw} + M_{extr}$$

(17.29)

$$M_d = \frac{M_{t,f}}{(1 - \omega_f/100)} + M_{dw} + M_{extr}$$

(17.30)

Where:

M_d = dry mass of the specimen (g)
M_{ds} = dry mass of the specimen excluding extruded material and washings (g)
M_{dw} = dry mass of washings (g)
M_{extr} = dry mass of extruded material (g)
$M_{t,f}$ = final moist mass of the specimen (g)
ω_f = final water content of the specimen (%)

2. Determine the initial water content (ω_N) using Equation 17.31:

$$\omega_N = \frac{(M_{srf,i} - M_{rf}) - M_d}{M_d} \cdot 100$$

(17.31)

Where:

ω_N = natural water content of the specimen (%)
$M_{srf,i}$ = initial mass of the specimen, ring, and filter (g)
M_{rf} = mass of the ring and filter (g)

3. Calculate the initial height (H_i) using Equation 17.32:

$$H_i = H_r - H_{fd} - H_{rt}$$

(17.32)

Where:

H_i = initial height of the specimen (cm)
H_r = height of the specimen ring (cm)
H_{fd} = height of one filter disc (cm)
H_{rt} = height of recess tool (cm)

4. Determine the initial void ratio (e_i) using Equation 17.33:

$$e_i = \frac{\left(\frac{\pi D_r^2}{4} \cdot H_i\right) - M_d\big/G_s \cdot \rho_w}{M_d\big/G_s \cdot \rho_w}$$

(17.33)

Where:

D_r = diameter of the specimen ring (cm)
G_s = specific gravity of solids (dimensionless)
ρ_w = mass density of water (g/cm^3)

5. Determine the initial degree of saturation (S_i) using Equation 17.34:

$$S_i = \frac{G_s \cdot \omega_N}{e_i}$$

(17.34)

Where:

S_i = initial saturation of the specimen (%)

6. Calculate the stress at each increment j ($\sigma'_{a,j}$) using Equation 17.35:

$$\sigma'_{a,j} = \frac{10 \cdot \left(F_j - F_{tr} + M_{tc} \cdot g \Big/ 1000 \right)}{A} \qquad (17.35)$$

Where:

$\sigma'_{a,j}$ = stress at increment j (kPa)
F_j = applied force at increment j (N)
F_{tr} = tare force of the assembly (N)
M_{tc} = mass of the top cap, top stone, and moment break (g)
g = acceleration due to gravity (m/s^2)
A = area of the specimen (cm^2)

Note that friction losses are ignored in this calculation. To account for them, estimate the friction losses using Equation 17.24 or Equation 17.25, and subtract the force from the numerator of Equation 17.35.

7. Convert the voltage readings from the displacement transducer to displacement for each reading i ($\delta_{j,i}$) using Equation 17.36. Record the initial and final displacement for increment j as $\delta_{j,0}$ and $\delta_{j,f}$, respectively.

$$\delta_{j,i} = \left(\frac{V_{disp,j,i}}{V_{in,j,i}} - \frac{V_{disp,0}}{V_{in,0}} \right) \cdot CF_{disp} \qquad (17.36)$$

Where:

$\delta_{j,i}$ = displacement at reading i (cm)
$V_{disp,j,i}$ = output of displacement transducer at reading i (V)
$V_{in,j,i}$ = input voltage to displacement transducer at reading i (V)
$V_{disp,0}$ = output of displacement transducer at zero reading (V)
$V_{in,0}$ = input voltage to displacement transducer at zero reading (V)
CF_{disp} = calibration factor of the displacement transducer (cm/V/V$_{in}$)

8. Calculate the deformation for each reading i during each increment j ($\Delta H_{j,i}$) using Equation 17.37:

$$\Delta H_{j,i} = \delta_{j,f} - \delta_{j,0} - \delta_{a,j} \qquad (17.37)$$

9. Calculate the strain at each reading i ($\varepsilon_{a,j,i}$) using Equation 17.38:

$$\varepsilon_{a,j,i} = \left(\frac{\Delta H_{j,i}}{H_i} \right) \times 100 \qquad (17.38)$$

10. Perform the constructions to determine the consolidation parameters for each increment where time curves were measured.

11. Calculate the coefficient of volume change for increment j ($m_{v,j}$) using Equation 17.39:

$$m_{v,j} = \frac{\varepsilon_{100,j} - \varepsilon_{100,j-1}}{100(\sigma_{a,j} - \sigma_{a,j-1})} \qquad (17.39)$$

Where:

$m_{v,j}$ = coefficient of volume change for increment j (kPa^{-1})
$\varepsilon_{100,j}$ = strain at the end of primary consolidation for increment j (%)

$\varepsilon_{100,j-1}$ = strain at the end of primary consolidation for increment j - 1 (%)
$\sigma'_{a,j}$ = applied axial effective stress for increment j (kPa)
$\sigma'_{a,j-1}$ = applied axial effective stress for increment j-1 (kPa)

12. Calculate the hydraulic conductivity for increment j ($k_{v,j}$) using Equation 17.40:

$$k_{v,j} = c_{v,j} \rho_w g \frac{\varepsilon_{s,j} - \varepsilon_{100,j}}{100(\sigma_{a,j} - \sigma_{a,j-1})} \tag{17.40}$$

Where:
$k_{v,j}$ = hydraulic conductivity for increment j (m/s)
$c_{v,j}$ = coefficient of consolidation for increment j (m^2/s)
$\varepsilon_{s,j}$ = strain at the start of consolidation for increment j (%)

    The coefficient of consolidation may be determined using the square root of time or the log of time construction, or the average of the two, depending on preferences.

13. Calculate the final height (H_f) using Equation 17.41:

$$H_f = H_{f,g} - H_{fd} \tag{17.41}$$

Where:
H_f = final height of the specimen (cm)
$H_{f,g}$ = final gage height of the specimen and one filter (cm)

14. For instructional testing or when trying to identify testing problems, perform an error analysis of height calculations (E_H) using Equation 17.42:

$$E_H = \frac{(H_f - (H_i - \Delta H_f))}{H_i} \cdot 100 \tag{17.42}$$

Where:
E_H = error in the height calculations (%)
ΔH_f = change in height of the specimen according to the dial gage (cm). This is the last test reading ($V_{disp,f}$)

15. Calculate the magnitude of error due to extrusion using Equation 17.27.

Report

Include the following in a report:

1. Plots of strain versus square root of time and strain versus log of time for each time increment with interpretive constructions.

2. A plot of the apparatus compressibility.

3. The magnitude of error due to extrusion.

4. The calculations of the error analysis of the height calculations.

5. A plot of strain vs. σ'_a (indicating both the end of primary and the end of increment) on a log scale and on a natural scale. Be careful the points are connected properly, especially at load reversals.

6. A plot of strain energy (work) vs. σ'_a.

7. Determine the preconsolidation stress using the Casagrande construction and the Strain Energy construction. Indicate these values and the method used on the plot of ε vs. σ'_a.

8. Determine the recompression ratio, the compression ratio, and the swelling ratio, and indicate these values on the plot of ε vs. σ'_a.

9. A plot of c_v (root t and log t) vs. log σ'_a, plotting the points at the average stress during each increment.

10. A plot of log k_v versus log σ'_a and versus e, plotting the points at the average stress and average void ratio during each increment.

11. A summary table for all increments including σ'_a, e_{50}, ε_s, t_{50}, ε_{50}, t_{90}, t_p, ε_{100}, t_f, ε_f, c_v, m_v, k_v, and $C_{\alpha\varepsilon}$.

12. A summary table for the test including CR, RR, SR, initial (and final if H_f measured) specimen state (water content, void ratio, saturation, density).

Criteria for judging the acceptability of test results obtained by this test method have not been determined by ASTM International.

PRECISION

Since the incremental loading consolidation test has so many stages, there are a multitude of aspects to check for errors. These have all been discussed in detail in the background section and important considerations section. A brief list of the most likely causes of error is reiterated in this section.

DETECTING PROBLEMS WITH RESULTS

Rounded curves are usually due to sampling disturbance, and the specimen quality should be reevaluated. The specimen preparation methods can also cause disturbance. The end surfaces must be flat and perpendicular to the loading axis to prevent softening of the ends and uneven loading. In some instances, the loading points may be located so that the most significant points of curvature are missed. If unfortunate loading points are believed to be the cause of a rounded curve, change the LIR around the preconsolidation pressure for subsequent tests.

If the preconsolidation stress is less than the estimated in-situ vertical effective stress, either the calculation of the in-situ stress is incorrect, or the interpreted preconsolidation pressure is wrong. Verify that the estimate of in-situ stress is reasonable, and evaluate the method of determining the preconsolidation stress. Another possible source of error is sample disturbance.

When the VCL is not easy to determine, verify that the EOP points are plotted (not the EOI) and that the loads have been maintained past the end of primary consolidation. If the interpreted values of CR or SR are lower than typical for the material, or the RR is higher than typical, the likely cause is due to sample disturbance.

Application of loads takes some practice. When the initial points on the time curves are rapidly increasing and decreasing, the likely cause is dropping the load onto the specimen. Make sure to apply the loads swiftly, but uniformly. Poor initial time readings lead to unreliable interpretation of the time curves and are best indicated as uninterpretable. Alternatively, if there is a large difference between ε_s and ε_0, the apparatus calibration may be calculated incorrectly and should be evaluated.

The hydraulic conductivity should have a log-linear relationship with void ratio. Scatter is expected and the unloading increments will generally have more scatter than the loading increments. When there are inconsistencies, review the interpretations of the time curves.

If the time curves show a trend of increasing or decreasing strain with time, especially during secondary compression, verify that the room temperature is stable. Thermal strain of both the soil and the machine will result in errors in the deformation measurement.

Saturation values should be within 2 percent of the actual value. If saturated soils are tested, then this measure provides a check on phase relationships.

REFERENCE PROCEDURES

ASTM D2435 One-Dimensional Consolidation Properties of Soils Using Incremental LoadingReferences

REFERENCES

Refer to this textbook's ancillary web site, www.wiley.com/college/germaine, for data sheets, spreadsheets, and example data sets.

Becker, D. E., J.H.A. Crooks, K. Been, and M. G. Jefferies. 1987. "Work as a Criterion for Determining In Situ and Yield Stresses in Clays." *Canadian Geotechnical Journal*, 24(4), November.

Casagrande, A. 1936. "The Determination of the Pre-Consolidation Load and Its Practical Significance." *Proceedings of the First International Conference on Soil Mechanics and Foundation Engineering*, Cambridge, Massachusetts.

Holtz, R. D., and W. D. Kovacs. 1981. *An Introduction to Geotechnical Engineering.* Prentice-Hall, Englewood Cliffs, New Jersey.

Ladd, C. C. 1973. "Settlement Analysis for Cohesive Soils," *Research Report R71-2*, No. 272, Department of Civil Engineering, Massachusetts Institute of Technology, Cambridge.

Ladd, C. C. 1991. "Stability Evaluation during Staged Construction," Twenty-Second Karl Terzaghi Lecture, *Journal of Geotechnical Engineering*, 117(4).

Ladd, C. C. and R. Foott. 1974. "New Design Procedure for Stability of Soft Clays," *Journal of the Geotechnical Engineering Division*, ASCE, vol. 100, GT7.

Ladd, C. C., R. Foott, K. Ishihara, F. Schlosser, and H. G. Poulos. 1977. "Stress-Deformation and Strength Characteristics." State of the Art Report, *Proceedings of the Ninth International Conference on Soil Mechanics and Foundation Engineering*, Tokyo.

Ladd, C. C., and U. Luscher. 1965. "Preliminary Report on the Engineering Properties of the Soils Underlying the MIT Campus," *MIT Research Report R65-58*, Cambridge, Massachusetts.

Ladd, C. C., J. S. Weaver, J. T. Germaine, and D. P. Sauls, 1985. "Strength-Deformation Properties of Arctic Silt," *Proceedings of the Conference Arctic '85*, San Francisco.

Lambe, T. W. 1951. *Soil Testing for Engineers,* John Wiley and Sons, New York.

Lambe, T. W. and R. V. Whitman. 1969. *Soil Mechanics*, John Wiley and Sons, New York.

Mesri, G., and A. Castro. 1987. "C_α/C_c Concept and K_0 during Secondary Compression," *Journal of Geotechnical Engineering*, ASCE, 113(3).

Mesri, G., and Y. K. Choi, 1985. "Settlement Analysis of Embankments on Soft Clay," *Journal of Geotechnical Engineering*, ASCE, Vol. 111, GT4.

Mesri, G., D.O.K. Lo, and T-W. Feng. 1994. "Settlement of Embankments on Soft Clays," *Vertical and Horizontal Deformations of Foundations and Embankments, ASCE Settlement '94*, GSP 40, College Station, TX.

Mitchell, J. K. and K. Soga. 2005. *Fundamentals of Soil Behavior*. John Wiley and Sons, Hoboken, NJ.

Saye, S. R., C. C. Ladd, P. C. Gerhart, J. Pilz, and J. C. Volk. 2001. "Embankment Construction in an Urban Environment: the Interstate 15 Experience." *Proceedings of the ASCE Foundations and Ground Improvement Conference*, Blacksburg, VA.

Schmertmann, J. H. 1955. "The Undisturbed Consolidation of Clay." *Transactions, American Society of Civil Engineers*, vol. 120.

Simons, N. E. 1974. "Normally Consolidated and Lightly Over-Consolidated Cohesive Materials," *Settlement of Structures*, British Geotechnical Society, Halsted Press, London.

Taylor, D. W. 1942. *Research on Consolidation of Clays,* Report Serial No. 82, Department of Civil and Sanitary Engineering, Massachusetts Institute of Technology, Cambridge.

Taylor, D. W. 1948. *Fundamentals of Soil Mechanics,* John Wiley and Sons, New York.

Terzaghi, K., and R. B. Peck. 1948. *Soil Mechanics in Engineering Practice,* John Wiley and Sons, New York.

Yin, Edward Yen-Pang. 1985. "Consolidation and Direct Simple Shear Behavior of Harrison Bay Arctic Silts." MS thesis, Department of Civil Engineering, Massachusetts Institute of Technology, Cambridge.

Appendix A

Constants and Unit Conversions

Table A.1 Constants.

Name	Symbol	Value	Units
Avogadro's Number	N_A	6.0221×10^{23}	Number of particles in one gram-mole
Gravity (Acceleration due to)	g	9.8067	m/s^2
		32.174	ft/s^2
Pi	π	3.14159	Dimensionless

Source: Mohr and Taylor (2004), except g, which was converted from m/s^2 to ft/s^2.

Name	Symbol	Factor
Nano	n	$0.000000001 = 10^{-9}$
Micro	μ	$0.000001 = 10^{-6}$
Milli	m	$0.001 = 10^{-3}$
Centi	c	$0.01 = 10^{-2}$
Deci	d	$0.1 = 10^{-1}$
—	—	1
Deca	da	10
Hecto	h	$100 = 10^{2}$
Kilo	k	$1,000 = 10^{3}$
Mega	M	$1,000,000 = 10^{6}$
Giga	G	$1,000,000,000 = 10^{9}$

Table A.3 Conversions

Property	Name	Symbol	Multiply By	To Get
Length	Angstrom	Å	10^{-10}	m
	Micron	μ	10^{-6}	m
	Meter	m	3.2808	ft
	Centimeter	cm	0.39370	in
	Millimeter	mm	0.03937	in
	Feet	ft	0.3048	m
	Inch	in	2.5400	cm
			25.400	mm
Area	Square meters	m^2	10.764	ft^2
	Square centimeters	cm^2	0.15500	in^2
	Square millimeters	mm^2	0.00155	in^2
	Square feet	ft^2	0.092903	m^2
	Square inches	in^2	6.4516	cm^2
			645.16	mm^2
Volume	Cubic meters	m^3	35.315	ft^3
	Cubic centimeters	cm^3	0.061024	in^3
	Cubic millimeters	mm^3	0.000061024	in^3
	Liter	L	1000	cm^3
			0.26417	US gallons
			0.035314	ft^3
	Milliliter	mL	1.0000	cm^3
			0.061024	in^3
	Cubic feet	ft^3	0.028317	m^3
			28.317	L
	Cubic inches	in^3	16.387	cm^3
			16.387	mL
			16387	mm^3

(*continued*)

Table A.3 (*continued*)

Property	Name	Symbol	Multiply By	To Get
Mass	Kilogram	kg	2.2046	lbm
	Gram	g	0.0022046	lbm
	Megagram	Mg	1.1023	ton (short)
	Pound (mass)	lbm	0.45359	kg
			453.59	g
	Ton (short)	T	2,000	lbm
Density (Mass Density)	Megagram per cubic meter	Mg/m^3	62.428	lbm/ft^3
	Grams per cubic centimeter	g/cm^3	62.428	lbm/ft^3
	Pounds per cubic foot	lbm/ft^3 (pcf)	0.016018	Mg/m^3
			0.016018	g/cm^3
Force	Newton	N	0.2248	lbf
	Pound (force)	lbf	4.4482	N
Pressure	Kilopascal	kPa	20.885	lbf/ft^2
			0.010443	ton/ft^2 (short ton)
			0.14504	lbf/in^2
			0.010197	kgf/cm^{2**}
			0.0098692	atm
	Pounds (force) per square foot	lbf/ft^2 (psf)	0.047880	kPa
	Kilopounds (force) per square foot	$klbf/ft^2$ (ksf)	47.880	kPa
	Tons (short) per square foot	ton/ft^2 (tsf)	95.760	kPa
	Pounds (force) per square inch	lbf/in^2 (psi)	6.8948	kPa
	Kilograms (force) per square centimeter	kgf/cm^2 (ksc)*	98.067	kPa
Hydraulic Conductivity	Centimeters per second	cm/s	0.032808	ft/s
		**	1.0354×10^6	ft/yr
	Meters per second	m/s**	1.0354×10^8	ft/yr
	Meters per year	m/yr**	3.2808	ft/yr
	Feet per second	ft/s	30.480	cm/s
	Feet per year	ft/yr**	0.30480	m/yr
		**	9.6585×10^{-7}	cm/s
Viscosity	Centipoise	cP	10^{-3}	Pa-s
	Pascal-second	Pa-s	1,000	cP
Coefficient Of Consolidation	Square centimeters per second	cm^2/s**	3155.8	m^2/yr
			9.3000	in^2/min
		**	3.3968×10^4	ft^2/yr
	Square meters per year	m^2/yr**	3.1688×10^{-4}	cm^2/s
	Square inches per minute	in^2/min	0.10753	cm^2/s
	Square feet per year	ft^2/yr**	2.9439×10^{-5}	cm^2/s

*Avoid using this unit if possible. The unit is included because older equipment may only display in these units, which mix force and mass.
**Assumed 1 year = 365.25 days = 8,766 hours = 525,960 minutes = 31,557,600 seconds.
Source: Taylor, 1995.

There are some conversion issues with the temperature conversion portion of this table. The current form has errors and is not usable. I will try to clearly indicate the changes necessary. To rectify the issues (for temperature conversions only), the words should be on one line, and the equation(s) should be on the following line(s) and indented. Where ever an "o" appears before a temperature variable (C or F), it should be superscripted to indicate degrees.

Table A.4 Other useful relationships

Conversions between Temperature in degrees Celsius (°C) and Temperature in degrees Fahrenheit (°F)

$(°C) = ((°F) − 32)/1.8$

$(°F) = (°C)*1.8) + 32$

Conversion between Temperature in Kelvin (K) and Temperature in degrees Celsius (°C)

$(K) = (°C) + 273.15$

$1 \text{ N} = 1 \text{ kg-m/s}^2$

$1 \text{ Pa} = 1 \text{ N/m}^2 = 1 \text{ kg/m-s}^2$

REFERENCES

Mohr, P. J., and Taylor, B. N. 2004. *The 2002 CODATA Recommended Values of the Fundamental Physical Constants*, web version 4.0. NIST Physical Data web site, http://physics.nist.gov/cuu/constants, December 2003; *Rev. Mod. Phys.*, 76, no. 4, October 2004.

Taylor, B. N. 1995. *Guide for the Use of the International System of Units (SI)*. NIST Special Publication 811, 1995 Edition, Superintendent of Documents, U. S. Government Printing Office, Washington, D.C.

Appendix B

Physical Properties of Pure Water

Table B.1 Mass density of water at various temperatures.

	0.0	0.1	0.2	0.3	0.4	0.5	0.6	0.7	0.8	0.9
	Mass Density of Pure Water, ρ_w, (g/cm^3)									
Temp (°C)	Leading Digits	Last Four Digits								
0	0.999*	8493	8558	8622	8683	8743	8801	8857	8912	8964
1	0.9999015	9065	9112	9158	9202	9244	9284	9323	9360	9395
2	0.9999429	9461	9491	9519	9546	9571	9595	9616	9636	9655
3	0.9999672	9687	9700	9712	9722	9731	9738	9743	9747	9749
4	0.9999750	9748	9746	9742	9736	9728	9719	9709	9696	9683
5	0.9999668	9651	9632	9612	9591	9568	9544	9518	9490	9461
6	0.9999430	9398	9365	9330	9293	9255	9216	9175	9132	9088
7	0.9999043	8996	8948	8898	8847	8794	8740	8684	8627	8569
8	0.9998509	8448	8385	8321	8256	8189	8121	8051	7980	7908
9	0.9997834	7759	7682	7604	7525	7444	7362	7279	7194	7108
10	0.9997021	6932	6842	6751	6658	6564	6468	6372	6274	6174

11	0.9996074	5972	5869	5764	5658	5551	5443	5333	5222	5110
12	0.9994996	4882	4766	4648	4530	4410	4289	4167	4043	3918
13	0.9993792	3665	3536	3407	3276	3143	3010	2875	2740	2602
14	0.9992464	2325	2184	2042	1899	1755	1609	1463	1315	1166
15	0.9991016	0864	0712	0558	0403	0247	0090	9932**	9772**	9612**
16	0.9989450	9287	9123	8957	8791	8623	8455	8285	8114	7942
17	0.9987769	7595	7419	7243	7065	6886	6706	6525	6343	6160
18	0.9985976	5790	5604	5416	5228	5038	4847	4655	4462	4268
19	0.9984073	3877	3680	3481	3282	3081	2880	2677	2474	2269
20	0.9982063	1856	1649	1440	1230	1019	0807	0594	0380	0164
21	0.9979948	9731	9513	9294	9073	8852	8630	8406	8182	7957
22	0.9977730	7503	7275	7045	6815	6584	6351	6118	5883	5648
23	0.9975412	5174	4936	4697	4456	4215	3973	3730	3485	3240
24	0.9972994	2747	2499	2250	2000	1749	1497	1244	0990	0735
25	0.9970480	0223	9965**	9707**	9447**	9186**	8925**	8663**	8399**	8135**
26	0.9967870	7604	7337	7069	6800	6530	6259	5987	5714	5441
27	0.9965166	4891	4615	4337	4059	3780	3500	3219	2938	2655
28	0.9962371	2087	1801	1515	1228	0940	0651	0361	0070	9778**

*Not applicable for 0.0° C
**Leading digit decreases by 0.001
Source: After Marsh, 1987.

Refer to Appendix C, "Calculation Adjustments for Salt," for the variation of the mass density of seawater with temperature and salinity.

Temp (°C)	Viscosity of Pure Water, μ_w, (mPa-s)	Temp (°C)	Viscosity of Pure Water, μ_w, (mPa-s)
0	—	15	1.145
1	1.732	16	1.116
2	1.674	17	1.088
3	1.619	18	1.060
4	1.568	19	1.034
5	1.519	20	1.009
6	1.473	21	0.984
7	1.429	22	0.961
8	1.387	23	0.938
9	1.348	24	0.916
10	1.310	25	0.895
11	1.274	26	0.875
12	1.239	27	0.855
13	1.206	28	0.836
14	1.175		

Table B.2 Viscosity of water at various temperatures.

Note: 1 mPa-s = 1 centipoise (cP)
Source: After Lambe, 1951.

REFERENCES

Marsh, K. N., Ed., 1987. *Recommended Reference Materials for the Realization of Physicochemical Properties*, Blackwell Scientific Publications, Oxford, England.

Lambe, T. W. 1951. *Soil Testing for Engineers*, John Wiley and Sons, New York.

Appendix C

Calculation Adjustments for Salt

ACCOUNTING FOR THE PRESENCE OF SALT IN PHASE RELATIONSHIPS

The presence of salt can have a significant effect on the calculation of phase relations and the physical properties of soil. At concentrations of about seawater (35 g/L), the measurement of specific gravity will be different by a value of the within-laboratory repeatability precision levels if the salt is ignored. Adjustments that account for the presence of salt can be made to the calculations of physical properties, such the void ratio, saturation, and specific gravity. The presence of salt particularly affects the phase relationships important to flow studies, such as void ratio.

The literature rarely discusses accounting for salt in geotechnical calculations. Further, there is disagreement about how exactly to account for the presence of salt in the calculations among those that do make correction. Refer to Appendix X.1 of ASTM D4542 Pore Water Extraction and Determination of the Soluble Salt Content of Soils by Refractometer for one method of performing the calculations, or the journal article by Iraj Noorany (1984) for a different method. The lecture notes of Charles C. Ladd provide yet another method (Ladd, 1998).

This appendix presents yet another method. The premise of these calculations is that the space available for flow is the volume of voids divided by the volume of soil solids, since the salt will be dissolved in water. Further, the saturation level of importance is the volume of water plus dissolved salts divided by the volume of voids. These concepts apply to concentrations below which salt will precipitate out of solution into the pore space. It is extremely important to note that the "$G_s \omega_C = Se$" relationship is not preserved in this method of calculation.

Separate measures of the water content and the salt concentration of the pore fluid are required to account for the presence of salt in the phase relationships for any method of correction. Refer to Chapter 2 and Chapter 6 for information on measuring the water content and salinity, respectively. For a determination of void ratio or saturation, a total volume of the soil specimen will be required, or 100 percent saturation can be assumed. This appendix details how to use those measurements to account for the presence of salt in geotechnical calculations.

Refer to Figure C.1 for a schematic diagram of the phase relationships of a soil element with salt present in the pore fluid.

The calculations presented herein measure the water content in the usual manner, and define the water content as the mass of pure water, M_w, divided by the dry mass, M_d. The dry mass includes the mass of soil plus the mass of salt. This preserves the water content definition as a simple measure of material characteristics. Equation C.1 presents this relationship:

$$\omega_C = \frac{M_w}{M_d} \cdot 100 \tag{C.1}$$

Where:

ω_C = measured water content of the specimen (%)
M_w = mass of water (g)
M_d = mass of dry solids (soil and salt) (g)

The mass of water is determined using Equation C.2:

$$M_w = M_t - M_d \tag{C.2}$$

Where:

M_t = total initial mass of specimen (g)

Equations C.3 through C.9 step through the derivation of the equation to determine the volume of seawater, V_{sw}, shown as Equation C.10. Equation C.3 and Equation C.4 are basic definitions:

$$\rho_{sw} = \frac{M_w + M_{sa}}{V_{sw}} \tag{C.3}$$

$$\frac{RSS}{1000} = \frac{M_{sa}}{V_{sw}} \tag{C.4}$$

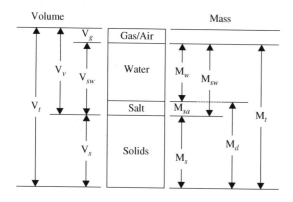

Figure C.1 Phase relationships of a soil element with salt present in the pore fluid.

$$M_{sa} = \frac{RSS}{1000} \cdot V_{sw} \tag{C.5}$$

$$\rho_{sw} = \frac{M_w + RSS/1000 \cdot V_{sw}}{V_{sw}} \tag{C.6}$$

$$\rho_{sw} \cdot V_{sw} = M_w + RSS/1000 \cdot V_{sw} \tag{C.7}$$

$$\rho_{sw} \cdot V_{sw} - RSS/1000 \cdot V_{sw} = M_w \tag{C.8}$$

$$V_{sw} \left(\rho_{sw} - RSS/1000 \right) = M_w \tag{C.9}$$

$$V_{sw} = \frac{M_w}{\left(\rho_{sw} - RSS/1000 \right)} \tag{C.10}$$

Where:

ρ_{sw} = density of seawater (g/cm^3)
M_{sa} = mass of salt (g)
V_{sw} = volume of seawater (cm^3)
RSS = salinity of the pore fluid (g/L)

Table C.1 presents the density of seawater as a function of temperature and salinity (in units of parts per thousand), s. To use the table, the salinity of the pore fluid in grams per liter should be converted to salinity in parts per thousand using Equation C.11:

$$RSS = s \cdot \rho_{sw} \tag{C.11}$$

Where:

s = salinity of the pore fluid (parts per thousand; i.e., g salt/1000 g seawater)

The conversion requires iterating for the density of seawater in Table C.1. However, since the differential in the mass of salt in a liter of seawater is relatively small, RSS can be considered equivalent to s for practical purposes.

Calculate the mass of soil, M_s, as shown in Equation C.12. The mass of salt is calculated using the volume of seawater and the salinity of the pore fluid, as shown above in Equation C.5.

$$M_s = M_d - M_{sa} \tag{C.12}$$

Where:

M_s = mass of dry soil (g)

The volume of soil solids, V_s, is calculated using Equation C.13:

$$V_s = \frac{M_s}{G_s \rho_{w,20}}$$ (C.13)

Where:

V_s = volume of soil solids (cm³)
G_s = specific gravity of soil (dimensionless)
$\rho_{w,20}$ = density of water at 20°C (g/cm³)

Table B.1 presents the density of water as a function of temperature. The specific gravity of soil can either be estimated or measured. Refer to Chapter 3 for procedures to measure specific gravity.

The void ratio, e, is calculated assuming that any salt is part of the void space since it is dissolved in the fluid, as shown in Equation C.14. This is the same equation and definition as used for pure water.

$$e = \frac{V_t - V_s}{V_s}$$ (C.14)

Where:

e = void ratio (dimensionless)
V_t = total volume of specimen (cm³)

The degree of saturation, S, is calculated assuming that any salt is dissolved in the fluid and therefore contributes to the saturation, as shown in Equation C.15:

$$S = \frac{V_{sw}}{V_v} \cdot 100$$ (C.15)

Where:

S = degree of saturation (%)

Table C.1 Mass density of seawater for various salt concentrations and temperatures.

	Mass Density of Seawater, ρ_{sw} (g/cm³)								
	Salinity (parts per thousand)								
Temperature (°C)	0	5	10	15	20	25	30	35	40
0	0.999843	1.003913	1.007955	1.011986	1.016014	1.020041	1.024072	1.028106	1.032147
5	0.999967	1.003949	1.007907	1.011858	1.015807	1.019758	1.023714	1.027675	1.031645
10	0.999702	1.003612	1.007501	1.011385	1.015269	1.019157	1.023051	1.026952	1.030862
15	0.999102	1.002952	1.006784	1.010613	1.014443	1.018279	1.022122	1.025973	1.029834
20	0.998206	1.002008	1.005793	1.009576	1.013362	1.017154	1.020954	1.024763	1.028583
25	0.997048	1.000809	1.004556	1.008301	1.012050	1.015806	1.019569	1.023343	1.027128
30	0.995651	0.999380	1.003095	1.006809	1.010527	1.014252	1.017985	1.021729	1.025483
35	0.994036	0.997740	1.001429	1.005118	1.008810	1.012509	1.016217	1.019934	1.023662
40	0.992220	0.995906	0.999575	1.003244	1.006915	1.010593	1.014278	1.017973	1.021679

Note: The values of density presented above are specifically for the chemical composition of the salt in seawater.
Source: Lide, 2007.

Usually, these calculations will be used directly to account for the presence of salt in some of the phase relationships for engineering tests. In the case of specific gravity testing, however, if salts are present the measured mass of solids at the end of the test will include salts and the density of the control fluid will be increased by the presence of salt. The mass of salt must be subtracted from the dry mass. In addition, the value for the density of water used in the calculations must be the density of the water after accounting for the salt from the pore fluid.

If salts are a concern, obtain two matching specimens. Obtain the salt concentration using one of the matching specimens and use the other to perform the specific gravity test. At the end of the specific gravity test, make sure to collect all of the soil, and oven-dry to determine the dry mass of solids. Use the salt concentration, the mass of water, and the dry mass of solids to obtain the mass of salt using Equation C.2, Equation C.10, and Equation C.5. Use Equation C.12 to obtain the mass of dry soil. The salinity of the resulting fluid used in the specific gravity test is calculated as the mass of salt divided by the volume of the bottle. Using Table C.1 (and Equation C.11, if desired), determine the density of the fluid in the volumetric flask.

After these two factors are accounted for, the specific gravity is calculated using Equation C.16. This is the same equation the appeared previously as Equation 3.8, except the mass density of water has been replaced by the mass density of seawater.

$$G_s = \frac{M_s}{\left\{ \left(M_B + V_B \rho_{sw} + M_s \right) - M_{B+W+S} \right\}}$$
(C.16)

REFERENCES

Lide, D. R., Ed. 2007. *CRC Handbook of Chemistry and Physics,* 88th edition. CRC Press, Boca Raton, FL.

Ladd, Charles C. 1998. "1-D Consolidation: Magnitude of Final Settlement." *1.322 Soil Behavior,* Lecture Notes.

Noorany, Iraj. 1984. "Phase Relations in Marine Soils," *Journal of Geotechnical Engineering,* ASCE, 110(4), pp. 539–543, April.

Index

A

activity 119
aggregates 16, 215
aggregation of particles 87, 88, 98, 123, 130
air bubbles 45, 75, 123, 129, 130, 134, 136, 232, 283
air compressor 5, 280
air dispersion 92, 98
air entry 283, 284
air/water interface 201, 202, 280
analog 205–206
"A" line 119–120, 142–143
analog/digital (A/D) converter 203–204
angle, friction 21, 239, 241, 244, 245, 247, 253–255, 280, 283, 284, 286
angularity 144, 145, 158
anisotropy 240, 242, 258, 263, 264, 284
AR
    See area ratio (AR)
area correction 277, 291
area ratio (AR) 166, 167
Atterberg limits 82, 117–139, 140, 142, 144, 153, 324
axial deformation 276, 280, 295
axial strain 31, 198, 260, 271, 277, 280, 290, 291, 303
axial stress 276, 304, 321

B

back pressure 229, 285, 322
base 278, 284, 312–314
base pedestal 284

B (continued)

bees' wax 168
bentonite 167
blending 9, 17–19, 88, 95, 98, 103, 107, 116, 218, 222
block sample 17, 164, 189
blocky structure 148
borderline group symbols 142, 144, 153, 159
boulders 84–86, 88, 144, 155, 158
bulk material 3, 17–19, 101, 104, 164, 184, 185, 189

C

calcite equivalent 60–66
calibration, transducer 164, 191–195
carbonate
    See calcite equivalent
Casagrande construction 309–310
Casagrande liquid limit cup 117, 122–128, 132–133
CD
    See Consolidated-Drained (CD) triaxial test
cell constant 75, 78
cell pressure 276, 279, 280, 285–288, 290, 292
cell, test 228, 229, 275, 276
cell, triaxial 4, 275, 276
cementation 60, 87, 88, 144, 147, 155, 158, 282, 298, 304
cemented soils 29, 98, 123, 130, 169, 283, 292, 322
chamber pressure
    See cell pressure
chamber, triaxial 278
classification, soil 25, 140–160
cobbles 85–86, 144, 155, 158

coefficient of consolidation 120, 243, 244, 246, 284, 294–296, 298, 300–306, 308, 313, 316, 318, 320, 322, 326, 330, 336
coefficient of curvature (C_c) 86, 144, 154
coefficient of uniformity (C_u) 86, 144, 154, 159, 232
coefficient of volume change (m_v) 303
cohesion 169, 239, 241, 254, 255, 280, 282
cohesionless 163, 169, 180, 223–238
color, soil 145–146, 148, 155, 156, 158, 159
combined sieve/sedimentation analysis 94, 100–103
compacted specimens 35, 185, 187, 188, 210–222, 229, 283, 284, 286
compaction 5, 12, 15 , 28–30, 35, 88, 163, 185–188, 210–222, 262, 294, 295
compaction method 217–28
compaction mold 30, 214, 218
compactive effort 210, 212
compliance 62
composite sieve analysis 94, 103–107
compression 13, 30, 31, 71, 211, 294–333
compression index (C_c) 303, 317, 318, 323
compression ratio (CR) 294, 303
computer automation 164, 189–209
computer control 312
computer-assisted testing 164, 190
cone penetrometer
    See fall cone
confining pressure 275–277, 280, 282–284, 292, 293
consistency 117–119, 147, 264, 265
consistency limits 117–119
Consolidated-Drained (CD) triaxial test 276
Consolidated-Undrained (CU) triaxial test 276, 285
consolidation 31, 32, 120, 163, 170, 172, 180, 188, 189, 203, 211, 226, 228, 242–244, 246, 247, 250, 283, 284, 294–333
consolidation, degree of 301–302
constant head hydraulic conductivity test 223–238
constant mass (drying to) 25, 32, 82
constant rate of strain (CRS) consolidation 228, 312
constant temperature bath 112
conversion factors 334–337
core-barrel sampler 165, 170
creep 211, 295
CRS
    See constant rate of strain (CRS) consolidation
crushing, particle 54, 214, 218, 298
CU
    See Consolidated-Undrained (CU) triaxial test

D

Darcy's Law 224, 225, 229, 233, 299
data acquisition (DAQ) 62, 164, 189–209, 249, 250, 260, 290, 325

data reduction 20–21, 208–209
deflocculate 98
degree of saturation
    See saturation
density
    See mass density
desaturation 231, 233, 238
description, soil 20, 25, 140–160, 292
desiccant 26, 33
deviator stress 277, 285, 291, 292
dial gage 189, 191, 195
digestion, acid 60, 62
dilatancy 144, 150, 151, 155, 292
dilation 247, 253, 283, 284
direct shear (DS) test 163, 239–255
disaggregate 94, 123
dispersion 92, 93, 98
dispersion agents 92, 95, 98
dispersion quick test 151–152, 158
displacement transducer 192, 195–197, 202
dissipation 243, 246, 295, 301, 308
distilled water 39, 42 , 70–73, 93, 95, 98, 111, 116, 123, 130, 214, 321
disturbance 6, 15, 36, 164, 166, 170–173, 177, 180, 182, 185, 257, 263, 264, 273, 282, 284, 285, 316–318, 321, 322, 331
double drainage 300, 301, 306
drilling mud 167, 170
drilling slurry 167
dry density 29, 210–222, 252
dry strength 144, 149–150, 155, 156, 159
drying temperature 32, 33, 43, 44
drying time 32
DS
    See direct shear (DS) test
dual symbol 120, 142–144, 159
dynamic compaction 185, 210–222

E

effective depth 109
effective diameter 86
effective stress 118, 189, 223, 240, 243, 282–285, 295–299, 303, 304, 309–311, 314, 315, 317–319, 321, 322, 330, 331
elastic deformation 297, 304
end of increment (EOI) 309, 318, 319, 330
end of primary (EOP) 296, 297, 305–309, 318, 319, 322, 323, 327, 329–331
energy level (effect on compaction) 217
EOI
    See end of increment
EOP
    See end of primary

equilibrated water 42, 43, 70, 232
excess pore pressure 240, 264, 284, 295, 300, 301
extrude (tube samples) 170, 172, 175, 177, 179
extrusion 245, 246, 315, 322, 323, 327, 328, 330

F

failure
   in liquid limit testing (Casagrande
     cup) 123, 126
   in Direct Shear testing 240, 242, 244–247, 254
   in strength index testing 258–260,
     262, 268, 273
   in Unconsolidated-Undrained triaxial
     testing 276, 277, 281, 282, 285, 290, 292
fall cone 122, 126–129, 256–274
falling head hydraulic conductivity test 223–238
filter design 86, 226
filter material 233, 238, 313–316, 322, 324, 325
fineness modulus 89
fissured structure 170
fitting methods
   See log time method or square root of
     time method
fixed ring 312–313, 321
fixed-piston sampler 165
floating ring 312–313, 321
flocculation 98
flocs 98
flow curve 125, 136
fluid state 118
fluidization limit 118
foil sampler 165
forced draft oven
   See oven
friction (equipment) 166, 186, 205, 246, 247,
    254, 265, 277, 278, 280, 285, 286, 291,
    312, 313, 321, 322, 329

G

gap graded 88
geostatic 189, 242, 246
gradient
   See hydraulic gradient (i)
grain size 54, 61, 62, 84–116, 142, 144
grain size analysis 84–116
grain size distribution 52, 54, 84–116, 140,
    142, 143, 159, 160, 185, 227
grooving tool 122–126, 132
group name 120, 141–143, 152–156, 159, 160
group symbol 16, 119–120, 123, 130, 141–144,
    152–156, 157, 159, 160

H

handheld shear vane 168, 175–177,
    256–274, 282, 292
hardness 144, 148, 156, 158
Harvard miniature compaction mold 185, 210, 212
Hazen's formula 227
head 46, 223–238, 322
head loss 228–230, 233–235, 238
heterogeneous deposit 165
homogeneous structure 18, 85, 130, 148,
    155, 165, 218
hydraulic conductivity 15, 163, 164, 172,
    180, 211, 212, 223–238, 243, 246, 295, 298–300,
    304, 305, 309, 318, 320, 322, 330, 331, 336
hydraulic gradient (i) 224, 229–233, 235–238, 312
hydrometer 84–116

I

ICR
   See inside clearance ratio (ICR)
identification, soil 140, 141, 143, 155
in situ 5, 6, 15–17, 27, 28, 30, 72, 123, 157,
    163, 164, 167, 189 , 283, 284, 304, 314,
    315, 317, 321, 331
incremental consolidation 163, 189, 294–332
indicator paper (pH) 70
inorganic soils 82, 120, 143, 144, 153, 155, 159, 324
inside clearance ratio (ICR) 166
in-situ earth pressure coefficient, K_0
   See lateral stress ratio (K0)
intact soil samples 6, 15, 16, 17, 143, 147–149, 157,
    163–168, 169–184, 189, 258–260, 262–264, 282,
    283, 314, 318
intergranular 28, 32
interparticle forces 118, 169, 211, 283, 284, 295, 320
interparticle structure 217, 283
iodine flask 40, 41, 43
isotropic 257, 258, 263, 264

K

K_0
   See lateral stress ratio (K0)
kneading compaction 185, 187, 212, 217
Kozeny-Carman equation 226

L

laboratory vane, miniature 256–274
laminar flow 91, 224, 231, 233, 238
laminated 0

lateral stress ratio (K_0) 189, 242, 244, 245, 284
length to diameter ratio 266, 276, 289
lensed 148, 155
LI
    See liquidity index (LI)
liquid limit (LL) 82, 117–139, 142–144, 149, 150, 189, 324
liquidity index (LI) 117–139
load cell 200, 207, 199–200, 207, 280, 285
load frame 246, 247, 260, 278, 280, 287, 296, 312–314, 324
Load Increment Ratio (LIR) 189, 315, 322, 331
loam 82
log time method 305, 307, 308, 313, 316
loss of fines 97, 229, 231–233, 238
loss on ignition (LOI) 80–83

M

major principal stress 242, 276, 277, 281
manometer 200, 228–230, 234, 237
mass density 24–38, 53, 56, 91, 93, 110, 133, 134, 167, 210–222, 224, 226, 236, 304, 323, 328, 338–339, 343
mass density, maximum 52–59, 186
mass density, minimum 52–59, 186
maximum dry density (MDD) 210–222
maximum particle size 17, 18, 26, 27, 52, 56, 85, 87, 94, 95, 100–104, 107, 113–115, 145, 155, 158, 186, 187, 213, 214, 239, 243, 252, 254, 286, 313
maximum past pressure
    See preconsolidation stress
mechanical analysis 84–116
membrane 277–279, 283, 285–288, 292
membrane correction 279, 285–287, 291
membrane stretcher 288
mercury 4, 122, 129
migration of fines
    See loss of fines
migration of particles
    See loss of fines
miniature laboratory vane 256–274
minor principal stress 276, 277, 281
MIT stress path space 275, 281, 282
miter box 29, 180, 182, 183
modified effort compaction 210, 212, 213, 215
modulus of elasticity 199, 240, 264, 286
Mohr-Coulomb 280, 281
Mohr's circle 239, 242, 254, 258, 260, 277, 278, 280–282
moisture conditioning
    See tempering
moisture content
    See water content

moisture/density relationship 185, 210–222
monofilament nylon
    See filter material
muck 28, 80
multiplexer 204
Munsell® soil color chart 146, 156, 158

N

natural water content (ω_N) 27–28, 77, 95, 119–121, 134, 137, 148, 220, 328
negative pore pressure 29, 169, 282, 284, 293, 314, 315, 322
nonplastic 144, 151, 155–157
normally consolidated (NC) 246, 254, 297, 310, 315, 318, 321, 323

O

odor 146, 155, 158, 159
oedometer 228, 294–333
oil/water interface 201, 202
one-dimensional consolidation 31, 188, 189, 294, 295, 297–299, 309–311, 324
optimum water content (ω_{opt}) 35, 211–214, 216, 218–221
organic content 80–83, 143
organic soil 88, 100, 120, 130, 132, 134, 142–144, 146, 153, 154, 159, 160, 265, 286, 324
o-rings 288
oven 4–5, 32–33
oven drying 25, 30, 32–34, 36, 38
overconsolidated (OC) 246, 247, 255, 297, 318, 321
overconsolidation ratio (OCR) 33, 189, 265, 286, 304, 315, 321, 324
oversize correction 215–216, 222

P

packing (of particles) 54, 186, 239
paraffin wax 168
partially saturated 281, 284, 286
particle shape 146, 155, 158
particle size
    See grain size
particle size distribution 84–116
peak condition 240–242, 244, 245, 247, 253, 254, 262, 271, 272, 277, 285
peak strength 242, 262
peat 28, 33, 39, 44, 68, 70, 81, 82, 142, 144, 146, 159, 324
pedestal 182, 278, 279, 283, 284, 288, 289, 292
permeability 226
permeameter 225, 228, 232, 233, 235
pestle 64, 65

petroleum jelly 168
pH 68–79
pH meter 70, 71, 74, 75, 78
phase relationships 24–38, 40, 340–344
PI
   See plasticity index (PI)
pipette 75, 76, 92
piston friction 278, 280, 285, 286, 288, 291, 292
piston seal 278, 280, 285–287, 291, 292
piston uplift 280, 285, 286, 291, 292
plastic deformation 298
plastic limit (PL) 117–139, 144, 149, 151
plastic state 118, 119
plasticity 76, 98, 117–139, 142–144, 149, 151, 153, 155, 156, 159, 173, 189, 214, 321
plasticity chart 119–120, 138, 142
plasticity index (PI) 12, 117–139, 144, 151, 173, 189
plug flow 235
pocket penetrometer 168, 256–274 , 292
poorly-graded soil 12, 56, 58, 87, 97, 115, 123, 144, 154, 159, 186, 219, 220
pore pressure 240, 243, 254, 275–277, 282–285, 292, 293, 295, 296, 300–302, 305, 308, 314, 315, 322
pore pressure parameter (B-value) 282
pore space 240
pore water 25
porosity 24–38, 225, 226, 227, 282
porous stones 246, 312–315, 322, 323
power source 4, 65, 191–193, 196, 199, 203, 204, 206–208, 312
preconsolidation pressure
   See preconsolidation stress
preconsolidation stress 294, 297, 298, 303, 304, 309–311, 314, 315, 317, 318, 320–322, 331
pressure gage 62, 191, 200
pressure gradient 167, 295
pressure head 232, 238
pressure transducer 62, 191, 195, 200–202, 280
primary consolidation 295–296, 304–309, 314–316, 318, 319, 322, 323, 327, 329, 331
principal stress 167, 242, 276, 277
Proctor compaction test
   See standard effort compaction
proving ring 189, 191, 200
pycnometer 40–43

Q

Q test 276
Q_c test 276
quartering 18–19, 104, 105, 107
quick clay 119, 149, 263
quick tests 143, 149–153, 155

quick triaxial test 276
quick, consolidated triaxial test 276

R

R test 276
radiography 148, 169–172, 314, 326
rate effects 284
rate of secondary compression 298, 304, 307, 308, 314, 318, 320, 322
reaction with HCl 60–67, 147, 155
recompression index (C_r) 303
recompression ratio (RR) 303
reconstitution, specimen 6, 163, 164, 184–189, 212, 246, 276
recovery 166, 168, 169
relative density (D_R) 52–53
remolded state 6, 149, 184, 258, 260, 262–265, 269, 271, 317, 318
representative sample/specimen 5, 15–19, 104, 107, 164, 170, 172, 219, 314, 321
resedimentation 188–189, 190
residual conditions 240–242, 244, 245, 247, 253, 254
residual soil 88, 100, 147
reuse of soil (effect on compaction) 218
riffle box 19, 20
ring shear 244–245
rotary core-barrel sampler 165

S

S test 276
safety 4–5
salinity 25, 28, 30, 68–79, 93, 100, 130, 131, 321, 324, 340–344
salt 25, 28, 30, 43, 68–79, 93, 100, 130, 131, 189, 321, 324, 340–344
salt water 43, 68–79, 321, 340–344
sample 5
sample tube 165–171, 177, 178, 183
saturation 214, 216, 232, 235, 238, 283, 292, 295, 299, 322
saturation line 214, 216
saturation, degree of 24–38, 211, 232, 233, 238, 283, 292, 295, 328, 332, 340, 341, 343
seawater
   See salt water
secondary compression 294–297, 299, 304, 305, 307–309, 314, 315, 318–320, 322, 327, 331
sedimentation 84–116, 151, 152, 156, 159
seepage 225, 226
semi-solid state 118
sensitivity (of measurement) 26, 93, 119, 195–197, 200–202, 205, 206

sensitivity (of soil) 69, 149, 157, 262–263, 272, 273, 324
SHANSEP
 See Stress History and Normalized Soil Engineering
  Properties (SHANSEP)
shear box 239–255
shear force 240–241, 244, 247, 252–254
shear induced pore pressure 246, 247, 282, 283
shear strain 242, 245, 262
shear strength 258, 260, 262–265, 284, 292
shear stress 118, 167, 189, 211, 241–243, 245, 246, 254,
  262, 270, 271, 276–278, 280, 281
shear zone 241–242
sheep's foot roller 185, 212
shrinkage limit (SL) 117–139
side friction 166, 186, 245, 247, 312, 313, 321, 322, 329
sieve diameter 96, 97
sieve shaker 89, 95–97, 101, 108, 113, 116
sieves 84–116
sign convention 13
significant digits 13–14
simple sedimentation 94, 98–100
simple sieve 94, 95–98
single drainage 300, 301, 306
slickensided 148, 155
slow triaxial test 276
sodium hexametaphosphate
 See dispersing agents
soil color 145–146, 148, 155, 156, 158, 159
soil fabric 6, 26, 27, 30, 35, 69, 118, 143, 148, 163, 170,
  185, 189, 227, 232, 240
soil structure 123, 143, 144, 148, 155, 157, 163, 184, 188,
  217, 245, 262, 295
solid state 118, 151
specific gravity 28, 30, 33, 36, 29–51, 58, 91, 93, 110,
  111, 214, 215, 226, 292, 323, 328, 340–344
specific surface of solids 226–227
specimen 5
splitting 17–19, 88, 94, 95, 103, 108, 116, 218, 219, 222
square root of time method 302, 305–308, 316, 322
standard effort compaction 210–222
static compaction 185–186, 211, 217
stockinette sampler 165
Stoke's Law 89, 91
storage (of soil samples) 165, 166, 169, 172, 177
strain 6, 31, 191–208, 239–242, 245, 247, 250 , 260, 262,
  263, 271–273, 275–277, 280, 284–286, 290–292,
  294–333
Strain Energy Method 309–311
stratified structure 148, 155
strength index tests 256–274
Stress History and Normalized Soil Engineering
  Properties (SHANSEP) 295
stress path 275, 281
stress-strain 239, 247, 250, 275, 277
structure, of soil

 See soil structure
surface tension 18, 29, 246, 282
suspension state 118
swell index (C_s) 303
swelling 166, 295, 298, 303, 309, 314–315, 317,
  320, 322, 323, 327, 331
swelling ratio (SR) 303

T

tamp 55, 98, 187–188, 218, 249
temperature (effect on consolidation) 314, 320–321, 331
tempering 112, 123, 135, 214, 218, 220
3-t method 308, 309
time factor 300–303, 305–308
time to failure 246
tooth test 151, 152
torsional ring shear 244–245
Torvane® 168, 256–274
total head 224, 230, 233, 237
total mass density 28, 33, 214
total stress 167, 243, 275–277, 280–285, 292, 295, 300
total stress path (TSP) 281
toughness 144, 151, 155, 156, 159
transducers 62, 164, 189–209, 285
transportation (of soil samples) 16–18, 68,
  164, 168–169
triaxial 163, 182, 185, 203, 207, 275–293
trimming methods 180–184
tube sample 164–184
turbulent flow 224, 225, 233, 299

U

UC
 See Unconfined Compression (UC) test
"U" line 119–120, 138
ultrasonification 98, 109
Unconfined Compression (UC) test 184, 256–274 , 277
unconfined compressive strength 258, 260,
  263, 265, 272
unconfined compressive stress 260
Unconsolidated-Undrained (UU) triaxial test 163,
  275–293
undercompaction 187–188
undisturbed soil samples
 See intact soil samples
undrained shear strength 35, 72, 127, 129, 257–259, 263,
  265, 270, 272, 275, 277, 280, 284, 285, 292
Unified Soil Classification System (USCS) 82, 85–87,
  117, 119, 123, 130, 134, 140–160, 218
units 12–13, 334–337
unsaturated 31, 283, 284, 292, 305, 322

USCS
  See Unified Soil Classification System (USCS)
UU
  See Unconsolidated-Undrained (UU) triaxial test

V

vacuum de-airing 44–47, 235
vertical stress 284
vibrating drum roller 212
viscosity 91, 224, 226, 227, 229, 320, 336, 339
visual-manual description 140–160, 292
void ratio 6, 24–38, 53, 54, 71, 72, 211, 227, 263, 295,
  303, 304, 317, 318, 322, 323, 328, 331, 340, 341,
  343
volume measuring device 195, 201, 202
volumetric strain 31, 242

W

water content 5, 6, 15, 16, 24–28
water content, natural
  See natural water content

water, physical properties of 338–339
wax 27, 29, 34, 117, 118, 122, 129, 130, 132, 134, 138,
  168, 169, 177
well-graded soil 87, 123, 144, 154, 220, 249
wet density
  See total mass density
Whatman 54 filter paper
  See filter material
work 310, 311

X

x-ray machine
  See radiography

Y

Young's modulus 264

Z

zero air voids line 214, 216